Hazardous Waste Management Handbook:
Technology, Perception, and Reality

Paul N. Cheremisinoff
Yeun C. Wu
Editors

PTR Prentice Hall
Englewood Cliffs, NJ 07632

Library of Congress Cataloging-in-Publication Data

Hazardous waste management handbook : technology, perception, and
    reality / Paul N. Cheremisinoff, Yeun C. Wu, editors
       p. cm,
    Includes bibliographical references and index.
    ISBN 0-13-138637-9
    1. Hazardous wastes--Management. I. Cheremisinoff, Paul N.
    II. Wu, Yeun C.
TD1030.H38 1993   93-15921
628.4'2--dc20   CIP

Editorial/production supervision: *Dit Mosco*
Cover design: *Bruce Kenselaar*
Cover photo: *Chris Collins, "The Stock Market"*
Manufacturing buyer: *Mary E. McCartney*
Acquisitions editor: *Michael Hays*

©1994 by P T R Prentice Hall
Prentice-Hall, Inc.
A Paramount Communications Company
Englewood Cliffs, New Jersey 07632

The publisher offers discounts on this book when ordered
in bulk quantities. For more information, contact:

    Corporate Sales Department
    P T R Prentice Hall
    113 Sylvan Avenue
    Englewood Cliffs, NJ 07632

    Phone: 201-592-2863
    Fax: 201-592-2249

All rights reserved. No part of this book may be
reproduced, in any form or by any means, without
permission in writing from the publisher.

ISBN 0-13-138637-9

Printed in the United States of America
10 9 8 7 6 5 4 3 2 1

Prentice-Hall International (UK) Limited, *London*
Prentice-Hall of Australia Pty. Limited, *Sydney*
Prentice-Hall Canada Inc., *Toronto*
Prentice-Hall Hispanoamericana, S.A., *Mexico*
Prentice-Hall of India Private Limited, *New Delhi*
Prentice-Hall of Japan, Inc., *Tokyo*
Simon & Schuster Asia Pte. Ltd., *Singapore*
Editora Prentice-Hall do Brasil, Ltda., *Rio de Janeiro*

CONTENTS

Preface vii

List of Contributors viii

1 SELECTION OF COST EFFECTIVE AND ENVIRONMENTALLY SOUND REMEDIAL ALTERNATIVES BASED ON RISK ASSESSMENT DERIVED CLEANUP LEVELS 1
Paul J. Gretsky

2 RULE-BASED SYSTEM FOR RISK ASSESSMENT 16
Allen C. Chao, Lee Kuo Hsiao, Kengo Watanabe, Seishu Tojo

3 A COMPREHENSIVE APPROACH TO DEVELOP AND IMPLEMENT AN ARMY OVERSEAS INSTALLATION HAZARDOUS WASTE MINIMIZATION PROGRAM 29
Byung J. Kim, Ernest Eddy, Naim Qasi, Jim Hartman

4 ENVIRONMENTAL IMPACT ANALYSIS IN THE EUROPEAN COMMUNITY: U.S. ARMY RESPONSE THROUGH THE ENVIRONMENTAL IMPACT ASSESSMENT PROGRAM 42
Wes Wheeler

5 ENVIROCAD: AN ADVANCED COMPUTING ENVIRONMENT IN SIMULATE, DESIGN AND EVALUATE INTEGRATED PROCESSES FOR WASTE RECOVERY, TREATMENT AND DISPOSAL 50
Demetri P. Petrides, Wei Li, Konstadinos Abeliotis

6 HAZARDOUS WASTE CLEANUP AT FORMER DEPARTMENT OF DEFENSE SITES 59
T. Julian Chu, Thomas J. Wash, Michael H. Fellows

Contents

7 A RAPID YET INEXPENSIVE METHOD TO MONITOR AIR QUALITY IN ASPHALT CONCRETE PLANTS — 70
Namunu J. Meegoda, Yaogin Cchen, Robert T. Mueller

8 CONTROL OF VOLATILE ORGANIC COMPOUND EMISSIONS FROM WASTE TREATMENT AND DISPOSAL FACILITIES — 81
Thomas T. Shen

9 EMISSIONS DATA FROM A PYROXIDIZER® OPERATING ON REGULATED MEDICAL WASTE — 90
Paul H. Kydd, H-M Chiang

10 MSW MASS BURN AND RDF INCINERATOR ASH LEACHATE CONSTITUENTS — 119
A. J. Perna, D. W. Sundstrom, H. E. Klei, B. A. Weir

11 INITIAL SCREENING OF THERMAL DESORPTION FOR SOIL REMEDIATION — 135
James J. Yezzi, Jr., Anthony N. Tafuri, Seymour Rosenthal, William L. Troxler

12 THERMAL DESORPTION OF ORGANIC CONTAMINANTS FROM SAND USING A CONTINUOUS FEED ROTARY KILN — 157
Samuel H. T. Chern, Anthony LaRosa, Joseph W. Bozzelli

13 ENERGY RECOVERY FROM SCRAP TIRES — 170
Hilary Davidson, Jy S. Wu, Ravi Vallabhapuram

14 FATE OF TOXIC METALS AND PHENOL IN THE BOTTOM, SEDIMENT OF WET DETENTION/RETENTION PONDS RECEIVING HIGHWAY RUNOFF — 180
L. Y. Lin, Y. A. Yousef, J. A. Feuerbacher

15 IN-SITU VOC REMOVAL FROM GROUNDWATER USING AIR STRIPPING WICKS — 193
Walter Konon

16 NEUTRON ACTIVATION ANALYSIS OF HEAVY METAL BINDING BY FUNGAL CELL WALLS — 204
Theodore C. Crusberg, John A. Mayer

17 **STRUCTURAL ANALYSIS OF A FUNGAL BIOTRAP FOR REMOVAL OF HEAVY METALS FROM WASTEWATERS** 213
T. C. Crusberg, G. Gudmundsson, S. C. Moore, P. J. Weathers, R. R. Biederman

18 **DETOXIFICATION TREATMENT OF CHROME SLUDGE** 225
Chin-Tson Liaw, Wen-Shung Ma Lin, Tzong-Tzeng Lin

19 **EFFECTS OF CHROMIUM CONTAMINATION ON CONCRETE** 232
Dorairaja Raghu, Hsin-Neng Hsieh

20 **REMOVAL OF CHROMIUM FROM A HIGHLY CONTAMINATED SOIL/SLAG MATRIX BY WASHING AT LOW pH** 247
Erez Gotlieb, Victor Ososkov, Joseph W. Bozzelli, Itzhak Gotlieb, Ed Stevenson

21 **BIODEGRADATION OF POLYCYCLIC AROMATIC HYDROCARBON DEPENDING UPON OXYGEN TENSION IN UNSATURATED SOIL** 257
Hun Seak Park, Ronald Sims

22 **APPLICABILITY OF BIODEGRADATION PRINCIPALS FOR TREATMENT OF SOILS AND GROUNDWATER** 274
Robert D. Norris, W. Wesley Eckenfelder, Jr.

23 **KINETIC EVALUATION OF ANAEROBIC BIODEGRADATION OF TRICHLOROETHYLENE USING ACTIVATED CARBON FLUIDIZED BED** 293
Suxuan Huang, Yeun C. Wu

24 **TREATMENT PROCESS EVALUATION FOR THE REMEDIATION OF HERBICIDE-CONTAMINATED GROUNDWATER: PERCEPTION AND REALITY** 308
Rong-Jin Leu, Chen-Yu Yen

25 **CONCEPT AND APPLICATION OF FIXED-FILM TECHNOLOGY FOR BIODEGRADATION OF HAZARDOUS ORGANIC WASTE** 321
Bruce E. Rittman

26 **EFFECTS OF INITIAL NITROGEN ADDITION ON DEEP-SOILS BIOVENTING AT A FUEL-CONTAMINATED SITE** 332
John W. Ratz, Peter R. Guest, Douglas C. Downey

27 DEMONSTRATION OF ENHANCED SOIL BIODEGRADATION
    OF TPH-CONTAINING SOILS AT A NEW JERSEY
    ECRA SITE 348
    Paul E. Levine, Donald R. Smallbeck

28 SUCCESSFUL IMPLEMENTATION OF CONTROLLED AEROBIC
    BIOREMEDIATION TECHNOLOGY AT HYDROCARBON
    CONTAMINATED SITES IN THE STATE OF DELAWARE 363
    C. Darrel Harmon, Anne V. Hiller, Judith B. Carberry

29 FULL SCALE UPFLOW ANAEROBIC SLUDGE BED
    PROCESS FOR INDUSTRIAL WASTEWATER TREATMENT 373
    Gwo-Dong Roam, Hsin Shao

30 KINETIC MEASUREMENT OF BIODECOMPOSITION
    OF ANILINE, 2,4-DICHLOROPHENOL AND
    1,2,4-TRICHLOROBENZENE USING A
    CALORIMETRIC METHOD 390
    Kuanchun Lee, Ming-Chin Chang, Yeun C. Wu, Lun-Wen Yao

31 SUCCESSFUL LEACHATE TREATMENT IN SBR-ADSORPTION
    SYSTEM 409
    Wei-chi Ying, John Wnukowski, David Wilde, Donald McLeod

32 ACHIEVEMENTS IN POLLUTION PREVENTION VIA
    PROCESS AND PRODUCT MODIFICATION 437
    Kevin F. Gashlin, Daniel J. Watts

33 PHOTOCATALYTIC OXIDATION OF DICHLORVOS IN WATER
    USING THIN FILMS OF TITANIUM DIOXIDE 446
    Ming-Chun Lu, Gwo-Dong Roam

34 USE OF FERRIC CHELATES FOR FENTON (Fe/H_2O_2)
    TREATMENT OF PESTICIDE CONTAMINATED WATER
    AND SOIL AT NEUTRAL pH 459
    Katharina Baehr, Joseph Pignatello

PREFACE

Hazardous Waste Management: Technology, Perception and Reality provides a forum for presentation of results and ideas on the practical application of various advanced methods for detoxification of hazardous wastes and restoration of contaminated sites. This volume presents data and information for professionals, engineers, scientists, and administrators worldwide with the latest technologies and innovations in solving the costly problems of hazardous and toxic waste management and disposal.

The chapters authored were originally presented as part of the International Conference on Hazardous Waste Management in 1992. The material originally presented was sponsored by:

- The Hazardous Substance Management Research Center and the Department of Civil and Environmental Engineering at New Jersey Institute of Technology

in conjunction with:

- New Jersey Water Pollution Control Association in cooperation with:
- U.S. Environmental Protection Agency
- U.S. Army Construction Engineering Research Laboratory
- U.S. Naval Civil Engineering Laboratory
- Kyoto University (Japan)
- Industrial Technology Research Institute (Taiwan)

The presentations covered in this volume include the following major areas:

Waste Minimization and Resource Recovery; Environmental Impact and Risk Assessment; Pollution Prevention; Heavy Metal Contamination and Treatment; Photo Oxidation and Physicochemical Detoxification; Biotechnology and Biotreatment Engineering; Combustion and Air Pollution Control; Contaminated Soil and Groundwater Remediation; and Site Investigation and Restoration.

Consulting firms, federal, state and local government employees, industrial representatives, engineers, researchers, scientists, educators and other professionals involved in hazardous waste management and disposal will benefit from this volume.

Paul N. Cheremisinoff
Yeun C. Wu

LIST OF CONTRIBUTORS

Konstadinos Abeliotis, Department of Chemical Engineering, New Jersey Institute of Technology, Newark, NJ

Katharina Baehr, Universität Hohenheim, Stuttgart, Germany

R. Biederman, Department of Biology and Biotechnology and Mechanical Engineering, Worcester Polytechnic Institute, Worcester, MA

Joseph W. Bozzelli, Department of Chemical Engineering, Chemistry and Environmental Science, New Jersey Institute of Technology, Newark, NJ

Judith Carberry, Department of Civil Engineering, University of Delaware, Newark, DE

Allen C. Chao, Department of Civil Engineering, North Carolina State University, Raleigh, NC

Hsien-Tsung Chern, Department of Chemical Engineering, Chemistry and Environmental Science, New Jersey Institute of Technology, Newark, NJ

T. Julian Chu, U.S. Army Corps of Engineers, Washington, D.C.

T. C. Crusberg, Department of Biology and Biotechnology and Mechanical Engineering, Worcester Polytechnic Institute, Worcester, MA

Douglas C. Downey, Engineering Science, Inc., Denver, CO

Ernest Eddy, Headquarters, Eighth U.S. Army Environment Program Office, Seoul, Korea

Michael H. Fellows, U.S. Army Corps of Engineers, Washington, D.C.

J. A. Feurbacher, Department of Civil and Environmental Engineering, University of South Florida, Orlando, FL

Erez Gotlieb, Department of Chemical Engineering, Chemistry and Environmental Science, New Jersey Institute of Technology, Newark, NJ

Itzhak Gotlieb, GHEA Associates, Roseland, NJ

Paul J. Gretsky, New Jersey Institute of Technology, Newark, NJ

G. Gudmundeson, Department of Biology and Biotechnology and Mechanical Engineering, Worcester Polytechnic Institute, Worcester, MA

Peter R. Gusest, Engineering Science, Inc., Denver, CO

C. Darrel Harmon, WIK Associates, Inc., New Castle, DE

Jim Hartman, Headquarters, Eighth U.S. Army Environment Program Office, Seoul, Korea

Anne V. Hiller, DNREC-UST Branch, State of Delaware, New Castle, DE

Lee Kuo Hsiao, Environmental Protection Administration, Republic of China, Taipei, Taiwan, Republic of China

Hsin-Neng Hsieh, Department of Civil and Environmental Engineering, New Jersey Institute of Technology, Newark, NJ

Suxuan Huang, Keystone Environmental Resources, Inc., Monroeville, PA

W. W. Eckenfelder, Jr., Eckenfelder, Inc., Nashville, TN

Byung J. Kim, U.S. Army Construction Engineering Research Laboratory, Champaign, IL

Walter Konon, Department of Civil and Environmental Engineering, New Jersey Institute of Technology, Newark, NJ

Paul Kydd, Envimed, Inc., Rocky Hill, NJ

Kuan-Chun Lee, Department of Civil and Environmental Engineering, New Jersey Institute of Technology, Newark, NJ

Rong-Jin Leu, Gannette Fleming, Inc., Baltimore, MD

Wei Li, Department of Chemical Engineering, New Jersey Institute of Technology, Newark, NJ

C.T. Liaw, Union Chemical Laboratory, Industrial Technology Research Institute, Hsin Chu, Taiwan

W. S. Ma Lin, Union Chemical Laboratory, Industrial Technology Research Institute, Hsin Chu, Taiwan

T. T. Lin, Union Chemical Laboratory, Industrial Technology Research Institute, Hsin Chu, Taiwan

L. Y. Lin, Christian Brothers University, Civil Engineering Department, Memphis TN

Ming-Jiunn Lu, Graduate Institute of Civil Engineering, National Chiao-Tung University, Hsin Chu, Taiwan

J. A. Mayer, Department of Biology and Biotechnology and Mechanical Engineering, Worcester Polytechnic Institute, Worcester, MA

David McLeod, Occidental Chemical Corporation, Technology Center, Grand Island, NY

Namunu J. Meegoda, Department of Civil and Environmental Engineering, New Jersey Institute of Technology, Newark, NJ

S. C. Moore, Department of Biology and Biotechnology and Mechanical Engineering, Worcester Polytechnic Institute, Worcester, MA

Victor Ososkov, Department of Chemical Engineering, Chemistry and Environmental Science, New Jersey Institute of Technology, Newark, NJ

Hun Seak Park, Washington State Pollution Liability Insurance Agency, Olympia, WA

A. J. Perna, Department of Chemical Engineering, New Jersey Institute of Technology, Newark, NJ

Demetri P. Petrides, Department of Chemical Engineering, New Jersey Institute of Technology, Newark, NJ

Joseph Pignatello, Connecticut Agricultural Experimental Station, New Haven, CT

Naim Qasi, Headquarters, Eighth U.S. Army Environment Program Office, Seoul, Korea

John W. Ratz, Engineering Science, Inc., Denver, CO

Dorairaja Raugh, Department of Civil and Environmental Engineering, New Jersey Institute of Technology, Newark, NJ

Bruce E. Rittmann, Department of Civil Engineering, University of Illinois, Urbana, IL

Gwo Dong Roam, Bureau of Environmental Sanitation and Toxic Substance Control, Environmental Protection Administration, Taipei, Taiwan

S. Rosenthal, U.S. Environmental Protection Agency, Edison, NJ

Jody M. Roud, Harding Lawson Associates, Princeton Junction, NJ

Hsin Shao, Union Chemical Laboratory, Industrial Technology Research Institute, Hsu Chu, Taiwan

Thomas T. Shen, Department of Environmental Conservation, State of New York, Albany, NY

Ronald C. Sims, Utah Water Research Laboratory, Utah State University, Logun, UT

Ed Stevenson, Division of Science and Research, New Jersey Department of Environmental Protection, Trenton, NJ

Anthony Tafuri, U.S. Environmental Protection Agency, Edison, NJ

Seishu Tojo, Department of Environmental Resources, Tokyo University of Agriculture & Technology, Fuchu, Tokyo, Japan

Bill Troxler, U.S. Environmental Protection Agency, Edison, NJ

Thomas J. Wash, U.S. Army Corps of Engineers, Washington, D.C.

Kengo Watanabe, Department of Environmental Resources, Tokyo University of Agriculture & Technology, Fuchu, Tokyo, Japan

P. J. Weathers, Department of Biology and Biotechnology and Mechanical Engineering, Worcester Polytechnic Institute, Worcester, MA

Wes Wheeler, USACE Construction Engineering, Champaign, IL

David Wilde, Occidental Chemical Corporation, Technology Center, Grand Island, NY

John Wnukowski, Occidental Chemical Corporation, Technology Center, Grand Island, NY

Yeun C. Wu, Department of Civil and Environmental Engineering, New Jersey Institute of Technology, Newark, NJ

Jy S. Wu, Department of Civil Engineering, University of North Carolina at Charlotte, Charlotte, NC

Lun-Wen Yao, Powel Duffryn Terminals, Inc., Bayonne, NJ

Chen Yaogin, Department of Civil and Environmental Engineering, New Jersey Institute of Technology, Newark, NJ

James Yezzi, U.S. Environmental Protection Agency, Edison, NJ

Wei-Chi Ying, Occidental Chemical Corporation, Technology Center, Grand Island, NY

Y. A. Yousef, Department of Civil and Environmental Engineering, University of South Florida, Orlando, FL

Chen-Yu Yen, Gannette Fleming, Inc., Baltimore, MD

1

SELECTION OF COST EFFECTIVE AND ENVIRONMENTALLY SOUND REMEDIAL ALTERNATIVES BASED ON RISK ASSESSMENT DERIVED CLEANUP LEVELS

Paul J. Gretsky
New Jersey Institute of Technology
Newark, NJ

INTRODUCTION

The Congress of the United States enacted in 1980 the Comprehensive, Environmental Response, Compensation and Liability Act (CERCLA). This act is more commonly known as "Superfund," a term coined from a comment made by a congressman in reference to the $1.6 billion budget, over a five year period, established to address the dangers posed to the nation's environment from the sudden or uncontrolled releases of hazardous constituents. Superfund established the basis for a national program to respond to the threat, created by inadequately managed or abandoned hazardous waste sites, to the nation's air, groundwater, land, and surface water. Superfund was subsequently amended and refunded in 1986 with the passage of the Superfund Amendment and Reauthorization Act (SARA). All following references to CERCLA are to be interpreted as "CERCLA as amended by SARA".

The United States Environmental Protection Agency (EPA) has been tasked by congress to manage and implement CERCLA with a mandated focus on permanent cleanups at Superfund sites. The remedy selected to provide the permanent cleanup, as stated in CERCLA section 121 (Cleanup Standards), should provide high reliability and provide long-term protection. The basis for the selection of the remedy/remedial action includes: that the remedial action be protective of human health and the environment; be cost effective; comply with other statutory laws; and utilize permanent solutions and alternative technologies and resource recovery alternatives to the maximum extent practical. Treatment, such as vapor extraction, soil washing, groundwater filtration/treatment, should be used as a principal element for the reduction of volume, mobility or toxicity of the hazardous substances.

The National Contingency Plan (NCP) details the organizational structure and procedures to respond to the releases and discharges of hazardous constituents into the environment. The Remedial Investigation Feasibility Study (RI/FS) is the procedural framework for identification, evaluation and selection of remedies at Superfund sites. This framework has been developed by EPA to determine the degree of protectiveness of a remedy based on a risk assessment approval. The risk assessment involves an evaluation of site specific data generated by the performance of the Remedial Investigation (RI), which includes the types of hazardous substances present and their concentrations, potential for exposure, and presence of sensitive populations.

In accordance with CERCLA, the degree of cleanup for a site or the acceptable exposure levels, are typically derived from applicable or relevant and appropriate regulations (ARARs). ARARs include federal and state environmental and public health laws and to be considered materials (TBCs), which are nonpromulgated adversaries or guidelines issued by regulatory authorities that are not legally binding.

ARARs are identified and evaluated on a site specific basis. The three types of ARARs that are considered pertinent to CERCLA sites include the following:

1. **Chemical—Specific or Ambient Requirements.** These establish health or risk based concentration limits or ranges in various environmental media for specific substances, pollutants, or contaminants, such as drinking water standards, slope factors, reference dose and reference concentration.

2. **Action—Specific, Performance, or Design Requirements.** These establish controls or restriction based on the types of actions to be employed in regards to the management of hazardous substances or contaminants. These ARARs are technology based restrictions corresponding to the remedies under consideration, and the action based requirements and associated limitations.

3. **Location—Specific Requirements.** These ARARs are based on the specific site locations or characteristics where restrictions are placed on the concentration of hazardous substances or the performance of remedies solely because they are situated in special locations (e.g. wetlands). ARARs will typically define the cleanup goals that a remedy must attain when set at acceptable levels with respect to site specific factors or risk derived acceptable exposure levels. In

defining the cleanup goals/acceptable exposure levels, the following factors are considered:

- Concentration levels for systematic toxicants to which the human population (including sensitive subgroups) could be exposed on a daily basis without appreciable risk of significant adverse effects during a lifetime (chronic daily intake);
- concentration levels of known or suspected carcinogens that represent an excess upperbound lifetime cancer risk to an individual of between 10^{-4} and 10^{-6};
- other factors related to exposure or to technical limitations, such as duration, and type.

The objective of the RI/FS is to secure data to a degree sufficient to support an informed risk management decision regarding which remedy has the potential for being the most appropriate for a given site, based on site-specific information. The RI/FS process is not restricted solely to CERCLA sites and the approach can be incorporated to address any site which has experienced a release of hazardous materials. The objective of the Remedial Investigation (RI) is to characterize the contaminants at the site and to obtain site specific information required to identify, evaluate and select cleanup alternatives. A critical component of the RI is the Baseline Risk Assessment.

A baseline risk assessment is intended to establish potential for an adverse effect to occur, either under current or future land use conditions, if no actions are taken to further control contamination at this site. If such risks are determined to be unacceptable by the regulatory agency, then the baseline risk assessment contributes valuable site characterization information for the selection of appropriate response alternatives.

The baseline risk assessment is composed of four phases. The first phase is data collection and evaluation to establish the Contaminants of Concern. It consists of analyzing relevant site data to identify potential contaminant pathways, release mechanisms, populations at risk, etc. The preliminary risk assessment starts this process and identifies initial data gaps and recommends activities to fill the data gaps.

The second phase is an Exposure Assessment. It consists of evaluating the magnitude of potential (or actual) human exposure, the frequency and duration of the exposures, and the exposure pathways. The results of the exposure assessment are models of pathway specific intakes of each contaminant, for both current and future land use.

Conducted in parallel to the exposure assessment is a Toxicity Assessment. The toxicity assessment determines:

1. the adverse health effects caused by exposure to the contaminant of concern,

2. the relationship between the magnitude of the exposure and subsequent adverse health effects, and
3. the uncertainties involved in determining 1 and 2.

Finally, the last phase of the baseline risk assessment is the Risk Characterization. The risk characterization summarizes the risk information and provides estimates (quantitative expressions, if available, or qualitative discussions) for both cancer and non-cancer risk potentials, as appropriate.

Included in this phase is an evaluation of uncertainty in the methods used to arrive at risk assessments.

The scope of the baseline risk assessment is determined by the complexity of the site. When a site is uncomplicated with limited numbers of contaminants and it is estimated that the site presents little, if any, threat to human health or the environment, the FS effort should be reduced to match the potential threat.

The objective of the Feasibility Study (FS) is to screen, evaluate and analyze the proposed remedy alternatives based on the nine NCP evaluation criteria. The nine NCP evaluation criteria are as follows:

Threshold Criteria

1. Overall protection of human health and the environment.
2. Compliance with ARARs

Balancing Criteria

3. Long-term effectiveness and permanence
4. Reduction of toxicity, mobility, or volume through treatment
5. Short-term effectiveness
6. Implementability
7. Cost

Modifying Criteria

8. State acceptance
9. Community acceptance

During the RI process, Preliminary Remediation Goals (PRGs) are established to provide long-term cleanup levels to be utilized during analysis and selection of remedy alternatives. If the PRGs can be achieved, they should comply with both the site specific ARARs and the requirements for the protection of human health and the environment. Chemical specific PRGs are individual contaminant concentration goals for specific media and land use combinations. Two general sources of chemical specific PRGs are:

1. Concentrations based on ARARs; and
2. Risk assessment derived concentrations.

Risk assessment based PRGs are derived from EPA data bases, (Integrated Risk Information System (IRIS) and Health Effects Assessment Summary Tables (HEAST), for toxicity values and exposure information. The site specific information required for the development of PRGs early in the RI/FS process include the following:

- media of potential concern;
- contaminants of potential concern; and
- current and probable future land use.

Upon the completion of the RI, including the Baseline Risk Assessment, the PRGs can be revised to incorporate site specific information through the review of the chemicals and media of potential concern, future land use, and exposure assumptions. The risk based PRGs may be recalculated using this newly required information and based on the addition or exclusion of contaminants from future considerations. PRGs that are modified, based on the results of the baseline risk assessment, are still required to meet the NCP evaluation "Threshold Criteria."

During the performance of the FS, identification, screening and analysis of remedy alternatives, an evaluation of the ARARs can be performed that focuses on certain ARARs overall applicability to the remedy of the site. This evaluation is performed through the recalculation of the PRGs based on site-specific data. This process requires direct interaction with the governing regulatory agency for concurrence. Through the results of this procedure, if applicable, a justification of ARAR non-applicability can be determined. The justification process relies on the site-specific data that documents the exposure pathways and routes, exposure parameters, and calculation equations that incorporate intake of a contaminant(s) from a given media and are based on the site unique exposure pathways and associated parameters. If the justification is approved by the governing regulatory agency, chemical and site-specific PRGs can be revised based upon risk. It should be noted that this procedure is strictly based on site-specific information and may not be implementable at all sites, but rather on a case-by-case basis. The target risk levels are based on concentration for contaminants of potential concern. The target risk levels for carcinogenic and non-carcinogenic effect are as follows:

- **Carcinogenic effects**—Concentration is calculated that corresponds to a lo-6 incremental risk of an individual developing cancer over a lifetime as a result of exposure to the potential carcinogen from all exposure pathways for a given medium.
- **Non-carcinogenic effects**—Concentration is calculated that corresponds to a Hazard Index (HI) of 1, which is the level of exposure to a chemical from all significant exposure pathways in a

given medium below which is unlikely for even sensitive populations to experience adverse health effects.

The PRGs are required to undergo an uncertainty assessment, which serves as a basis for recommending further modifications to the PRGs prior to establishing final goals for site cleanup. The uncertainty assessment focuses on all aspects of the RI and baseline risk assessment. The objective of this assessment is to qualify elements of the study that may present a degree of uncertainty in accuracy, comprehensiveness, or finality. These issues need to be brought forth and discussed to determine if modifications are required to ensure protectiveness of the remedy during the FS process.

The FS systematic process of remedial alternative evaluation involves the screening of alternatives followed by a detailed analysis of those alternatives that pass the screening phase. The remedial alternatives are evaluated based on the nine NCP evaluation criteria. At this junction in the RI/FS process, the risk based PRGs (cleanup levels) and the Best Demonstrated Available Technologies (BDATs) can be directly brought together and integrated into a formal evaluation process. The focused objective of the evaluation of remedial alternatives is to select a BDAT that provides protection to human health and the environment, is implementable and cost effective. This evaluation process requires "balancing" of the documented remedial effectiveness of alternatives with attaining the PRGs; any residual risk which would remain at the site; the alternatives, implementability; and cost to implement. In cases where the alternative that represents the best balance factor is unable to achieve cancer risks within the risk range of 10^{-6} or an HI of 1, institutional controls may be used to supplement treatment to ensure protection of human health and the environment.

The approach to establishing balancing factors, given specific site characteristics, is to evaluate the remedial alternative (BDAT) to determine the following:

- **Performance Criteria** —The level of effectiveness a BDAT can achieve for the given media and contaminant and the degree of permanence (short-term and long-term).
- **Cost**—The cost, based on present worth, that would be incurred to achieve the performance criteria for the given media and contaminant.
- **Implementability**—A determination, based on historical data, site characteristics, public and local government acceptance, of the constructability and operational limitations.
- **Reduction of Toxicity, Mobility or Volume**—The resultant that the BDAT can achieve to address one or a combination of these three elements. The actual methodology utilized to perform the balancing evaluation can vary by establishing a ranking system where numerical values are assigned to each of the criteria.

The BDAT's achievement of the goal is rated, and a scoring is developed where weighted values are established for each of the criteria based on site specific knowledge and a scoring exercise is then performed. It is critical to understand that the selected remedy(ies)/remedial alternatives are based on documented site characteristics and must protect human health and the environment.

The BDATs that are most frequently utilized as remedial alternatives have demonstrated their ability to achieve given goals for specific chemicals in certain medias. The balancing evaluation becomes additionally qualitative for those BDATs that have not been "field proven" or are still in the treatment testing stages. In certain site-specific cases and under a case-by-case basis, the balancing evaluation has resulted in the selection of a remedial alternative that does not fully achieve the risk based PRGs due to the chemicals or site-specific characteristics. The remedial alternative's level of performance, degree of cleanup that can be achieved for a chemical concentration, can be evaluated in respect to the resultant risk that would remain if the alternative is implemented. Through this determination, the balancing process can select a remedial alternative that provides the greatest degree of protection and is cost effective and implementable. Through the inclusion of institutional controls and/or the addition of additional remedial elements (capping of the site to prevent direct contact with contaminated soils) the degree of resultant risk can be reduced. The degree that the remedial alternative has achieved for protectiveness (risk) can be derived from formulation developed by the EPA and presented in the documents, "EPA December 1991. Risk Assessment Guidance for Superfund: Volume 1 - Human Health Evaluation Manual (Part B, Development of Risk-Based Preliminary Remediation Goals), Interim, Publication 9285.7-01B", and, "EPA December 1991, Risk Assessment Guidance for Superfund: Volume 1 - Human Health Evaluation Manual (Part C, Risk Evaluation of Remedial Alternatives), Interim, Publication 9285.7-01C".

A rationale and summary of the procedures and equations for calculating risk based PRGs are presented for residential land and water use scenarios as an example. The aforementioned EPA publications also address additional issues such as industrial land use, volatilization and particulate emission factors, radiation, and individual pathways, and should be referenced prior to calculating risk based PRGs to ensure that proper assumptions and modifications are incorporated. It should be noted that the following equations utilize standard default exposure parameters consistent with Oswer Directive 9285.6-03 (EPA 1991b).

These equations should be modified based on site specific conditions and/or additional equations should be developed.

Risk based PRG equations have been derived in order to reflect the potential risk from exposure to a chemical, given a specific media, pathway, and land use combination.

The ability to modify and/or develop alternate equations, in accordance with EPA protocols and procedures, enables the balancing evaluations for remedial alternatives to be compared on a risk basis. By setting the total risk level of 10^{-6} for carcinogenic effects, the equation can be solved for the concentration value (risk based PRG). In the case where the remedial alternative's effectiveness is documented, through treatability studies/pilot projects or historical data, the equation can be set to the effective cleanup concentration of the remedial alternative to determine the resultant risk that could potentially remain after the remediation has been completed. This same process can be implemented for total non-carcinogenic effects by setting an HI of 1 for each chemical in a stated media. This process enables the remedial alternatives to be evaluated on a risk basis.

The risk based PRG for each chemical should be calculated by considering all of the relevant exposure pathways and "totaled" to combine risk for an individual contaminant from all exposure pathways for a given media and for both carcinogenic and noncarcinogenic effects. Note that through this process of "totaling," the calculated concentration level could fall below the analytical detection limit for the contaminant.

Based on the specific-site characteristics, typically if both non-carcinogenic and carcinogenic hazard risk based PRG are calculated for an individual contaminant, then the lower of the two calculated values is considered as the PRG for that chemical.

In conclusion, risk hazard based PRGs can be developed based on site-specific information which justifies the utilization of risk assessment derived cleanup levels rather than ARARS. Through the performance of the FS remedial alternative evaluation, these risk based PRG can be integrated into the "balancing" of the alternative process, with the nine NCP evaluation criteria, to select a remedial alternative that is most protective of human health and the environment, implementable and cost effective.

Table 1. Residential Water–Carcinogenic Effects

$$TR = \frac{SF_o \times C \times IR_o \times EF \times ED}{BW \times AT \times 365 \text{ days/yr}} + \frac{SF_I \times C \times K \times IR_o \times EF \times ED}{BW \times AT \times 365 \text{ days/yr}}$$

$$= \frac{EF \times ED \times C \times [(SF_o \times IR_w) + (SF_I \times K \times IR_o)]}{BW \times AT \times 365 \text{ days/yr}}$$

$$C(\text{mg/L}) \text{ risk based} = \frac{TR \times BW \times AT \times 365 \text{ days/yr}}{EF \times ED \times [(SF_I \times K \times IR_o) + (SF_o \times IR_w)]}$$

where:

Parameters	Definition (units)	Default Value
C	Chemical concentration in water (mg/L)	–
TR	Target excess individual lifetime cancer risk (unitless)	1×10^{-4}
SF_I	Inhalation cancer slope factor [(mg/kg day)$^{-1}$]	Chemical specific
SF_o	Oral cancer slope factor [(mg/kg day)$^{-1}$]	Chemical specific
BW	Adult body weight (kg)	70 kg
AT	Averaging time (yr)	70 yr
EF	Exposure frequency (days/yr)	350 days/yr
ED	Exposure duration (yr)	30 yr
IR_I	Daily indoor inhalation rate (m³/day)	15 m³/day
IR_w	Daily water ingestion rate (L/day)	2 L day
K	Volatilization factor (unitless)	0.0005 × 1,000 L/m³ (Andelman 1990)

Table 2. Residential Water–Noncarcinogenic Effects

$$THI = \frac{C \times IR_o \times EF \times ED}{R\!\int\! D^o \times BW \times AT \times 365 \text{ days/yr}} + \frac{C \times K \times IR_o \times EF \times ED}{BW \times AT \times 365 \text{ days/yr}}$$

$$= \frac{EF \times ED \times C \times \left[\left(\dfrac{1}{R\!\int\! D_o} \times IR_o\right) + \left(\dfrac{1}{R\!\int\! D_1} K \times IR_o\right)\right]}{BW \times AT \times 365 \text{ days/yr}}$$

$$C \text{(mg/L) risk based} = \frac{THI \times BW \times AT \times 365 \text{ days/yr}}{EF \times ED \times \left[\left(1R\!\int\! D_I \times K \times IR_o\right) + \left(\dfrac{1}{R\!\int\! D_o} \times IR_w\right)\right]}$$

where:

Parameters	Definition (units)	Default Value
C	Chemical concentration in water (mg/L)	–
THI	Target hazard index (unitless)	1
$R\!\int\! D_o$	Oral chronic reference dose [(mg/kg day)$^{-1}$]	Chemical specific
$R\!\int\! D_o$	Inhalation chronic reference dose [(mg/kg day)$^{-1}$]	Chemical specific
BW	Adult body weight (kg)	70 kg
AT	Averaging time (yr)	30 yr (for noncarcinogens, equal to ED)
EF	Exposure frequency (days/yr)	350 days/yr
ED	Exposure duration (yr)	30 yr
IR_I	Daily indoor inhalation rate (m^3/day)	15 m^3/day
IR_w	Daily water ingestion rate (L/day)	2 L day
K	Volatilization factor (unitless)	0.0005 × 1,000 L/m^3 (Andelman 1990)

Table 3. Age Adjusted Soil Ingestion Factor

$IF_{soil/adj}$ (mg-yr/kg-day)

$$= [IR_{soil/age1-6}] \times \left[\frac{ED_{age1-6}}{BW_{age1-6}}\right] + [IR_{soil/age7-31}] \times \left[\frac{ED_{age7-31}}{BW_{age7-31}}\right]$$

Parameters	Definition (units)	Default Value
$IF_{soil/adj}$	Age adjusted soil ingestion factor (mg-yr/kg-day)	114 mg-yr/kg-day
BW_{age1-6}	Average body weight from ages 1–6 (kg)	15 kg
$BW_{age7-31}$	Average body weight from ages 7–31 (kg)	70 kg
ED_{age1-6}	Exposure duration during ages 1–6 (yr)	6 yr
$ED_{age\,7-31}$	Exposure duration during ages 7–31 (yr)	24 yr
IR_{age1-6}	Ingestion rate of soil age 1–6 (mg/day)	200 mg/day
$IR_{age7-31}$	Ingestion rate of soil all other ages (mg/day)	100 mg/day

Table 4. Residential Soil–Carcinogenic Effects

$$TR = \frac{SF_o \times C \times 10^4 \text{kg/mg} \times EF \times IF_{\text{soil/adj}}}{AT \times 365 \text{ days/yr}}$$

$$C(\text{mg/kg, risk based}) = \frac{TR \times AT \times 365 \text{ days/yr}}{SF_o \times 10^4 \text{kg/mg} \times EF \times IF_{\text{soil/adj}}}$$

where:

Parameters	Definition (units)	Default Value
C	Chemical concentration in soil (mg/kg)	–
TR	Target excess individual lifetime cancer risk (unitless)	1×10^{-4}
SF_I	Inhalation cancer slope factor [(mg/kg day)$^{-1}$]	Chemical specific
SF_o	Oral cancer slope factor [(mg/kg day)$^{-1}$]	Chemical specific
BW	Adult body weight (kg)	70 kg
AT	Averaging time (yr)	70 yr
EF	Exposure frequency (days/yr)	350 days/yr
$IF_{\text{soil/adj}}$	Age adjusted ingestion factor (mg/yr/kg-day)	114 mg/yr/kg-day (see Equation 3)

Table 5. Residential Soil–Noncarcinogenic Effects

$$THI = \frac{C \times 10^{-4} \text{ kg/mg} \times EF \times IF_{\text{soil/adj}}}{R\!\int\! D_o \times AT \times 365 \text{ days/yr}}$$

$$C(\text{mg/kg, risk based}) = \frac{THI \times AT \times 365 \text{ days/yr}}{\dfrac{1}{R\!\int\! D_o} \times 10^{-4} \text{ kg/mg} \times EF \times IF_{\text{soil/adj}}}$$

where:

Parameters	Definition (units)	Default Value
C	Chemical concentration in soil (mg/kg)	–
THI	Target hazard index (unitless)	1
$R\!\int\! D_o$	Oral chronic reference dose (mg/kg-day)	Chemical specific
AT	Averaging time (yr)	30 yr [for noncarcinogens, equal to ED (which is incorporated in $IF_{\text{soil/adj}}$)]
EF	Exposure frequency (days/yr)	350 days/yr
ED	Exposure duration (yr)	30 yr
$IF_{\text{soil/adj}}$	Age adjusted ingestion factor (mg/yr/kg-day)	114 mg-yr/kg-day (see Equation 3)

REFERENCES

EPA. 1988a. *"CERCLA Compliance With Other Laws Manual,"* Part (Interim Final). Office of Emergency and Remedial Response. EPA/540/G-89/006 (OSWER Directive #9234.1-01).

EPA. 1988c. *"Guidance for Conducting Remedial Investigations and Feasibility Studies Under CERCLA."* Interim Final. Office of Emergency and Remedial Response. EPA/540/G-89/004 (OSWER Directive #9355.3-01).

EPA. 1988d. *"Guidance on Remedial Actions for Contaminated Ground Water at Superfund Sites."* Interim Final. Office of Emergency and Remedial Response. EPA/540/G-88/003 (OSWER Directive #9283.1-2).

EPA. 1988f. *"Superfund Exposure Assessment Manual."* Office of Emergency and Remedial Response. EPA/540/1-88/001 (OSWER Directive #9285.5-1).

EPA. 1988. *"Availability of the Integrated Risk Information System (IRIS)."* 53 Federal Register 20162.

EPA. 1989a. *"CERCLA Compliance with Other Laws Manual,"* Part II: Clean Air Act and Other Environmental Statutes and State Requirements. Office of Emergency and Remedial Response. EPA/540/G-89/009 (OSWER Directive #9234.1-01).

EPA. 1989b. *"Interim Final Guidance on Preparing Superfund Decision Documents."* Office of Emergency and Remedial Response. (OSWER Directive #9355.3-02).

EPA. 1989c. *"Methods for Evaluating the Attainment of Cleanup Standards,"* (Volume 1: Soils and Solid Waste). Statistical Policy Branch. NTIS #PB89-234-959/AS.

EPA. 1989d. *"Risk Assessment Guidance for Superfund"*: Volume 1 Human Health Evaluation Manual (Part A, Baseline Risk Assessment). Interim Final. Office of Emergency and Remedial Response. EPA/540/1-89/002.

EPA. 1990b. *"Guidance for Data Useability in Risk Assessment."* Office of Solid Waste and Emergency Response. EPA/540/G-90/008 (OSWER Directive #9285.7-05).

EPA. 1990d. *"National Oil and Hazardous Substances Pollution Contingency Plan (Final Rule),"* 40 CFR Part 300.55 Federal Register 8666.

EPA. 1991a. *"Conducting Remedial Investigations/Feasibility Studies for CERCLA Municipal Landfill Sites."* Office of Emergency and Remedial Response. EPA/540/P-91/001 (OSWER Directive #9355.3-11).

EPA. 1991b. *"Risk Assessment Guidance for Superfund"*: Volume 1 Human Health Evaluation Manual (Part B, Development of Risk Based Preliminary Remediation Goals). Interim. Office of Emergency and Remedial Response. (OSWER Directive #9285.7-01B).

EPA. 1991c. *"Role of the Baseline Risk Assessment in Superfund Remedy Selection Decisions."* Office of Solid Waste and Emergency Response. (OSWER Directive #9355.0-30).

EPA. 1991d. *"Risk Assessment Guidance for Superfund"*: Volume 1 Human Health Evaluation Manual (Part C, Risk Evaluation of Remedial Alternatives). Interim. Office of Emergency and Remedial Response. (OSWER Directive #9285.7-01C).

EPA. *"Health Effects Assessment Summary Tables (HEAST)."* Published quarterly by the Office of Research and Development and Office of Solid Waste and Emergency Response. NTIS #PB 91-921100.

EPA. 1990d. *"National Oil and Hazardous Substances Pollution Contingency Plan (Final Rule)."* 40 CFR Part 300.55 Federal Register 8666.

EPA. 1991a. *"Conducting Remedial Investigations/Feasibility Studies for CERCLA Municipal Landfill Sites."* Office of Emergency and Remedial Response. EPA/540/P-91/001 (OSWER Directive #9355.3-11).

EPA. 1991b. *"Risk Assessment Guidance for Superfund"*: Volume 1 Human Health Evaluation Manual (Part B, Development of Risk Based Preliminary Remediation Goals). Interim. Office of Emergency and Remedial Response. (OSWER Directive #9285.7-01B).

EPA. 1991c. *"Role of the Baseline Risk Assessment in Superfund Remedy Selection Decisions."* Office of Solid Waste and Emergency Response. (OSWER Directive #9355.0-30).

EPA. 1991d. *"Risk Assessment Guidance for Superfund"*: Volume 1 Human Health Evaluation Manual (Part C, Risk Evaluation of Remedial Alternatives). Interim. Office of Emergency and Remedial Response. (OSWER Directive #9285.7-01C).

EPA. *"Health Effects Assessment Summary Tables (HEAST)."* Published quarterly by the Office of Research and Development and Office of Solid Waste and Emergency Response. NTIS #PB 91-921100.

2

RULE-BASED SYSTEM FOR RISK ASSESSMENT

Allen C. Chao
Department of Civil Engineering
North Carolina State University
Raleigh, N. C.

Lee Kuo Hsiao
Environmental Protection Administration
Republic of China
Taipei, Taiwan, Republic of China

Kengo Watanabe and Seishu Tojo
Department of Environmental Resources
Tokyo University of Agriculture & Technology
Fuchu, Tokyo, Japan

INTRODUCTION

Risk analysis is of great importance for establishing the priority in risk management that concerns human health environmental quality. When there is a potential health hazard, a quantity question such as "how much the risk is involved" is often of primary concern to those involved. Thus a quantitative measurement of the risk is a fundamental requirement of all risk analyses. Since uncertainty is always introduced to the calculation of risk, the results obtained by including the uncertainty involved is more significant than a single numerical estimate of the risk.

Computer-based models have been widely used for environmental applications. Complicated numerical calculations can be easily carried out with computers. In this regard, conventional computer models have been designed for solving problems using a procedural algorithm. These models may offer numerical solutions but do not offer much insight into the procedures of making risk decisions. Thus, using these models may not produce significant results to satisfy the need of policy makers. In some cases, the numerical results may generate serious controversy and thus hinder the risk assessment.

Uncertainty of Risk Analysis

Although the term "Risk" has been extensively used in recent years, it is often vaguely defined in literature. Hansson (1989) defines "risk" as "a concept which includes both the probability and the character of the undesirable event." Thus, the concept of risk may include both quantitative and quality phases or senses. The quantity sense includes the magnitude of consequence and the probability of risk while the qualitative sense may include voluntariness, equality, catastrophic potential and other characters that relate to the perception, consequence and nature of the risk (Slovic, 1987; Cohrssen, 1989).

Previous researchers often ignored the qualitative phase of risk and defined the risk entirely based on quantitative consideration. For example, some risk assessors may define "risk" as the expected value of loss by using the following formula:

$$\text{Risk} = \Sigma \ P_{(i)} \times C_{(i)}$$

Where: $P_{(i)}$ = probability of event i.
$C_{(i)}$ = consequence of event i.

Such a definition does not show any qualitative consideration of the risk and hence provides no information about the uncertainty of the risk. For example, the risk of "50,000 persons having a probability of death rate of 0.001%" and that of "one particular individual having a death probability of 50%," obviously are not qualitatively the same event even though both have the same numerical value. To alleviate this problem, some risk analyses include a probability distribution of the consequence in order to express the quantitative uncertainty. But this approach does not deal with the qualitative phase of the risk.

The quality phase of risk is important but often ignored. For example, the risk perception is affected by many factors such as observability, dread, catastrophic potential, etc. (Slovic, 1987). Additionally, the quality of the risk information is often affected by the assumptions and procedures used in the analysis. The qualitative but not the quantitative information is, in most cases, the major source of disagreement among the parties involved. These quality characteristics cannot not be adequately handled by the quantitative risk analysis.

Definition of Probability

The probability is often defined as the long term frequency of an event. Since most risk analyses deal with rare events, the long term frequency of these events is usually difficult to determine because the information about its frequency does not exist. Thus, the probability is rather defined as "a numerical measure of a state of knowledge, a state of confidence" that may be influenced by statistical data and frequency measurement, if available (Kaplan, 1981; Apostolakist, 1978).

Sources of Uncertainty in Risk Analysis

The risk analysis involves the analytical processes of evaluating, ordering and structuring our knowledge of the risk so that decisions can be made with limited knowledge. For a typical analysis, there may be several sets of experiences available, and the analyst may not be sure which set of experiences is the most relevant to the event being studied. Even if one set of experiences is selected, there may be several theories to account for theses experiences. Additionally, some theories may be converted into several versions of mathematical models, and each model may agree with the experience and theory to a different extent. A decision that must be made for each of the aforementioned steps will contribute uncertainty to the final result but in various degrees. Therefore, documentation and explanation of the decision-making processes are of great importance for future reference and review.

There are two major types of uncertainty associated with risk analyses: the uncertainty of empirical quantities and the uncertainty of selected models, according to Morgan and Henrion (1990).

Uncertainty of the selected model is the second concern of risk analyses. Mathematical models are used to analyze and simulate the events and their consequences. These models are constructed based on existing theories. If several theories exist, uncertainty about the selected theory, and hence the model, is likely to have a substantial effect on the risk analysis results. Unfortunately, there is little information that can be found in literature for comparing the uncertainty of selecting different models.

Iterative Refinement in Risk Analysis

Risk analyses are carried out in an iterative mode. During the process of analysis, new data of knowledge about the problem being evaluated may be generated. The new information is then used to modify or change the comprehension of the issue. This leads to an iterative process and dynamical loops of refinement in risk analysis. In some cases, due to limited resources or lack of professional training, many risk analyses are conducted in a linear mode.

The iteractive approach of risk analysis is much more complex than the simple linear approach. It is often considered as a process of learning, discovery and refinement. Due to the iterative character of risk analysis, detailed descriptions of the procedure are of great importance and should be documented for future reference and modification.

Process Documentation

The above discussion leads to the importance of recording and documenting the evaluation process of risk analyses. There are other reasons for carrying out process documentation. Risk analyses are often seen by some as a black box, especially for those who are not involved in the evaluation process. Recipients of the risk information often interpret the result according to personal perception. This may lead to a misunderstanding of the risk or even raising questions between analysts and other concerned parties. A detailed description or record of the procedure will greatly alleviate this problem. When evaluating the uncertainty, detailed descriptions of the measurement procedure, assumptions and implication of the model, the judgement of experts as well as the information sources should be documented (US EPA, 1986).

Preferred Computer Programs for Risk Assessment

Conventional computer programs use an algorithm to generate numerical results. Using these programs, if the results generated are considered accurate, there is no need for the user to examine the procedures. In most cases the user does not have access to examine the procedures. In risk analyses the uncertainty that is often introduced in the various stages is one of the major concerns for making policy decisions.

If conventional computer programs are used for risk analyses, the analyst can only conduct the analysis using the procedures implemented by the program, without the ability to test different theories, assumptions and models. Thus, the conventional program has a serious limitation of its flexibility to deal with dynamic procedures. In other words, a good risk analysis program should allow the user to carry out iterative procedures and redefine the procedures.

There are several types of computer programs available for risk analyses or decision making. Some are based on a decision tree theory, others use a non-procedural modeling for financial planning or even spread sheets (Henrion, 1985). The programs using a definite procedure may not be appropriate for risk analyses. None of these programs allows the user to perform iterative analyses. To overcome the limitations of conventional models, a computer-aided system, DEMOS, was developed (Henrion, 1985; Morgan, 1990). This program is capable of achieving most of the requirements of analysis and has been applied in many policy modeling projects. However, it lacks the ability to combine different sets of information.

There are several basic requirements for a computer program developed for handling risk assessment:

1. Ability to handle risk quantitatively and qualitatively.
2. Ability to test sensitivity of different sources of information.
3. Ability to combine different pieces of message.

4. Flexibility of analytical approach.
5. Capable of iterative refinement.
6. Capable of improving risk communication.

These requirements are discussed in the following sections.

Ability to handle risk quantitatively and qualitatively—The ability of a program to perform quantitative calculations is essential but qualitative characterization of the risk is also of great importance. A computer program should be sufficiently self-expressed to describe or express any message dealing with the scope, limitation, assumption, methodology and other relevant procedure of the model (Henrion, 1990).

Ability to test sensitivity of different information—Risk analysts may sometimes have to choose a set of information from among several available sets of competitive information. If the results achieved by using a particular set of information are significantly different from those achieved by using a different set of information, the user must perform sensitivity studies to check whether the difference will significantly influence the final result. One of several alternatives can be taken pending the outcome of the sensitivity studies. The alternatives may include those, to convert the difference into the uncertainty of a parameter, to combine all the available information, or simply to ignore the difference. If a computer program is used, it should have the capability to allow the user to perform a sensitivity analysis of all the information from different sources.

Ability to combine different pieces of message—In most cases, the message from a certain source may provide some useful information that is not reveled by another piece of information from a different source. Thus, a better decision can be made if it is based on the combination of all the available information. Although some researchers may argue against the concept of combining the probability of information from different sources (Genest 1986), such a practice can provide a comprehensive understanding of the whole problem and others consider that the benefit is worth the effort (Newbold, 1974). Using a computer program, combining the information from different sources can be made effortless.

Flexibility of analytical approach—One of the objectives of decision analyses is to divide a complex issue into simpler problems, solve these sub-problems separately, and then logically link all the solutions in a final conclusion (Raiffa, 1968). The judgement process can be significantly improved with analytical estimation instead of performing a direct assessment (Gettys, 1973; Amstrong, 1975). In some cases, some empirical quantity is critical for a decision but is difficult or impossible to measure. Thus the user must try to use an analytical model to estimate the quantity. For example, if the ambient concentration of pollutant is important but there is no quantitative data available, the user can use data of emission rate and a dispersion model to estimate its concentration. Further, the results of risk analyses are used by people of diversified backgrounds for different purposes. Some users may consider the analytical work adequate while others my regard the same work as too rough or too detailed. Hence there is little doubt that a computer program should offer the flexibility of problem solving using different levels of analytical strategies.

Capability of performing iterative refinement—Most of the variables used in risk analysis models are not independent variables but dependent variables that are mathematically defined by different mathematical models. For example, a dependent variable A is related to variables B and C by $A = f1(B, C)$. The variables B, and C may be independent variables or they may be related to variables E, F and G, H by $B = f2(E, F)$ and $f3 = h(G, H)$, respectively. Because of a rapidly changing world, the knowledge base of an independent variable expands. It is likely that new models may become available for calculating the value of a variable that used to be treated as an independent variable. The appropriate and necessary way for solving a problem is dynamically changing. Hence, the computer program should be made flexible and capable of progressively refining its models and continuously modifying its knowledge base.

Improvement of risk communication—Risk communication is a crucial point for making successful policy decisions (Covello, 1986). The scientific complexity of the information and the uncertainty inherent in the technology make risk communication extremely difficult. This problem is further augmented by use of computer programs because conventional computer programs lack the ability to express qualitative information.

Trust and credibility are the most important factors in risk information communication (Covello, 1987; Weterings, 1989). The results of an analysis should be expressed thoroughly, honestly, clearly and consistently. Hence the information and judgement of the analyst should be saved and recorded and later be opened for public criticism in order to maintain trust and credibility. In this regard, a computer program should be capable of documenting the qualitative information along with performing quantitative calculations to assist users in risk communication.

Rule-Based Artificial Intelligence for Risk Analysis

The field of Artificial Intelligence (AI) is defined as "the part of computer science concerned with designing intelligent computer systems that exhibit the characteristics we associate with intelligence in human behavior" (Barr, 1981). Among all the AI research and application developments, knowledge representation that is a major focus is used to capture the essential features of a problem domain and make that information accessible to a problem-solving procedure (Luger, 1989). AI computer programs offer many advantages over the conventional computer programs for risk analyses (Fiksel, 1987). These advantages are:

    a. Repository for consensus knowledge.
    b. Clarity of logic and assumptions.

c. Ability to process qualitative information.
d. Handling of uncertainty.

Repository for consensus knowledge—Using AI-based programs allows the user to examine and criticize the procedures used for conducting risk analyses. The reasoning underlying the procedures can be made known or understood to the users. New knowledge can be easily integrated into existing knowledge base.

Clarity of logic and assumptions—In AI-based programs, the assumptions, reasoning, logic and other inferential procedures can be clearly recorded and examined. Different assumptions about risk can be used to explore the sensitivity of the conclusion. This feature is the most important for the policy maker to gain an insight about the uncertainties of the risk analysis.

Ability to process qualitative information—Qualitative knowledge of risk is related to the physiological and conceptual status of human perception. This type of information is very difficult to express using traditional procedural calculating skills. AI-based programs are capable of providing verbal descriptions so that both quantitative and qualitative considerations of risks can be included.

Handling of uncertainty—Several decision making strategies or methods have been developed by AI researchers to handle uncertainty of human knowledge. These include Bayesian rule, Fuzzy logic rule, certainty factors and Belief function, etc. Although there are still many debates about some of these methods, they can be applied for solving the uncertainty problems of risk analyses (Mumpower, 1986).

Rule-Based ESIRA

ESIRA (Expert System In Risk Analysis) is a rule-based prototype of a computer program developed in PROLOG using the concept of artificial intelligence as a modeling tool for risk analysis. The PROLOG language uses rules to process knowledge and to describe problem solutions. Unlike conventional computer programs, a PROLOG-based program consists of many rules and each of these rules has a goal and conditions that must be satisfied to achieve the specified goal. If the knowledge given does not satisfy the conditions for a specific goal, another goal will be tested and this procedure repeated until all goals are satisfied. The computer program developed in PROLOG is capable of handling both quantitative and qualitative information regarding risk in an iterative mode.

The prototype of ESIRA is design to demonstrate that both the quantitative and qualitative information for risk analysis can be handled. It is used to manipulate information of empirical quantities (variables) and the relationships among these variables. Each variable consists of three parts: (1) identification name, (2) qualitative information, and (3) quantitative information. The probability distribution of a variable concerned can be directly obtained by field measurement as independent variables. It may also be calculated as dependent variables using appropriate mathematical models to operate on other dependent variables. The field measurements can be in the form of a set of field data or a method that can be used to estimate the probability of the variable directly. Main features of ESIRA are:

1. Quantitative and qualitative approach to risk analysis.
2. Analytical and dynamic approach to modeling.
3. Combine different information using expert's judgement.
4. Clarify the reasoning using expert's explanation.
5. Graphics and documentation to facilitate communication.
6. User friendly menu driving, status line, and on-line help.

Quantitative and qualitative approach to risk analysis—In ESIRA, the quantitative information about a variable includes mean, median, 95% confidence range, probability distribution and qualitative weight. The qualitative weight is provided by the user and all the other values are calculated by combining all the measurements and models that are available to the problem. The probability of the variable is translated into discrete distribution and the final distribution is calculated using the Monte Carlo Simulation.

Besides the quantitative assessment of the variable, the scope, source, assumptions, limitations, judgements and other relevant qualitative information concerning the characteristics of a variable are included in its qualitative descriptions. These descriptions are provided and keyed in by the user and can be examined, and modified later. Similarly, the comments and opinions of other users can also be included and edited as an integrated part of the qualitative discussion of the variable in question.

Analytical and dynamic approach to modeling—When the relationships among the different variables are known, ESIRA is capable of using all the variables to create a new variable. Unlike the procedural approach used by conventional programs that uses a static and fixed formula, the procedures used in ESIRA to manipulate variables are dynamically defined by the user. Once the relationship of variables is defined by the user, an inference diagram is constructed. Using the backtrack feature of the PROLOG language, ESIRA will calculate the value of each variable by first calculating the values of its offsprings in an iterate mode.

Using multiplication as the prescribed relationship between two variable as in many risk analyses (Chrostowski, 1989; Crawford-Broen, 1987; US EPA, 1987) to calculate the Excess Cancer Rate (ECR):

ECR = Exposure Dose (EXPOSURE) × Cancer Potency Factor (POTENCY)

In fact, a more general mathematical function can be included to define the relationship of variables. In this example, ELCR can be measured or it can be calculated using the model. When the model is used, values of both EXPOSURE and POTENCY are shown to be obtained by two sets of measurement each.

ESIRA is designed to accommodate a dynamic change of the inference diagram so that new measurements or models become available for calculating the values of ELCR, EXPOSURE or POTENCY, the inference diagram can be easily modified to reflect such changes of knowledge. The user may also define different sets of inference diagram to compare and study the results obtained with different models.

Combine different information provided by expert's judgement—Expert's probability judgement is considered better calibrated than laymen's opinion (Wallsten, 1980; Keren, 1987). Hence, in ESIRA, the user is asked to examine and compare different information and then assign a weight to each piece of information to reflect his confidence in the accuracy of the information being evaluated. The final result is a weighed average of all the information combined.

In some cases, the information from different sources is contradictory to one another such as in the study of the health effects of sulfur air pollution (Morgan, 1984). A "1" can be assigned to the weighing factor of one piece of information and "0" to another one. The result thus obtained can be used to evaluate the sensitivity of the information from different sources.

Clarify the reasoning using expert's explanation—The expert's personal bias of the is of great concern when a subjective judgement is offered. In order to reduce the possible personal bias, the user must provide some explanation about his reasoning for the weighing process when the weighing factor is being offered. This explanation is then recorded for public examination and criticism that may help to further refine the weighing factor. This feature will help the expert to provide less biased judgments (Slovic, 1977; Koriat, 1980).

Graphics and documentation to ease communication—Graphic information is considered to be an effective means for risk communication (Ibrekk, 1987). ESIRA provides a graphic display of probability density function (PDF), cumulative distribution function (CDF) and the mean and median of a variable in question.

User friendly menu driving, status line and On-Line Help—All the functions provided by ESIRA are grouped according to their functions with main and pull-down menus. Selection of the function is done by highlighting or typing the first letter of a menu item. During execution of the program, a line of status is provided to inform the user about the current status, as well as when text of documentation is shown on the screen. The user may use the F1 hot key to display detailed information of a highlighted key word. These features allow the user to get educated about unfamiliar facts and procedures.

Demonstration of ESIRA

The Excess Lifetime Cancer Rate (ELCR) of the residents who live in the vicinity of a hazardous waste incinerator X is estimated. The excess cancers are thought to be caused by dioxins (TCDD), and the only suspected source of TCDD in this area is from the air emissions of the incinerator. For TCDD, its ELCR is related to the multiplication of the exposure and potency (Chrostowski, 1989; US EPA, 1985: 1987):

$$ELCR = EXPOSURE \times POTENCY$$

For exposure, there are two measurements and no model is currently available for calculating its value. One of the two sets of measurement is from biological monitoring data and another set from the intake rate from inhalation and fish ingestion. As for the potency of TCDD, there are two sets of measurements from laboratory animal tests. One set was carried out by using rat data from a multi-stage model and another uses rabbit data from a one-hit model.

Input of Data

Input of the quantitative data is done by selecting the "INPUT" item of the main menu. A pull-down menu is then displayed allowing the user to input a measurement or a model definition. The knowledge about the direct measurements includes both the quantitative information (distribution type and parameters) and qualitative information (explanation, sources, date, sampling, procedure, instruments, assumptions, application history, and any other factors that are related to the quality of measurement). All the quantitative and qualitative knowledge is recorded and saved for later risk analyses.

If there is any confusion about the procedures to be used in entering a field, the user can obtain immediate help by pressing the F1 key. Should the user hit the Carriage Return key to select the menu item "type," a menu of the available probability distributions is displayed for the user to make a selection. After the input of measurement 1 for ELCR is completed, the user may define the model that can be used to calculate the value of ELCR based on EXPOSURE and POTENCY. Since these two variables were not in the available knowledge base, the program will add them to the variable list and request more information from the user about these two new variables. Using the iterate procedures, all the information (measurements and models) needed for estimating the distribution value of ELCR have been acquired and stored in the knowledge base.

Analysis Procedure

Once all the relevant models have been selected, the user may select the appropriate menu item to examine all the measurements and models of any number of variables. The user may compare the quality of all the existing knowledge and input the qualitative weight factors to the different measurements and models to reflect his confidence of each item along with his explanations and reasoning. Such a qualitative knowledge is of great importance for discussions on the equality, voluntariness of risk or the scopes and causality of the analysis.

The analysis can be repeated for other variables. After analyzing all the relevant variables, the user may select the "CALCULATE" function from the menu to calculate the combined probability. Since ELCR is on top of the influence structure, ESIRA will first calculate the combined probability of EXPOSURE × POTENCY by use of the Monte Carlo Simulation method. The calculated results are combined with the measured probability to derive the final probability distribution of ELCR.

Results and Graphic Display

The final quantitative and qualitative results of all the relevant variables will be displayed by selecting the "RESULT" function from the main menu. These displays can be text reports or graphical probability distribution curves. Further statements or discussions can be added to the final display.

CONCLUSIONS

The rule-based computer program developed in PROLOG allows the user to manipulate both quantitative and qualitative information for risk analyses. With this program, the user may structure and carry out the analyses with greater flexibility and efficiency. Additionally, the user may create, examine and refine the model in order to easily conduct a sensitivity test or combine the various pieces of information. Because both qualitative and quantitative knowledge can be manipulated in this program, the user can gain more insight and comprehension of the uncertainty of the analysis in question.

Another advantage of this program is that the user analyst is asked to re-examine and explain his reasoning. The qualitative documentation of his reasoning will be made available for others to review and criticize. Under such provisions, the personal bias can be reduced to a minimal leading to improving the judgment made by expert analyst. The flexibility of editing and modifying the document allow a better risk communication among all those involved in the risk analysis process.

All the aforementioned features are included in the prototype ESIRA developed using the rule-based PROLOG language. However, further additions and modifications to develop the prototype into a working model include the following features:

1. The capability of calculating the probability using dependent variables.
2. More complicated mathematical models.
3. Graphic display of the influence diagram.
4. Automatic checking unit and weighing factor for inconsistent inputs.
5. Formatted input and output.

These features are expected to be included in the future version of ESIRA.

REFERENCES

1. Apostolakist, G. "Probability and Risk Assessment: The Subjectivistic Viewpoint and Some Suggestions," *Nuclear Safety*, 19, 3, 305-315, 1978.

2. Armstrong, J. S., Deniston, W. B., & Gordon, M. M., "The Use of the Decomposition Principle in Making Judgments," *Organizational Behavior and Human Performance*, 14, 2, 257-263, 1975.

3. Barr, A., & Feigebaum, E., *The Handbook of Artificial Intelligence, Vol 1*, William Kaufman, Los Altos, CA, 1981.

4. Chrostowski, P. C., Foster, S. S., & Fogg, A., "Assessing the Risks of Incinerating Dioxin-Contaminated Soil," *Hazardous Materials Control*, 2, 4, July-Aug., 1989.

5. Cohrssen, J. J., & Covello, V. T., "Risk Analysis: A Guide to Principles and Methods for Analyzing Health and Environmental Risks," Council on Environmental Quality, 1989.

6. Covello, V. T., Winterfeldt, D. & Slvic, P., "Risk Communication: A Review of the Literature," *Risk Abstract*, 3, 171-182, 1986.

7. Covello, V. T., Winterfeldt, D. & Pavlova, M. T., "Principles and Guidelines for Improving Risk Communication, Effective Risk Communication," *Proceedings of the Workshop on the Role of Governmentin Health Risk Communication and Public Education*, 1989, Plenum Press, 1989.

8. Fiksel, J., "The Impact of Artificial Intelligence on the Risk Analysis Profession," *Risk Analysis*, 7, 3, 277-280.

9. Genest, C., Michel, C., Seiger, J. H., Kelly, C. W., & Peterson, C. R., "Research on Human Multiplestage Probabilistic Inference Process," *Organizational Behavior and Human Performance*, 10, 3, 318-343, 1973.

10. Hansson, S. O., "Dimensions of Risk," *Risk Analysis*, 9, 107-112, 1989.

11. Henrion, M., "Predict! Uncertainty Analysis in a Spread Sheet," *Risk Analysis*, 5, 4, 539-541, 1987.

12. Henrion, M., "Computer Tools for Environmental Policy Analysis," *The CPSR Newsletter*, 8, 3, 1-5, 1990.

13. Ibrekk, H. & Morgan, G., "Graphical Communication of Uncertain Quantities to Nontechnical People," *Risk Analysis*, 7, 4, 519-529, 1987.

14. Karen, G., "Facing Uncertain in the Game of Bridge: A Calibration Study, "*Organizational Behavior and Human Decision Processes,*" *38*, 1, 98-114, 1987.

15. Kaplan, S. & Garrick, B. J., "On the Quantitative Definition of Risk," *Risk Analysis*, *1*, 1, 11-27, 1981.

16. Luger, G. F., & Stubblefield, W. A., *Artificial Intelligence and the Design of Expert System*, Benjamin Cummings, Redwood City, CA, 1989.

17. Morgan, M. G., & Henrion, M., *Uncertainty: A Guide to Dealing with Uncertainty in Quantitative Risk and Policy Analysis*, Cambridge University Press, Cambridge, 1990.

18. Mumpower, J. L., Phillips, L. D., Ren, O., & Uppuluri, V. R. R., "Expert Judgment and Expert Systems," *Computer and Systems Sciences, V 35*, NATO Scientific Affairs Division, 1986.

19. Newbold, P., & Granger, C. W. J., "Experience with Forecasting Univariate Time Series and the Combination of Forecasts," *Journal of Royal Statistic Society, 137*, 131-165, 1974.

20. Slovic, P., & Fischhoff, B., "On the Psychology of Experimental Surprises," *Journal of Experimental Psychology: Human Perception and Performance, 3*, 4, 544-551, 1977.

21. US EPA, Health Assessment Document for Polychlorinated Dibenzo-p-dioxins, EPA 600/8-84-014F, 1985.

22. US EPA, Health Assessment Document for Polychlorinated Dibenzo-p-dioxins, EPA 600/8-84-014F, 1986.

23. US EPA, Human Exposure Estimation for 2,3,4,8-TCDD, EPA 600/8-84-014F, 1986.

24. Wallsten, T. S., & Budescu, D. V., "Encoding Subjective Probabilities: A Psychological and Psychometric Review," Report to the Strategies and Standards Division of the US EPA, Research Triangle Park, NC 1980.

25. Weterings, R. A. P. M., & Van Eijndhoven, J. C. M., "Informing the Public about Uncertain Risks," *Risk Analysis, 9*, 4, 473-482, 1989.

3

A COMPREHENSIVE APPROACH TO DEVELOP AND IMPLEMENT AN ARMY OVERSEAS INSTALLATION HAZARDOUS WASTE MINIMIZATION PROGRAM

Byung J. Kim
U.S. Army Construction Engineering Research Laboratory
Champaign, IL

Ernest Eddy, Naim Qasi and Jim Hartman
Headquarters, Eighth U.S. Army Environment Program Office
Seoul, Korea

INTRODUCTION

The Hazardous and Solid Waste Amendments (HSWA) of 1984 declare hazardous waste reduction to be a national U.S. policy. Hazardous waste reduction, which is often used as synonym for hazardous waste minimization or pollution prevention, is in reality a new approach to environmental management because conventional pollution abatement or control technologies often simply shift pollutants from one medium to another.

In recognition of the long- and short-term liabilities of hazardous waste generation in economic costs and effective mission performance, the Army has established a hazardous waste minimization goal—to achieve a 50 percent reduction in the quantity of hazardous wastes generated, by 1992, as compared to the baseline calendar year 1985. The Army's hazardous waste minimization program includes source reduction, recycling, on-site treatment, reuse and treatment to reduce the quantity or volume and toxicity of hazardous wastes (Army Regulation [AR] 200-1, 1990).

Different socioeconomic, regulatory, and technical requirements of host countries create hazardous waste minimization priorities and options for overseas installations that differ from those for installations in the Continental United

States. Therefore, the Army's hazardous waste minimization goal is not always mandatory for overseas installations. However, the Eighth U.S. Army (EUSA) in Korea still makes its best effort to minimize hazardous waste and will be able to meet the Army goal.

This paper summarizes the technical assistance provided by the U.S. Army Construction Engineering Research Laboratory (USACERL) to successfully develop and begin to implement a hazardous waste minimization program for the EUSA. Emphasis will be on its experience in developing the program and its efforts to coordinate the program with host country regulations.

A Comprehensive Approach

An effective hazardous waste minimization program is impossible without the full support of the entire target organization. Figure 1 shows a flow diagram for hazardous materials and waste, in which:

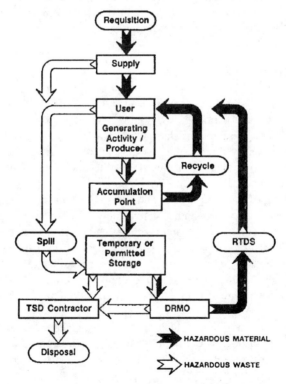

Figure 1 Flow of hazardous material and waste.

- Hazardous materials enter an Army installation when the Army Logistics / Supply Office purchases them. It is the Army's policy to avoid the purchase of hazardous material as much as possible. However, it is sometimes difficult to control local purchase of hazardous materials.
- By definition, users become hazardous waste generators when wastes are generated. The users can reduce the generation volume by material substitution, process change, or recycling. Hazardous waste is stored at the generator's accumulation points, or in temporary or permitted storage areas.
- By definition, used or excess hazardous materials do not become hazardous waste until the Defense Reutilization and Marketing Office (DRMO) considers reutilization, transfer, donation, or sales (RTDS) opportunities and declares that the materials should be disposed of. Spill-related debris is considered as waste. All RTDS or recycled items are regarded as hazardous materials.

The EUSA hazardous minimization program required concerted efforts from many different units. In fiscal year 1991, the EUSA tasked USACERL to develop the EUSA hazardous waste minimization program from "point zero." The EUSA's hazardous waste minimization program was developed in steps:

1. A method was developed to inventory hazardous materials using existing computer data. Hazardous material inventory by an installation was provided.
2. The Korean environmental regulations were translated into English so that EUSA installations could understand and comply with the Korean regulations. Korean Ministry of Environment (MOE) officials were contacted and environmental requirements in Korea were thoroughly discussed.
3. The EUSA Hazardous Waste Management and Minimization Program Guidance was developed. The EUSA Supplement 1 to AR 200-1, *Environmental Protection and Enhancement* was prepared.
4. EUSA employees were trained in the methods and procedures of hazardous waste minimization and safety in hazardous waste management.
5. A hazardous waste minimization study was conducted at two typical EUSA installations (Yongsan Garrison and Camp Carroll). A hazardous minimization strategy for each waste stream was recommended.
6. Based on recommendations from the hazardous waste minimization study, equipment was provided to EUSA installations for concept validation and implementation. An environmental audit program was developed to reflect the EUSA's unique requirements.

7. Installation hazardous waste management and minimization plan; installation spill prevention, control and countermeasure plan; and installation spill contingency plan were prepared for each installation.

Discussion

A New Method for Hazardous Material Inventory (Kim, B., et al., September 1991)

In a previous study, USACERL had developed a hazardous material identification system (HMID). The HMID identifies hazardous materials from the Logistics Intelligence File (LIF) and Central Demand Database (CDDB) maintained by the Logistics Control activity (LCA), Presidio, CA using the following criteria:

- materials with the codes listed in Appendix G of AR 55-355, *Defense Traffic Management Regulation*
- all class III supply petroleum, oils, and lubricants (POLs)
- Federal Supply Class (FSC) 6505 materials (hospital, infectious).

Because the HMID defined hazardous materials more broadly than required for this inventory, the HMID data was compared with the Material Safety Data Sheet (MSDS) in the Department of Defense's (DOD's) Hazardous Material Information System (HMIS); only common elements between two data bases were reported. The hazardous material inventory was prepared by installation or DOD Activity Address Code (DODAAC).

EUSA Hazardous Waste Management and Minimization Program Guidance

The guidance included responsibilities within EUSA organizations: organizational structures for effective hazardous waste management and minimization; hazardous waste minimization guide; inspection requirements; storage, handling, transportation requirements; and specific waste disposal instructions (Kim, B., et al., May 1991).

EUSA Hazardous Waste Minimization Study

First, the U.S. hazardous waste management system was compared with its counterpart in Korea. The Resource Conservation and Recovery Act (RCRA) definition of hazardous waste is based on characteristics and specific sources of hazardous waste, while Korean definitions are based on the type of waste-generating industry, specifically designated hazardous substances, and specific disposal standards and methods. In other words, Korean hazardous wastes are defined in terms of how safely industrial wastes are managed and disposed of (Kim, B., et al., March 1991).

Based on the hazardous material and wastes survey data and discussions with EUSA hazardous waste managers, the following hazardous waste streams of most concern were identified (Kim, B., April 1991):

- PCP ammunition boxes
- waste batteries and acids
- PCB and PCB-contaminated items
- antifreeze
- waste solvents
- used oil.

To meet the RCRA, DOD, and DA requirements, the EUSA will be able to formulate its own strategies to minimize hazardous waste disposal quantity and disposal expenses. An appropriate hazardous waste minimization strategy was proposed:

1. PCP-treated ammunition boxes:

    a. Background: In the past, the EUSA has returned PCP ammunition boxes to CONUS for disposal (an expensive alterative). By definition, however, PCP ammunition boxes are not hazardous waste. 40 CFR Part 261 (March 1990) defines the regulatory level of PCP as 100 mg/L by the toxicity characteristics leaching procedure (TCLP). USACERL contracted an Illinois Environmental Protection Agency-certified laboratory to analyze ammunition boxes randomly sampled from EUSA installations. The analysis results showed PCP concentrations ranging from 0.15 to 1.2 mg/l, with median value of 0.52 mg--well within safe limitations. These values clearly showed that the ammunition boxes were not hazardous waste.

At the time of study, the Korean Waste Management Law defined PCP items with 0.003 parts per million (ppm) or more in extracted solution as a special hazardous industrial waste (Article 2. MOE *Order To Implement the Waste Management Law*). This 0.003 mg/L limitation seemed unreasonably low compared with the 100 mg/L standard in the United States. It appeared that the limitation for wastewater effluent was misapplied to PCP-treated wood. After thorough discussion on the limitations, the Korean MOE eliminated the requirements.

    b. Proposed strategy

        (1) Coordinate negotiations between the appropriate Army agencies and host country authorities to officially acknowledge that PCP ammunition boxes are not a hazardous waste.

(2) DRMO-Bupyong and the EUSA will evaluate alternatives to dispose of PCP ammunition boxes as nonhazardous items. Reutilization, transfer, donation and sales (RTDS) by DRMO is a priority consideration.

(3) The following provisions should be made before RTDS activities: (1) an inspection program must be implemented to certify that the ammunition boxes are not contaminated with explosives, and that no crystallization is observed on the boxes; and (2) open burning of wood is prohibited.

(4) Technical committees will be established to resolve technical questions.

2. PCB and PCB-contaminated items:

a. Background: In 1976, the USEPA issued 40 CFR Part 761 to implement the Toxic Substance Control Act (TSCA), prohibiting the manufacture, processing, distribution, and use of PCB items. Two exceptions for continued use were: (1) use in a totally enclosed manner, and (2) use without presenting an unreasonable risk of injury to health or the environment. The USEPA provided three concentration ranges: PCB free, which is less than 50 parts per million (ppm), PCB-contaminated, which contains from 50 to 500 ppm, and PCB, which contains more than 500 ppm.

DLA, DRMS, DRMOs, and the Logistics and the Environmental staff of EUSA could not find any solution to the problems of PCB transformers. For more than 10 years, attempts to coordinate hazardous waste management with host countries have met with difficulty. Important unresolved issues include: (1) whether the host country can help with PCB transformer disposal, and (2) how host countries dispose of their own PCB transformers. Commonly, the U.S. Army has been refused such help on the basis that host countries often store spent transformers without treatment, and on the perception that the U.S. Army is better equipped than the host country to handle its own wastes.

It should be noted that Korean Governments do not regulate PCB items in the same manner as does the USEPA. In Korea, Article 2 of the MOE *Order To Implement the Waste Management Law* designates PCB concentrated to 0.003 mg/L or more in extracted solution, as a special hazardous industrial waste. This standard, which is considerably lower than the USEPA's 50 mg/L cutoff, was apparently the minimum detection level when the law was written. In fact, the Korean MOE does not have any implementation regulations to control PCB transformers. Both Korean Electric Power Corporation (KEPCO) and Korean MOE representatives have reported that they were not aware of any problems in the storage or disposal of PCBs. Recently, Korean MOE changed the PCB limit in dielectric fluid from 0.003 mg/L to 50mg/L.

b. Minimization and treatment technologies

   (1) Reuse: The TSCA does not regulate PCB items with a concentration of less than 50 ppm. This allows mineral oils with less than 50 ppm PCB concentrations to be reused and recycled. However, any waste oil containing any detectable concentration of PCBs cannot be used as a sealant, coating, dust-control agent, pesticide or herbicide carrier, or rust preventative.

   (2) Retrofill and reclassification: A PCB transformer is energized allowing PCB leach-out from the core after PCB oil is replaced with a specially prepared PCB-free fluid. The replacement is repeated until the oil PCB concentration becomes less than 50 mg/L. The USEPA requires a 90-day reclassification period after which the transformer may be filled with permanent dielectric fluid. Initial treatment takes about 12 to 24 hours and subsequent treatment requires 8 to 10 hours. Sometimes a processor is attached to the transformer after initial replacement to destroy residual PCB.

   (3) Chemical treatment: Many chemical processes have been successfully used to destroy PCBs in dielectric fluids with concentrations up to 10,000 ppm. Those processes also allow reuse of dielectric fluids. The chemical processes attack PCB molecules at their chlorine-carbon bonds, stripping the chlorine from the biphenyl molecules. The biphenyls are then polymerized to form an insoluble sludge.

   (4) Incineration: Only USEPA-approved high-efficiency incinerators may be used to destroy PCBs at concentrations greater than 500 ppm. Incinerators that burn PCB liquids must meet the following criteria: 2-second dwell time at $1200 \pm 100\,°C$ and 3 percent excess oxygen in the stack gas; or 1.5-second dwell time at $1600 \pm 100\,°C$ and 2 percent excess oxygen. Combustion efficiency must be at least 99.9 percent. Once operation conditions have been met, the destruction and removal efficiency must be no less than 99.9999 percent. Commonly available incineration technologies in the United States include the use of liquid injection, rotary kiln, multiple hearth, fluidized bed, and co-incineration. (Detailed discussion can be found in Freeman 1988.)

PCBs in concentrations between 50 and 500 ppm may be burned in a high-efficiency boiler that meets the following criteria:

- It must be rated at a minimum of 14.6 MW (50 million Btu/h).

- For gas- or oil-fired boilers, CO concentration in the flue gas must be less than 50 ppm.
- For coal-fired boilers, the CO concentration in the stack cannot exceed 100 ppm.
- Excess oxygen must be at least 3 percent.
- The waste cannot exceed 10 percent by volume of the total fuel fed to the boiler.
- Waste can only be fed into the boiler when it is at operating temperature.
- Specific process-monitoring and operating procedures must be followed (Freeman 1988).

(5) Landfill: Landfilling wastes with concentrations of PCB less than 50 mg/L are not regulated, but hazardous waste landfill is recommended. PCBs in concentrations of more than 500 mg/L may not be landfilled. Landfilling PCBs between 50 and 500 ppm requires USEPA approval with stringent technical provisions to prevent migration and leaching.

(6) Other methods: Innovative treatment methods include microbiological treatment, ultrasonic radiation, plasma system, catalytic oxidation, catalytic hydrogenation, hydrothermal decomposition, and photodegradation. (Detailed discussion of these technologies is beyond the scope of this report.)

c. Proposed strategy

(1) Discussion: The concept of PCB management differs in Korea from that in the United States. The USEPA distinguishes three different concentration categories and prohibits mixing of the three types of waste. Korean standards are based on the detection capability limitation at the time of the enactment of the standards. The Korean Government does not have implementation regulations for the use of PCB transformers.

Based on this preliminary evaluation, the most feasible in-country disposal methods are incineration (for PCB more than 500 ppm), and use of a high-efficiency boiler (for PCBs with a concentration of 500 ppm or less).

(2) Recommendations: Korean PCB regulations should be complied with. Since neither the USEPA nor host governments require subsequent PCB testing for locally manufactured transformers, new transformers bought from Korean manufacturers should be used in a totally enclosed manner. In-house rebuilding and interior repair of

transformers should be prohibited. PCB transformers should be strictly monitored. The DRMO should dispose of unserviceable transformers in accordance with Korean procedures. One option is to lease transformers from the Korean Government or a Korean Company.

A technical committee should be formed to establish a strategy for the disposal of existing PCB transformers. Based on the available information, the committees will assess the risk of in-country disposal and make recommendations on the best methods of compliance with host country regulations.

3. Waste Solvents

   a. Background: Solvent cleaning and degreasing operations can be categorized as: (1) cold cleaning (solvent application either by brush or dipping), (2) vapor degreasing (used for the high flashpoint chlorinated hydrocarbons such as 1,1,1-trichloroethane or methylene chloride, in the vapor phase, to clean metallic surfaces, and (3) precision cleaning (cleaning of work pieces prior to application of final surface coatings). Solvent has a great recycling potential with little capital and operation costs.
   b. Minimization technologies

      (1) Segregation: Without segregation, no effective recycling and management of wastes are expected. Especially in EUSA installations, the need for segregation cannot be exaggerated. Before Korean environmental regulations are enforced, the current practice of mixing wastes with slop oil must be stopped.
      (2) Substitution: Alternatives to chlorinated solvents include the use of alkaline cleaners and high-pressure hot water washers. Flammable petroleum distillate (PD) Type I can be substituted with PD Type II.
      (3) Reuse and recycling: High quality solvents once used for precision cleaning can be reused for cleaning other items requiring less purity. The use of distillation and condensation is a very effective process to recycle solvents.
      (4) Process and equipment modification and good housekeeping: Small modifications can result in hazardous waste minimization. Techniques include: (1) covering all solvent cleaning units, and (2) using refrigerated freeboard on vapor degreaser units. Randolph (1989) details solvent minimization methods.

c. Proposed strategy: The EUSA has not actively participated in the Army's Used Solvent Elimination (USE) program because Defense Reutilization and Marketing Office (DRMO) was able to sell used solvents to a Korean contractor under its slop oil contract. However, it is more economical for the EUSA to recycle solvents. Furthermore, proposed, more stringent Korean Environmental regulations will require the EUSA to pay a disposal fee. Also, since it is more economical to recover solvents on site, it may be a good opportunity for the EUSA to start a solvent recovery program. Recommended locations for solvent recovery are the consolidation points for Camp Carroll (Bldg 327 [Major Assembly Plant]; Bldg 405 [Direct Support Maintenance]; and Bldg 658 [Tactical Vehicle Maintenance]), and the consolidation point for vehicle maintenance facilities in Camp Coiner, Yongsan, Seoul. USACERL purchased eight solvent recovery systems for EUSA for field evaluation of solvent recovery.

4. Waste oil

   a. Background: Waste oil refers to lubricating oils that have gone through their intended use cycle. In the EUSA, used oil is released to Korean contractors under a sales contract.
   b. Minimization and treatment:

   (1) Combustion: 40 CFR part 266, subpart E, defines regulations for used oil burned for energy recovery.

Used-oil space heaters are considered to be a reasonably efficient means of disposing of waste oil generated on site. The USEPA has provided a conditioned exemption from the prohibition of burning of off-specification fuels in used oil space heaters. The referenced conditions are: (1) that the heater burn the oil generated by the facility, (2) that the heater be designed with a maximum capacity of not more than 0.5 Mbtu/h, and (3) that the combustion gases vent to the ambient air.

   (2) Re-refining: Currently available technologies have the capability to yield a recycled oil product with characteristics comparable to virgin lube oil. Refiners do not currently operate at system capacity due to the expense of refining costs and the demand for waste oil as fuel.

   c. Proposed strategy:

   (1) The experience at Camp Humphrey shows that a Korea-wide waste oil collection may not be economical. However, small

systems that burn near the waste oil generation site will be economical.
(2) Waste oil should be used as fuel within the waste-generating installation. Used-oil space heaters have been provided at several locations for demonstration.

5. Waste batteries and antifreeze solution

   a. Background: The USEPA has reported a proven concept for recycling battery acids (USEPA 1990). Although commercial systems for recycling antifreeze are available, none are currently used on EUSA installations. In the absence of an antifreeze recycling system, after coordination with the Utilities Division or the Sanitation Branch Chief, antifreeze solution may be discharged into a sewer system instead of being collected into used-solvent containers.
   b. Recommendations:

   (1) Recycling systems including both distillation units and oxidation-filtration units have been provided to EUSA. The recycled anti-freeze from an oxidation-filtration unit should not be used for tactical vehicles or vehicles under warranty. Oxidation-filtration units should not be used more than twice in a row.
   (2) Battery acid recycling should also be considered.

OTHER CONSIDERATIONS

Because each Army Installation had unique conditions, hazardous waste generation quantities were reported on differing bases. It was recommended hazardous waste quantity reporting should be divided into RCRA wastes, state-unique wastes, nonrecurring wastes, and nation-unique wastes (Kim, B.J., et al., April 1990).

Although hazardous waste minimization may not be driven by cost, disposal cost information is important to measure the no-action alterative costs. Department of Defense hazardous waste disposal costs vary widely. More research may be needed to reduce the disposal costs (Kim, B.J., et al., April 1991).

CONCLUSION

Minimizing hazardous waste at overseas installations is a complex task of keeping up with the rapid changes at distant CONUS installations while still meeting host country requirements. The U.S. Army's comprehensive approach to hazardous waste minimization promises to accomplish this goal and to serve as a valuable model for other overseas installations now establishing hazardous waste minimization programs. Through a comprehensive approach, the Army's overseas hazardous waste minimization program can be successfully developed and effectively implemented in a relatively short period.

It is critical to gain a mutual understanding of the host country's and the U.S. Army's requirements, and to maintain concerted efforts to protect the host country environment based on scientific foundation is also critical. These, in concert with a spirit of cooperation, will make even such difficult goals obtainable.

Real success will come when each installation and hazardous waste generator sets hazardous waste minimization as an operational priority. Such efforts will ensure that the EUSA's program will continue to be known as the "Vanguard of Overseas Hazardous Waste Minimization Programs."

REFERENCES

Army Regulation AR 200-1, *Environmental Protection and Enhancement*, Headquarters, Department of the Army (HQDA), 23 April 1990.

Freeman, H.M., ed., "Standard Handbook of Hazardous Waste Treatment and Disposal," (McGraw-Hill, New York, 1988).

Kim, Byung J., "Hazardous Waste Minimization and Treatment Opportunities in the Eighth U.S. Army and the U.S. Army, Japan," TR N-91/16 (U.S. Army Construction Engineering Research Laboratory [USACERL], April 1991).

Kim, Byung J., et al., "Validation of the U.S. Army's Current Hazardous Waste Data," TR N-90/10 (USACERL, April 1990).

Kim, Byung J., et al., "Korean Waste Management Law and Waste Disposal Forms," TR N-91/19 (USACERL, March 1991).

Kim, Byung J., et al., "An Analysis of Army Hazardous Waste Disposal Cost Data," TR N-91/17 (USACERL, April 1991).

Kim, Byung J., et al., "Hazardous Waste Management and Minimization Guidance," (U.S. Forces, Korea/Eighth U.S. Army [USFK/EUSA] Environmental Program Office, May 1991).

Kim, Byung J., et al., "A New Approach to Inventorying Army Hazardous Materials, A Study Done for the Eighth U.S. Army, Korea," Technical Report (TR) N-91/33 (USACERL, September 1991).

Randolph, E.R., "Solvent Waste Reduction Alternatives," EPA/625/4-89/021 (U.S. Environmental Protection Agency [USEPA], 1989) ch. 9.

Waste Minimization Opportunity Assessment: Fort Riley Kansas, EPA/600/S2-90/03 1 (USEPA, August 1990).

4

ENVIRONMENTAL IMPACT ANALYSIS IN THE EUROPEAN COMMUNITY: U.S. ARMY RESPONSE THROUGH THE ENVIRONMENTAL IMPACT ASSESSMENT PROGRAM

Wes Wheeler
USACE Construction Engineering
Research Laboratory - Environmental

INTRODUCTION

In recent years, the ability of the U.S. Army Europe (USAREUR) to perform its mission has been challenged by increasing pressure from the Federal Republic of Germany (FRG) government and the general public to reduce environmental impacts resulting from U.S. military activities. On several occasions, costly delays in USAREUR projects have occurred when the German governmental agency that approves U.S. military construction projects or *Oberfinanzdirektion*, at the state level, *Bauamt* at the local level, or Military Construction (MILCON), determined that environmental considerations had not been adequately addressed. Recent environmental disasters and chronic environmental problems in Europe (e.g., the Chernobyl explosion, contamination of the Rhine, deforestation from acid rain, pollution in the North Sea) have led to a heightened environmental concern in the general public as well as in Germany's political parties. In addition, more stringent environmental regulations are expected to be adopted by the FRG as the result of the European Community's (EC) Environmental Directives of 1985. The EC Directives require member countries to adopt legislation by 1992 to require environmental assessments of all projects and activities that may significantly affect the environment.

Headquarters USAREUR has been responsive to the need for integrating environmental risk assessment and minimization into project planning early on so that operational and mission capabilities will not be jeopardized. As a result, the

U.S. Army Construction Engineering Laboratory (USACERL) was tasked by USAREUR to develop an Environmental Review Guide (ERG) that will evaluate and minimize the environmental risk of USAREUR projects and activities[1]. This ERG will be responsive to EC Directives requiring environmental assessments.

PURPOSE

The purpose of the ERG is to provide environmental guidance to USAREUR environmental coordinators and planners during the early stages of project/activity planning. The ERG is a tool to assist planners in both forecasting and mitigating potentially adverse environmental effects. The ERG provides specific guidance on minimizing the environmental risk of proposed projects and actions.

It is also intended to act as an "early warning" system, identifying sensitive environmental issues in the early stages of a proposed action, thereby allowing for avoidance of impact and a reduction in the chances that project approval will be delayed due to inadequate environmental consideration.

The ERG procedure involves environmental risk identification and evaluation. Risk identification simply involves the recognition that a hazard exists. Risk evaluation refers to the process of determining the acceptability of that risk. The objective of environmental risk assessment is to forecast and evaluate potential impacts throughout a project's life cycle. The ERG facilitates risk evaluation by using a thorough procedure to review and anticipate possible impacts of a proposed project or activity.

After an environmental risk has been identified, the appropriate action to either intervene, delay, or proceed with an action can then be determined, based on the predicted results and the environmental risk. The ERG is intended to identify the potential environmental risk for activities in four functional areas that are typically performed in the host nation (see Table 1). These activities are described in Volume 1 of the ERG and include: 1) construction; 2) training; 3) stationing; and 4) operations, maintenance, and repair. Environmental impact is measured in eleven technical areas, such as health and safety, air quality, surface water, etc. (see Table 2). Risk assessment questionnaires and associated listings of ramifications and mitigations for each of these functional areas are addressed in the subsequent volumes of the ERG Manual.

PROCEDURE

The ERG procedure is based on the Environmental Impact Computerized System (EICS) developed by USACERL. EICS identifies types of potential impacts resulting from changes in activities and guides baseline investigation to help in the preparation of environmental documentation. The basic assumption of the EICS system is that certain environmental qualities are inherently affected by specific Army activities. Based on this assumption, EICS develops cause and effect relationships between specific Army activities and their corresponding effect on the environment.

The environmental evaluation standards within the EICS system were developed by a panel of engineers and scientists representing various environmentally-related disciplines (see Appendix 1 of Volume 1, ERG). This panel of experts identified nine broad types of Army activities that could affect the environment. A comprehensive list of army activities was established along with a complete list of biological/physical attributes that can be affected by these activities. The experts also identified environmental ramifications and mitigation strategies for activities in each functional area. From this research, three primary components of the EICS were developed:

1. Functional Areas (broad types of army activity);
2. Technical Areas (broad categories of environmental attributes which can be affected by army activity);
3. Ramifications and Mitigations (descriptions of specific environmental consequences and mitigative solutions).

The ERG utilizes the EICS definition of Army activities and their described relationship to biological/physical attributes, but within a broader, overview context, without the use of computerization. The ERG utilizes four of the nine functional areas found in EICS that are most pertinent for USAREUR (i.e., construction; training; stationing; and operations, maintenance, and repair). The ERG's use of the EICS framework allows the environmental consequences of specific Army activities in the FRG to be forecasted and for applicable mitigating solutions to be incorporated within a simplified process. Other non-automated environmental assessment methods reviewed during preparation of the ERG were found to be either too elementary to provide useful guidance, or lacked a recognition of activities unique to the Army, or required the use of computerized systems.

The ERG is intended to serve as the first step in providing environmental guidance for proposed USAREUR projects and actions. The ERG is useful as an initial indicator of environmental risk and for guiding site-specific analysis. However, as additional resources for environmental analysis become available, more precise site-specific methods of environmental analysis can be implemented for USAREUR. For instance, the creation of computerized environmental data bases and interactive geographic information systems (GIS) can allow quick, precise quantification of impacts for proposed projects/actions at specific locations. Data bases and GIS systems do have a moderately high initial cost, although they generally prove to be cost effective in the long-term, if adequately developed and maintained.

INTEGRATION WITH PROJECT PLANNING

In order for the ERG to yield the greatest net benefit (i.e., reduce costly delays and avoid adverse environmental impacts), the process must be employed during the early stages of project planning and programming (i.e., within the guidance year for construction projects) and well in advance of the execution phase of the project. Completion of all of the possible project design details are not necessary in order to apply a sound, conscientious environmental analysis. Rather, the environmental analysis should focus on the results of the proposed project/action along with those of several alternatives to the project/action, which may be implemented at either a particular location or in one of a series of locations. In addition, early analysis of potential environmental effects is desirable for identifying suitable mitigating strategies and incorporating these strategies into the project/action.

Environmental planning should be seen as an important element of project planning and, to be most effective, it must be proactive in nature. Whether the project involves construction, stationing, training, or operations and maintenance, the ability to properly identify environmental consequences reduces the chance that weak documentation or changes in scope will delay or remove a project from the funding cycle. Integration of the ERG into the USAREUR master planning process will help to augment the Phase II & III environmental assessment of future development as outlined in the USAREUR Master Planning Technical Manual.[2] In addition, the ERG process can be completed along with the DD Form 1391 report (the Department of Defense uses the DD Form 1391 to submit document funding requests for military construction made to Congress) to minimize the chances that the *Bauamt* (the German equivalent of a county or city government construction agency) would delay approval of USAREUR construction projects due to inadequate consideration of environmental issues. The Army goal to conduct environmental reviews at the same time as other Army planning and decision-making actions will also help to avoid delays in mission completion.

CONCLUSION

The net effect of the ERG process should be that proposed projects/actions are studied more thoroughly. Operational requirements need to be reconciled along with economic, political, and environmental considerations. Risk assessment in decision-making though the ERG environmental impact analysis process will play an important part in lowering project costs, reducing and mitigating environmental impacts, and facilitating project approval by the host nation.

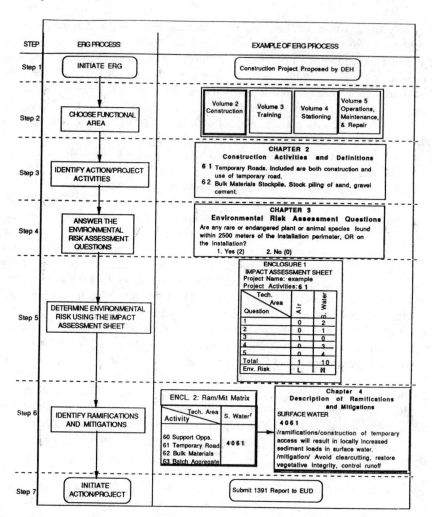

Figure 1 Environmental review guide process.

Table 1 Description of Functional Areas

Functional Area	Related Activities
Construction	Building, housing, and road construction and their associated activities (e.g., painting, welding, electrical work, plumbing, excavation, utilities, drainage).
Training	Any expansion, addition, or termination of training activities (e.g., troop exercises, firing ranges, tanks, artillery, air operations).
Stationing	Troop movement in or out of the installation (temporary or permanent), change in installation mission, major vehicle movement.
Operations	Any expansion, addition, or termination of daily activities or maintenance, programs within the installation (e.g., land management, utility and repair services, recreation, medical facilities, electronics, communications, post supply).

Table 2 Technical Areas and Affected Attributes

Technical Area	Affected Attributes
Ecology	Ecosystem stability, terrestrial and aquatic plant and animal life
Health and Safety	Human safety and health hazards
Air Quality	Particulates, gases, and vapors
Surface Water	Sedimentation, chemical point and nonpoint input, biocides, thermal impacts
Groundwater	Accidental spills, waste disposal, infiltration, recharge rate, depletion of groundwater
Earth Science	Soils, geology, landforms, biota
Land Use	Nuisances, natural landscape, conflicting land use
Noise	Task interference, invasion of privacy, sleep interference, hearing loss, mental stress
Transportation	Congestion, safety hazards, wear on streets
Aesthetics	Quality of visual landscape, natural landscape diversity, unique cultural or natural features
Energy & Resource Conservation	Fuel resources, renewable resource degradation

REFERENCES

1. Forrest, Russell, Fittipaldi, Erb, Clint, Tyler, Elizabeth, and Wheeler, Wes, *Environmental Review Guide for United States Army Europe*. Draft USAERL Special Report, May 1992.

2. Harland Bartholomew & Associates, Inc. *USAREUR Comprehensive Planning Technical Manual*, January 1990.

5

ENVIROCAD: AN ADVANCED COMPUTING ENVIRONMENT IN SIMULATE, DESIGN AND EVALUATE INTEGRATED PROCESSES FOR WASTE RECOVERY, TREATMENT AND DISPOSAL

Demetri P. Petrides, Wei Li, and Konstadinos Abeliotis
Department of Chemical Engineering
New Jersey Institute of Technology
Newark, NJ

INTRODUCTION

The successful design and evaluation of integrated waste recovery, treatment and disposal processes is a difficult task that can be facilitated by the use of expert systems and advanced computer-aided process design tools. Such a software tool, called EnviroCAD, is under development at the New Jersey Institute of Technology in collaboration with a number of industrial partners. EnviroCAD is being designed to enable scientists and engineers to evaluate the environmental impact of new processes and minimize waste generation during the early stages of process development when it is easy to make process modifications. EnviroCAD takes as input data the waste streams of a process plant and based on their flowrates and compositions it recommends alternatives for waste recovery and recycling. For whatever cannot be recovered, it recommends alternatives for treatment and disposal. In addition to generation of feasible flowsheets for waste recovery and treatment, EnviroCAD provides the facilities to analyze and evaluate those alternatives and select the best among them. EnviroCAD's user interface makes use of advanced graphics to improve the human/computer communication and reduce the learning period.

In the chemical and other process industries, most R&D scientists and engineers, during process development, are mainly interested in product yield and product quality and pay little attention to the impact that a new process may have on the environment. By the time the environmental issue is raised, the process has usually undergone significant development and it is difficult and/or costly to make major modifications (e.g., change an extraction solvent, add a recycle stream to minimize waste generation, etc.). Furthermore, for pharmaceutical processes, the FDA limits the improvements or retrofits that may be made after the product has been approved as "safe and effective." This attitude used to work in the past when the cost of end-of-pipe treatment constituted a very small fraction of the operating cost. With rapidly rising waste treatment costs, however, the industrial world is realizing that something else ought to be done that would eliminate the formation of waste at the first place.

The solution to the problem is to use a software design tool that enables research and development engineers to consider the environmental impact of new processes during the early stages of process development (Venkataramani et al., 1990). EnviroCAD is such a software design tool currently under development at the New Jersey Institute of Technology in collaboration with a number of industrial partners. The development of EnviroCAD is to a large degree an extension of BioDesigner, a program that was developed at the Massachusetts Institute of Technology to facilitate the design and evaluation of integrated biochemical processes (Petrides, 1990).

The formulation and development of process simulation tools for environmental applications, however, is a difficult problem to process because: 1) many processes are poorly understood and predictive mathematical models that could be used for design do not exist; 2) bulk stream properties (BOD, COD, TSS, TKN) are used to characterize material streams in end-of-pipe treatment processes instead of full composition and as a result unit operation models in environmental simulators must be able to deal with those properties; 3) a number of compounds are present at very low concentrations requiring special models that can deal with trace contaminant levels.

DESCRIPTION OF ENVIROCAD

Figure 1 shows the scope of EnviroCAD from a process point of view. EnviroCAD takes as input data the waste streams that a plant generates and as a first step it recommends alternatives for waste recovery and recycling. For whatever cannot be recycled or recovered, EnviroCAD recommends alternatives for waste treatment and disposal. In addition to recommendation of alternatives for waste recovery, treatment and disposal, EnviroCAD provides the facilities to analyze and evaluate the various alternatives and select the best feasible solution.

EnviroCAD runs on the Apple Macintosh Computer and closely follows the graphical interface guidelines set by Apple Computer. It combines technologies of process simulation and knowledge based expert systems (KBES). The architecture of the program is shown in Figure 2. EnviroCAD consists of three main components, the graphical user interface, the simulation (analytic) component, and the expert system modules.

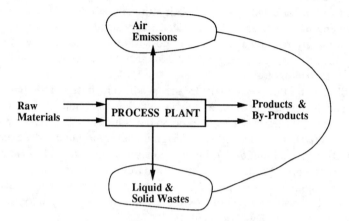

Figure 1 The scope of EnviroCAD from a process point of view.

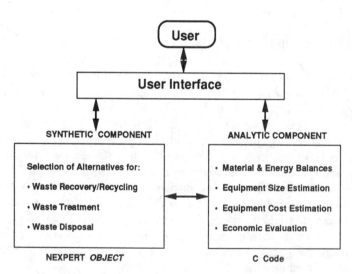

Figure 2 System architecture of EnviroCAD.

The interface and the analytic component are written in THINK C (a C compiler available from Symantec, Inc.). C was the programming language of choice because it supports advanced data structures and runtime memory allocation. Advanced data structures are needed to store and efficiently manipulate information about various objects, such as unit operations, chemical components, material streams, etc. The runtime memory allocation features allows the program to handle flowsheets of any number of unit operations, streams, etc. with the actual computer memory (RAM) as the only limitation.

The synthetic component consists of three knowledge bases and the inference engine of the Nexpert *Object* expert system shell (Nexpert *Object is* available from Neuron Data, Inc.). Nexpert *Object* was selected among several expert system shells because: 1) it provides an object-oriented environment that facilitates representation of declarative knowledge, 2) it can be fairly easily interfaced to programs that are written in the C programming language, and 3) its development version has a user friendly interface that facilitates development of knowledge bases. Nexpert *Object is* used in its dynamic library version (NDL) and is invoked from within EnviroCAD as an external dynamic subroutine. This architecture allows the analytic component of EnviroCAD to be used, if desired, as an independent application with lower memory requirements.

The Graphical Interface: EnviroCAD features a fully graphical and interactive interface that results in a tool that is simple to use and easy to learn even for occasional users who have limited process design and environmental expertise. All input-output information is provided or/and displayed through appropriate dialogue windows. Figures 3 and 4 show how information about a flowsheet is displayed on the screen of EnviroCAD. The typical user of EnviroCAD is expected to be a chemical or environmental engineer who has some process design and environmental engineering experience but not necessarily programming experience. EnviroCAD provides an advanced "Help" facility that minimizes the need for written manuals. The user has quick access to information on "How to Use EnviroCAD," information on the various unit operations, raw material prices, etc. Several items of the help facility are editable and the user can document his/her own information.

Process Synthesis in EnviroCAD: Process synthesis deals with the selection and assembly of a set of unit operations that are capable of meeting certain process constraints. The number of alternative flowsheets for waste recovery, treatment and disposal is usually very large and if one wants to consider and evaluate all alternatives the problem becomes combinatorial. The number of alternatives can be reduced significantly by using domain-specific experiential design knowledge. Experiential knowledge in the form of heuristics and rules of thumb is knowledge accumulated and refined by experts over years of problem-solving experience in a specific domain. Experiential design knowledge can be captured and documented on the computer in an active and easily accessible and modifiable way. Designing a computer program to do this activity is the objective of Knowledge Based Expert Systems (KBES). KBES are appropriate for dealing with difficult, ill-structured

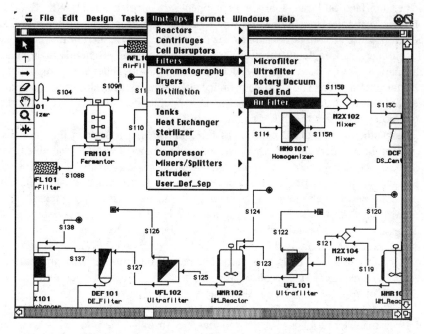

Figure 3 EnviroCAD makes use of advanced graphics to facilitate the interaction of the user with the computer.

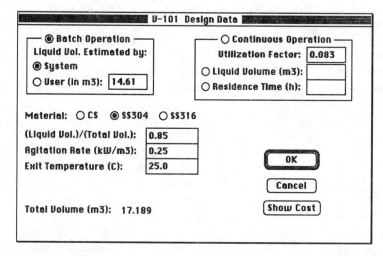

Figure 4 When the user clicks on a unit operation picture using the mouse, EnviroCAD displays information about that unit through dialogue windows like the one above.

problems in complex domains, such as the process synthesis problem, typically not amenable to purely algorithmic solutions. EnviroCAD uses a KBES to carry out synthesis. It is composed of three knowledge bases and the inference engine of Nexpert *Object*. The first knowledge base contains information to carry out synthesis of waste recovery processes, the second of waste treatment processes, and the third of waste disposal processes. Modification of the knowledge bases in EnviroCAD can be done easily using the development version of Nexpert *Object* that provides advanced editors and a friendly user interface.

Waste Recovery Module. This KBES analyzes the waste streams of a process plant and recommends alternatives for recycling and recovery of valuable components and treatment and disposal of everything that cannot be recycled or recovered. Decisions are based on the relative amounts and physical/chemical properties of the various compounds. For instance, if a certain solvent is present in a waste stream at a high concentration, distillation may be recommended to recover and reuse it. If on the other hand a solvent is present at a low concentration, then steam stripping may be recommended to recover the solvent and reduce VOC emissions in the waste treatment facility.

Waste Treatment Module. This KBES recommends alternatives for waste treatment for streams from which all valuable components have been recovered. Before a stream is sent to an activated sludge treatment plant, this module checks for the presence of any hazardous or toxic compounds, biocides, and odor causing compounds. If there are any, it recommends alternatives for their removal.

Waste Disposal Module. This KBES recommends alternatives for waste disposal for whatever cannot be recovered or treated. Disposal options include various types of incineration as well as landfill and land application of sludge. The results of a synthesis session are a number of feasible flowsheets, that can be directly transferred to the simulation component for further analysis and evaluation.

Process Simulation in EnviroCAD: The analytic component of EnviroCAD is an interactive process simulator that enables engineers to quickly develop, analyze, and evaluate integrated processes for waste recycling, recovery, treatment and disposal. Both dynamic (e.g., batch distillation) and steady-state unit operation models are included in the program because both modes of operation are practiced in such processes. The program carries out detailed material and energy balances and estimates concentrations of contaminants down to sub-ppm levels. Table 1 reported in the stream report. Table 2 shows a partial list of unit operation models that will be included in the final version of EnviroCAD. EnviroCAD is being interfaced to a database module that stores physical and environmental properties for a large number of chemical compounds (including many that are priority pollutants).

EnviroCAD enables the user to ask and readily answer "what-if" questions and carry out sensitivity analysis with respect to key design variables. Using EnviroCAD, the engineer can easily investigate the effect of the addition of a

Table 1 Environmental stream properties in EnviroCAD.

BOD_5	Total Solids (TS)
BOD_L	Dissolved (TDS)
COD	Volatile
	Volatile
TOC	
ThOC	Suspended (SS)
	Fixed
TKN	Volatile
Free NH3	pH
Nitrites	
Nitrates	Ash
Phosphorus	Heavy Metals
Organic	
Inorganic	
Sulfate	
Chlorides	
Organic	
Inorganic	

recycle stream or the change in operating conditions of a unit operation on the economics and the performance of an entire process.

For an integrated waste recovery, treatment and disposal process, EnviroCAD estimates the purchase cost of equipment, the fixed capital investment, the annual operating cost, and carries out a detailed economic evaluation. EnviroCAD is intended to provide cost estimates for preliminary economic evaluation with a maximum error of $\pm 25\%$. Equipment cost is estimated as a function of equipment capacity, material of construction, and certain design characteristics. The fixed capital investment is estimated based on the total purchase cost of equipment using multipliers (Valle-Riestra, 1983). The user can modify any multipliers that are used in the economic evaluation.

CONCLUSIONS

EnviroCAD constitutes an attempt to introduce the benefits of process simulation and computer-aided process design into environmental problems. Such tools have been widely used in the chemical/petrochemical industry in the last three decades

Table 2 Partial list of process models in EnviroCAD.

Mechanical Separators	Phase Separators
Plate and Frame Filtration	Batch Distillation
Membrane Microfiltration	Continuous Distillation
Membrane Ultrafiltration	Condenser
Gas Separation Membranes	Flash
Reverse Osmosis	Evaporation
Rotary Vacuum Filtration	Drying
Basket Filter Centrifugation	Extraction
Disc-Stack Centrifugation	Steam Stripping
Decanter Centrifugation	Air Stripping
Clarification	Wet Scrubbing
Flocculation	Decanter Tank
Granular Media Filtration	
Adsorption Separators	**Biological Treatment**
Activated Carbon	Activated Sludge
Ion Exchange	Trickling Filter
	Anaerobic (various designs)
	Lagoons
Sterilization	**General Reactors**
Continuous Heat Sterilization	Stoichiometric Reactor
Batch Heat Sterilization	Well Mixed Reactor
Chemical Sterilization	Plug Flow Reactor
General Unit Ops	**Other Models**
Liquid and Slurry Pump	Flaring
Compressor	Liquid Injection Incineration
Shell and Tube Heat Exchanger	Fluid-Bed Incineration
Plate and Frame Heat Exchanger	Rotary Kiln Incineration
Spiral Heat Exchanger	Supercritical Oxidation
General Mixer	Oxidation in Molten Metal
Flow Splitter	Electrostatic Precipitators
Component Splitter	
Storage Tanks	
Blending Tanks	

resulting in significant cost savings from improved utilization of raw materials and decades resulting in significant cost savings from improved utilization of raw materials and energy. We feel the time has come for this technology to start playing a major role in cleaning up the environment and designing industrial processes that are environmentally friendlier. The ultimate objective of this effort is the development of more efficient production and pollution control technologies that result in manufacturing facilities with almost "zero discharge" of pollutants to the environment.

The use of EnviroCAD in industry will enable engineers to consider the environmental issues during the early stages of process development when there is still room for process modifications. Further, the estimation of a dollar value for the cost of waste recovery/treatment/disposal early on will set up flags to warn scientists and managers that process modifications to reduce waste generation may make more sense than end-of-pipe treatment.

The detailed material balances for a waste recovery/treatment/disposal process estimated by EnviroCAD will satisfy the requirements of all regulatory agencies concerning the fate of waste compounds. Efforts will be made, through EnviroCAD, to introduce a common language of communication between the corporate world and the regulatory agencies by making sure that the format of the reports generated by EnviroCAD are accepted by those agencies.

Last but not least, EnviroCAD has the potential to become a successful educational tool that will train students and engineers how to consider environmental constraints during the design of new processes. Its user friendly and interactive interface will stimulate a dialogue between the user and the computer resulting in effective training and refined final designs.

ACKNOWLEDGEMENTS

Financial support for this project by the Commission on Science and Technology of the State of New Jersey and a number of corporations (Envirogen, Merck, Pfizer, Schering Plough, and SmithKline Beecham) is greatfully acknowledged.

REFERENCES

1. Petrides, D., "Computer-Aided Design of Integrated Biochemical Processes; Development of BioDesigner," Ph.D. Thesis, Department of Chemical Engineering, Massachusetts Institute of Technology, Cambridge, Massachusetts, 1990.

2. Valle-Riestra, J. Frank, *Project Evaluation in the Chemical Process Industries*, McGraw-Hill Book Company, 1983.

3. Venkataramani, E.S., House, M.J. and Bacher, S., "AN EXPERT SYSTEM BASED ENVIRONMENTAL ASSESSMENT SYSTEM (EASY)," Merck & Co., Inc., P.O. Box 2000, Rahway, NJ 07065-0900, 1990.

6

HAZARDOUS WASTE CLEANUP AT FORMER DEPARTMENT OF DEFENSE SITES

T. Julian Chu, Thomas J. Wash, and
Col. Michael H. Fellows
U.S. Army Corps of Engineers
Washington, D.C.

INTRODUCTION

Protecting the environment and natural resources for present and future generations is an integral part of all Department of Defense (DOD) missions. The Defense Environmental Restoration Program (DERP) reflects DOD's commitment to correcting environmental damage that resulted from past practices. This program currently includes: (1) the Installation Restoration Program (IRP), where potential contamination at DOD installations and formerly owned or used properties are investigated and, as necessary, site cleanups are conducted; and (2) Other Hazardous Waste Operations, through which research, development, and demonstration programs aimed at improving remediation technology and reducing DOD waste generation rates are conducted. The focus of this paper is to provide an overview of the Formerly Used Defense Sites (FUDS) Program (also known as DERP-FUDS), which is a subprogram of IRP under DERP.

PROGRAM DEVELOPMENT

The antecedents of the DERP program evolved out of agency decisions within DOD in the early seventies. In 1974, DOD directed the U.S. Army Corps of Engineers to conduct a study to determine the extent of environmental consequences of abandoned military debris on Federal lands in Alaska. A draft environmental impact statement was prepared for a proposal to remove and dispose of debris and obsolete buildings. In 1975, DOD launched a pilot IRP to respond to known environmental contamination at several Army installations. The Army IRP was subsequently extended throughout DOD in 1976.

These activities were originally designed to primarily prevent contamination from leaving the boundaries of DOD sites. Cleanup was not required unless a site was going to leave DOD ownership or be used in another DOD mission.

In the early eighties, Congressional concern over abandoned military buildings and debris in Alaska dovetailed with its concern over releases of hazardous substances from Federal facilities, and the foundation of the DOD environmental restoration program was laid. Soon after the passage of the Comprehensive Environmental Response, Compensation, and Liability Act (CERCLA) in December 1980, the President delegated to DOD the authority to clean up hazardous substances released from DOD facilities. Thus, actions of both the President and Congress changed the emphasis of DOD's environmental restoration program. In December 1983, the Defense Appropriations Act (Public Law 98-212) provided one-year funding for cleanup of hazardous substances released from DOD sites as well as removal of unsafe or unsightly DOD buildings and debris. The Act also initiated environmental restoration activities at sites formerly used by DOD.

The line-item appropriations were continued in 1985 and 1986; however, buildings and debris eligible for removal were restricted from "unsightly" to "unsafe" and such removal activities were limited to former DOD properties currently owned by state or local governments or native corporations in Alaska. In October 1986, Congress passed the Superfund Amendments and Reauthorization Act (SARA) which authorized the Secretary of Defense to carry out the DERP under his jurisdiction and established a new transfer account to be known as the Defense Environmental Restoration Account (DERA).

The DOD role in DERP is to provide centralized policy, provide consistency, and manage the overall program. Execution of the program at each active installation is left to each DOD component. At former installations or formerly used defense sites, execution of the program has been delegated to the U.S. Army Corps of Engineers. Therefore, the Corps has become the chief executor for environmental restoration activities at Army installations and former DOD sites. Additionally, the Corps has been supporting the DERP-IRP activities of other DOD components such as the Air Force and the Defense Logistic Agency.

Funding appropriated to DERP has increased steadily since the program's inception and it continues to be the largest component of the Corps environmental restoration mission. Beginning in 1989, the Corps has been assigned the responsibility for environmental assessment and restoration activities at Army sites scheduled for base realignment and closure. The specific experience of real estate, engineering, and construction in various environmental programs across the nation has enabled the Corps to provide support to the U.S. Environmental Protection Agency (EPA) Superfund program and to other Federal agencies in meeting the

Hazardous Waste Cleanup 61

challenges of protecting and restoring our natural and cultural resources. The Corps' overall environmental restoration program has grown from $250 million in 1988 to more than $1 billion in 1992.

PROGRAM OBJECTIVES AND EXECUTION PROCESS

Section 211 of SARA lists three "goals" of DERP: (1) identification, investigation, research and development, and cleanup of contamination from hazardous substances, pollutants, and contaminants; (2) correction of other environmental damage, such as detection and disposal of unexploded ordnance, which creates an eminent and substantial endangerment to the public health or welfare or to the environment; and (3) demolition and removal of unsafe buildings and structures, including buildings and structures of DOD at sites formerly used by or under the jurisdiction of the Secretary of Defense. Because these are program goals rather than requirements, DOD retains discretion to prioritize and carry out activities among the three categories of environmental damage.

Environmental restoration activities at FUDS conform to the requirements of the National Oil and Hazardous Substances Pollution Contingency Plan of CERCLA. The DERP-FUDS program has three major phases: inventory, study, and removal/remediation. The inventory phase consists of site identification; real estate search to verify previous DOD usage; and preliminary assessment (PA) to determine site and project eligibilities. The study phase includes site inspection (SI), if necessary, to confirm the contamination; engineering evaluation and cost analysis (EE/CA) for a removal project; remedial investigation and feasibility study (RI/FS) for a remedial project; and/or litigation, negotiation, and settlement with other parties relative to defining and resolving the DOD liability for a potentially responsible party (PRP) site. The removal/remedial phase consists of engineering design and removal/remedial action. All program activities will be executed consistent with CERCLA and SARA as amended, and with applicable requirements of the Resource Conservation and Recovery Act.

DETERMINATION OF SITE ELIGIBILITY

SARA generally requires that a site must have been formerly owned by, leased to, possessed by, or otherwise under the jurisdiction of the Secretary of Defense at the time of actions leading to contamination by hazardous substances. SARA also requires that former sites, which were transferred by DOD components to the U.S. General Services Administration (GSA) after October 17, 1986, be cleaned up prior to transfer. However, if it is discovered later that a cleanup was incomplete, response activities may be conducted either by the DOD component or under the DERP-FUDS program.

In practice, it is necessary for DOD to provide more specific eligibility criteria. Under the DOD policy, eligible sites may include: (1) sites for which real property accountability previously rested with DOD irrespective of current ownership or responsibility for accountability within the Federal government; (2) sites used by DOD components under leases or other agreements; (3) sites occupied by DOD components over which significant controls were exercised without the benefit of formal real estate instruments or other agreements; (4) manufacturing facilities which were owned by DOD components and real property accountable to DOD but operated by contractors; (5) National Guard and reserve facilities where property accountability at one time rested with DOD; and (6) sites which were used for the disposal of DOD materials or wastes where installations responsible for the materials or wastes are inactive and no longer under DOD.

Former DOD sites under any of the following conditions are not considered eligible: (1) sites outside the 50 states and outside those districts, territories, commonwealths, and possessions over which the U.S. has jurisdiction; or (2) sites that were excess to the holding agency's requirements which have not been formally transferred to another Federal agency or disposed of by GSA.

After a possible FUDS is placed on the inventory list, the local Corps district will undertake a preliminary assessment to gather information on the extent of DOD ownership or use of the site and the subsequent use of the site by non-DOD entities. The findings and determination of eligibility will be included in an Inventory Project Report (INPR) prepared by the district.

DETERMINATION OF PROJECT ELIGIBILITY

For an eligible site, the INPR also recommends eligible projects, if any, for possible DERP-FUDS funding. An eligible project is one which DOD has or shares potential CERCLA responsibility for the hazardous condition on an eligible site. The eligible project must also meet the DOD policy considerations. There may be none, a single, or multiple projects proposed for an eligible site. Proposed projects are classified into three categories, stemming from the language of the Defense Appropriations Acts and the DERP statutory goals: (1) hazardous, toxic, and radioactive waste (HTRW) cleanup or containment activities; (2) ordnance and explosive waste (OEW) removal; and (3) building demolition and debris removal (BD/DR).

It is DOD policy that potential projects under any of the following conditions cannot not be proposed: (1) projects identified on sites where the current property owner refuses right of entry; (2) projects to remedy hazards which resulted from civil works activities rather than military activities; (3) projects initiated or completed by past or current owners; and (4) asbestos containment, removal, and/or disposal, except when part of and incidental to other eligible projects.

Ordinance and explosive waste (OEW), by policy, is addressed as a safety hazard. Clauses restricting property usage in real estate deeds, leases, and other legal instruments may absolve or limit DOD responsibility. Containerized hazardous and toxic waste (CON/HTRW) projects involving underground and/or aboveground storage tanks, which were beneficially used by owners subsequent to DOD usage, cannot be proposed unless there is evidence of a release from previous DOD operations. The BD/DR projects are only applicable to former DOD properties currently owned by state or local governments or native corporations in Alaska, and the buildings or structures were never beneficially used by owners subsequent to DOD usage. Former DOD buildings and structures must have been unsafe as a result of DOD usage and must have been inherently unsafe when the property was transferred or disposed of by GSA. Buildings and structures where the hazard is a result of neglect by an owner/grantee subsequent to DOD usage are ineligible.

PROGRAM EXECUTION STATUS

The initial emphasis of the DERP-FUDS program was placed on execution of the inventory phase and BD/DR, particularly in Alaska. The funding for the DERP-FUDS program, as indicated in Figure 1, has shown a steady growth since 1990, consistent with increasing national interest in the environment. The BD/DR activities were temporarily suspended between 1988 and 1990 because of higher priority HTRW and OEW projects, and then resumed in 1991.

Over time, a total of 7,150 former DOD sites have been identified through inventory efforts. The distribution of these sites among 50 states and U.S. territories is indicated in Table 1. The top five states are California (847 sites), Alaska (547 sites), Florida (518 sites), Hawaii (378 sites), and Texas (323 sites). As of August 1992, the Corps have initiated preliminary assessments (PAs) for 5,217 sites; of which, 4,111 (58% of the total) have been completed and 1,106 are underway. Of the completed 4,111 PAs, 2,865 sites or 70% have been determined to be eligible under the DERP-FUDS program and the remaining 1,246 sites are ineligible.

As previously explained, at each eligible site there may be none, a single, or multiple projects eligible for cleanup. Of the 2,865 eligible sites, environmental restoration projects are being conducted at 983 sites. The remaining 1,882 eligible sites either have all cleanup completed or require no cleanup actions. In total, 1,785 eligible projects have been identified through the PA process. These projects consist of 28% for hazardous, toxic, and radioactive waste (HTRW) cleanup; 37% for removal of underground and/or aboveground storage tanks (CON/HTRW); 13% for removal of ordnance and explosive wastes (OEW); 19% for building demolition and debris removal (BD/DR); and 3% combined projects.

As indicated in Table 2, eleven FUDS have been placed on EPA's National Priorities List (NPL). DOD shares the CERCLA liability with other potentially responsible parties at seven of these sites. Consistent with the DOD worst-first policy, the NPL sites have the highest priority for environmental restoration under the DERP-FUDS program. To help expedite cleanups, many sites are divided into Operable Units (OUs) and individual OUs at a site are allowed to progress independently through the study and remedial phases. At present, RI/FS activities are either underway or completed at all eleven NPL FUDS. In addition, remedial design has been initiated at six sites and remedial action is underway at one site.

Table 1. Inventory of Formerly Used Defense Sites

State	Number of Sites	State	Number of Sites
Alabama	130	Montana	106
Alaska	503	Nebraska	101
Arizona	180	Nevada	37
Arkansas	83	New Hampshire	26
California	783	New Jersey	119
Colorado	87	New Mexico	226
Connecticut	35	New York	261
Delaware	29	North Carolina	94
District of Columbia	21	North Dakota	66
Florida	393	Ohio	83
Georgia	101	Oklahoma	80
Hawaii	375	Oregon	109
Idaho	64	Pennsylvania	111
Illinois	74	Rhode Island	54
Indiana	68	South Carolina	98
Iowa	34	South Dakota	92
Kansas	121	Tennessee	63
Kentucky	21	Texas	325
Louisiana	66	Utah	30
Maine	90	Vermont	13
Maryland	74	Virginia	172
Massachusetts	221	Washington	282
Michigan	136	West Virginia	21
Minnesota	64	Wisconsin	60
Mississippi	132	Wyoming	69
Missouri	86	Territories and Others	102
		Total	6,771 sites

Table 2. Formally Used Defense Sites on National Priorities List

Site	State	Current Phase
Crab Orchard National Wildlife Refuge/Sangamo Electric Dump* (Illinois Ordance Plant)	IL	RD (OU1 & OU2) RI/FS (OU3)
Fisher-Calo** (Kingsbury Ordance Plant)	IN	RD
Hastings Groundwater Contamination (Blaine Naval Ammunition Depot)	NE	RD (OU1) RI/FS (OUs 2 to 4)
Malta Rocket Fuel Area* (Multa Test Station)	NY	RI/FS
Marathon Battery Corporation* (Cold Spring Battery Plant)	NY	RD
Nebrasa Ordnance Plant	NE	RI/FS
New Hanover County Airport Burn Pit*	NC	RI/FS
Morgantown Ordnance Works*	WV	RD (OU1) RI/FS (OU2)
Phoenix-Goodyear Airport* (Litchfield Park Naval Air Field)	AZ	RD
Weldon Spring Ordnance Works*	MO	RI/FS
West Virginia Ordnance Works	WV	RA

* Principally Responsible Party project
** No significant contribution of DoD contaminants
RI/FS: Remedial design
RA: Remedial Action

Table 3. Special Contracts Used for Environmental Restoration Activities

Contract Name	Contract Type	Pricing
Large-Dollar Architect-Engineer Indefinite Delivery (IDT)	Fixed Price or Cost Plus Award/Fixed Fee Delivery Orders (DOs)	More than $0.4M per year (one basic year with two option years)
Rapid Response IDT	Fixed Price or Cost Plus Award Fee DOs	$50M per year (one basic year with four option years)
Preplaced Remedial Action IDT	Fixed Price or Cost Plus Award Fee DOs	$50M per year (One basic year with four option years)
Underground Storage Tank Removal IDT	Fixed Unit Price	$1.5M per year (one year)
Professional Service IDT Requirements	Fixed Price or Cost Plus Award/Fixed Fee DOs	$5M per year (one basic year with one option year)

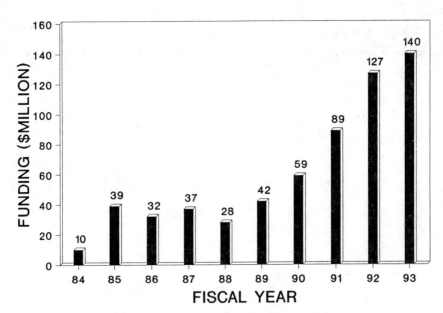

Figure 1 DERP-FUDS program funding.

Since 1990 when program funding began to significantly increase, the Corps has been accelerating the inventory phase or PA activities. The current goal is to complete all PAs by September 1994, two years ahead of the original schedule established in 1986.

MANAGEMENT APPROACH

To cope with the growing environmental restoration workload, the Corps has been constantly enhancing its capability to better manage and execute all programs including DERP-FUDS. Every geographic division of the Corps now has established at least one HTRW design district capable of executing the study phase and remedial design of HTRW projects. In addition, there are 32 local districts across the nation who are able to perform CON/HTRW and BD/DR projects as well as removal and remedial actions.

To further strengthen the quality assurance and quality control of projects executed by various divisions and districts, the Corps has established the HTRW Mandatory Center of Expertise (MCX) at the Missouri River Division and the OEW MCX and Design Center at the Huntsville Division. The MCX roles are to maintain state-of-the-art technical capabilities and to provide technical assistance to other Corps divisions and districts. Additionally, as the OEW design center, the Huntsville Division executes site inspection through removal design of OEW projects for geographic divisions.

RELIANCE ON PRIVATE CONTRACTORS

Although highly specialized environmental expertise and experience exist in many divisions and districts, the Corps conducts less than 20 percent of the DERP-FUDS workload with in-house assets. The vast majority of work has been and will continue to be accomplished through contracts with the private sector. Because of the complex and urgent nature of contracting requirements for many DERP-FUDS projects, the Corps has to consider both traditional and non-traditional contracts. The traditional firm-fixed-price contract is normally used when reasonable design certainty and prices can be established and the perceived financial risks are minimal. Non-traditional contracts are used when project requirements are not fully defined and/or response requirements are of such urgency that there is no sufficient time to acquire the needed activity through standard contracting procedures. Table 3 indicates those non-traditional contracts established and used at the Corps districts for the DERP-FUDS program.

To address contractor liability issues and to ensure the quality of services, DOD is evaluating measures for providing equitable risk sharing between the government and private contractors.

COORDINATION WITH REGULATORY AGENCIES

The Corps works closely with Federal, state, and local regulatory agencies in the execution of all projects. The coordination begins at the inventory phase. For instance, as soon as an INPR is approved, the Corps district notifies all current landowners and regulatory agencies of proposed actions. The approved INPR is also available, upon request, to the general public. Regulatory agencies are provided opportunities to fully participate in the evaluation and oversight of the progress of the subsequent study and removal/remedial phases.

In 1988, DOD and EPA completed negotiation of the Interagency Agreement (IAG) model language for NPL sites. Subsequent guidance was issued concerning the state role in the IAG progress. The Corps routinely enters into IAGs during the RI/FS phase to establish good working relationships with both EPA and state agencies.

To further involve state regulatory authorities in the DERP-FUDS program, DOD has recently included FUDS in the Defense and State Memorandum of Agreement (DSMOA) program. Based on the criteria established under the DSMOA program, DOD reimburses states and territories for regulatory services in direct support of DERP-FUDS projects conducted by DOD. It is believed that the DOD initiative will expedite the regulatory review process and develop a new partnership with states and territories for addressing problems at eligible FUDS.

FUTURE PERSPECTIVE

The total cost for the DERP-FUDS program, estimated in 1991, is about $3 billion. As of the end of fiscal year 1992, $460 million has been expended with emphasis on the inventory phase and initiation of remedial activities at NPL sites. Future program funding will support the completion of all program phases, from preliminary assessment through removal/remedial actions as well as operation and maintenance of remedial systems for up to ten years. It is expected that the annual funding for the DERP-FUDS program will continue to increase in the nineties, because of the transition from the inventory and study phases to the more costly removal/remedial phase. DOD and the U.S. Army Corps of Engineers will review the total program cost estimate periodically as the program matures and more information becomes available.

Increasing the pace at which FUDS cleanups are conducted entails many new challenges. DOD and the Corps are committed to working closely with Congress, environmental organizations and regulatory authorities, private industries, and the general public in meeting these challenges.

REFERENCES

1. Public Law 98-212, "Defense Appropriations Act," 1983.

2. Public Law 98-473, "Defense Appropriations Act," 1984.

3. Public Law 99-190, "Defense Appropriations Act," 1985.

4. Public Law 96-510, "Comprehensive Environmental Response, Compensation, and Liability Act," Section 120, 1980.

5. Public Law 99-499, "Superfund Amendments Reauthorization Act," Section 211, 1986.

6. Public Law 94-580, "Resource Conservation and Recovery Act," Sections 3004, 3008, and 9001 through 9010 (Subtitle I), 1976, as amended.

7. U.S. Department of Defense, "Annual Defense Environmental Restoration Program Report to Congress for Fiscal Year 1991," February 1992.

8. McCabe, M. A., "An Introduction to the Defense Environmental Restoration Program," Master Thesis, the National Law Center, George Washington University, Washington, DC, 1990.

7

A RAPID YET INEXPENSIVE METHOD TO MONITOR AIR QUALITY IN ASPHALT CONCRETE PLANTS

Namunu J. Meegoda and Yaogin Cchen
Department of Civil and Environmental Engineering
New Jersey Institute of Technology
Newark, NJ

Robert T. Mueller
State of New Jersey
Department of Environmental Protection and Energy
Trenton, NJ

INTRODUCTION

The higher emphasis on recycling of contaminated soils and waste products has forced the asphalt industry to incorporate such material in the production of asphalt concrete. When such materials are used to produce asphalt concrete, due to partial incineration and volatilization of organic compounds, there is higher VOC emissions. However, to comply with the new federal air quality regulations asphalt concrete plants are under constant scrutiny by the enforcing agencies to reduce the VOC emissions. Currently the VOC emissions in asphalt plants are monitored by drawing exhaust gases into field gas chromatographs mounted in mobile trucks. For each asphalt plant two technicians and the mobile truck need to be hired for two days, and hence this method is expensive.

In this paper a rapid, yet inexpensive air sampling and analysis system for the monitoring of volatile organic compounds, emitting from stacks of asphalt plants, is proposed. It is based on solid absorbent sampling tubes with thermal desorption followed by gas chromatographic analysis, where capillary column separations with cryogenic refocusing techniques are used. Air samples are collected using the solid absorbent, Tenax, in stainless gas collection tubes. Stripping of the adsorbed analytes from the sampling tube is accomplished by thermal desorption followed by gas chromatographic anaoClysis. This paper also describes the VOC emission test results from a regular asphalt concrete plant, and a plant which incorporates contaminated soils in the production of hot mix asphalt concrete.

The emission of volatile organic compounds (VOCs) from asphalt plants has been a real concern for USEPA. The bulk of the VOC emissions are due to the improper combustion of fuel used to heat the aggregates. Back in 1980, USEPA industrial environmental research laboratory (USEPA, 1981) developed a quantitative method to estimate emission of VOCs from drum-mix asphalt plants. A total of 67 VOC samples were collected in 26 test runs from five drum-mix asphalt plants in this study (USEPA, 1981). The sampling method was the proposed EPA method 25, modified to filter out particulate emissions. Results showed that VOC emission factors for drum-mix plants were of the order 0.1 to 0.4 pounds of VOCs (as carbon) per ton of asphalt concrete. It was also found that VOC emissions were independent of plant operating parameters, over the normal range of plant operations and within the limited scope of the statistical testing employed. It was also shown that wet scrubber reduced VOC emissions. Based on the above study the nationwide emission of VOCs from all drum-mix asphalt plants was estimated to be about 20,600 tons per year (USEPA, 1981).

1990 Amendments to the Clean Air Act

The 1990 amendments to the Clean Air Act, which is a major milestone in the evolution of environmental protection in the United States with a goal to reduce the air pollution by 56 billion pounds a year, i. e., 224 pounds for every man, woman, and child in the US. The reduction will come from cutting the emissions from several principal sources including the emissions from asphalt industry. Projected cost of implemention may run as high as $25 billion a year since the 1990 amendments significantly toughen Clean Air Act by mandating regulations to reduce acid rain, urban smog, air toxins and ozone-depleting chemicals over the next decade. Following are some of the major implications of the above act.

Reduction of urban smog: New provisions of the bill call for reductions in urban smog by areas that have not attained health-based ambient air quality standards. Controls will be imposed on emitters of 10 tons of hydrocarbons and nitrogen oxides in areas which the law defines to be extremely polluted such as Los Angles; 25 tons in areas which the law defines as severely polluted such as New York; and 50 tons in areas that the law defines as seriously polluted, such as Washington, D.C.

New standards for automobile tailpipes: New tailpipe standards will require automobiles to reduce nitrogen oxides by 60% and hydrocarbons by 30% from currently accepted standards. Target dates for accomplishment are between 1994 and 1998. The nine smoggiest cities in the U.S. will be required to sell reformulated gasolines in 1995, and cities with CO problems will be required to sell oxygenated fuels. Auto manufacturers will also be required to design and build vehicles for the California market that are capable of running on non-gasoline fuels.

Industrial emissions of toxic compounds: The new law will require industrial sources of 189 listed air toxins to install Maximum Achievable Control Technologies (MACT) by the year 2003. However utility emissions are exempted until, EPA determines the need for regulation. Municipal incinerators will be subject to these air emission controls and other requirements for monitoring emissions, training operators and issuing operating permits. Ash disposal provisions and waste recycling requirements were deleted from the law pending reauthorization of the RCRA, up for consideration in '91.

Emissions from utility industries: Acid rain controls include a cap on utility emissions of sulfur dioxide of 8.9 million tons a year by the year 2000, representing a 10 million ton reduction from 1980 levels. 111 of the dirtiest utility plants must account for the largest cuts during the first 5 years, and cleaner plants are responsible for later emission reductions. The law also sets up a system to provide pollution credits to "dirty" utilities in exchange for cuts in sulfur dioxide below their required limits.

One of the casualties of this new air quality act is the asphalt industry. On one hand the federal and state governments force the asphalt industry to recycle contaminated soils and waste products (i.e., tires, reclaimed asphalt pavements (RAP), etc.). When such material is used to produce asphalt concrete, due to partial incineration and volatilization of organic compounds, there is higher VOC emissions. However, to comply with the new federal air quality regulations, asphalt concrete plants are under constant scrutiny by the enforcing agencies to reduce the VOC emissions. Currently to comply with the air quality act the NJDEPE has proposed a 250ppm limit for VOC emissions from asphalt plants.

Furthermore, to continuously monitor the emissions NJDEPE suggest the installation of online monitoring devices at a cost of $50,000.00 and to interpret the data regular technicians. The on-line monitoring devices provide information on total VOCs, amount of water vapor and the solid emissions. Measuring of total VOCs is not good enough as it will not indicate the source of pollution (i.e., from the burner, from asphalt cement or from waste products) and also it will not indicate the toxic effect on individual compounds. Therefore, to address the above, a rapid but yet inexpensive method based on a broad spectrum approach using thermal desorption techniques was developed to monitor organic compounds emitted from the stacks of asphalt plants. The air sampling and analysis system is based on solid adsorbent sampling tubes and thermal desorption gas chromatographic analysis. After collection, the sampling tubes were put into a desorber, then the VOCs were twice condensed in cryogenic trap, to improve capillary column resolution. Before cryogenic refocusing, a purge step serves to remove any oxygen remaining in the tube. This eliminates the problem of the solid adsorbent reacting with the oxygen when heated, it also removes traces of water from the tube.

Experimental Procedure

Air samples from stacks of hot mixing asphalt plants were collected by drawing a volume of air through a 5/8" stainless steel tube packed with 1.5 g of 60/80 mesh Tenax TA using a air sampling pump. Samples were drawn at 500 ml/min. Figure 1 shows the schematics of extracting air samples from stacks. Prior to sample correction, the sampling flow rate was calibrated over a range including the rate to be used for sampling. The flow rate was checked before and after each sample collection. The Tenax cartridges were prepared for tests by baking them for 24 hours at 250°C under an inert gas purge (50 -100 ml/min). The collected samples in Tenax cartridges were stored under 4°C temperature and analyzed within 1 week after sampling.

Preparation of Tenax Cartridges

The following Procedure was used to prepare a 5/8" tube containing Tenax cartridges:

Prior to using the Tenax resin it was subjected to a series of solvent extraction and thermal treatment steps. The operation was conducted in an area where levels of volatile organic compounds (other than the extraction solvents used) were minimal. All glassware used in Tenax purification as well as cartridge materials were thoroughly cleaned by rinsing with water followed by rinsing with acetone and drying in an oven at 250°C. The bulk Tenax was placed in a glass extraction thimble and held in place with a plug of clean glasswool. The resin was then placed in the soxhlet extraction apparatus and sequentially extracted with methanol and then with pentane for 16-24 hours for each solvent for approximately 6 cycles/hour. Glasswool used for cartridge was cleaned in the same manner as that for Tenax. The extracted Tenax was placed immediately in an open glass dish and heated under an infrared lamp for two hours inside a hood. Care was exercised to avoid over heating of the Tenax by the infrared lamp. The Tenax was then placed in a vacuum oven (evacuated using a water aspirator) without heating for one hour. Then it was purged with an inert gas (helium or nitrogen) at a rate of 2-3 ml/minute to aid the removal of solvent vapors. Then the oven temperature was increased to 110°C, while maintaining the flow of inert gas for one hour. The oven temperature control was shut off and the oven was allowed to cool to room temperature. Before opening the oven, it was slightly pressurized with nitrogen to prevent contamination with ambient air. The Tenax was removed from the oven and sieved through a 40/60 mesh sieve (acetone rinsed and oven dried) into a clean glass vessel. If the Tenax is to be used later, then it was stored in a clean glass jar having a Teflon-lined screw cap and placed inside a desiccator.

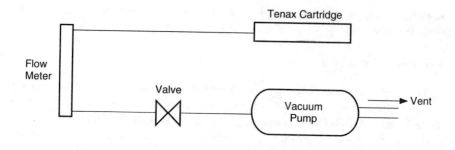

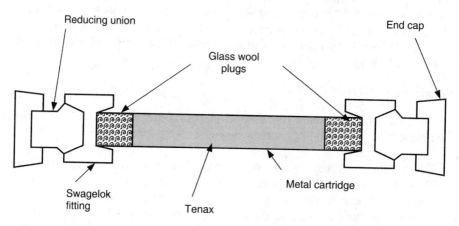

Figure 1   Schematics of the Air Sample Monitoring Setup.

The cartridge used for the monitoring of air was packed by placing a 0.5 to 1 cm long glasswool plug at the bottom of the cartridge and then filling the cartridge with Tenax up to approximately 1 cm from the top. Then a 0.5 to 1 cm long glasswool plug was placed on the top of the Tenax. Before each air quality monitoring test, all Tenax cartridges were thermally conditioned by heating for four hours at 270°C while purging with an inert gas (helium at 100-200 ml/min).

Calibration and Quantification

Collected air samples in Tenax traps were first desorbed using a Tekmar modal 5010 Automatic Desorber connected to a Varian 3400 GC with a flame ionization detector. Both systems were interfaced so as to automate the entire analysis. The desorption/detection conditions:

Prepurge:	5 min, at 10 ml/min
Desorb:	8 min 210°C, 10 ml/min
Cryotrap 1 :	-150°C
Cryotrap 2 :	-150°C
Transfer :	10 min, 210°C
Inject :	0.75 min, 210°C

The column used for GC/FID analysis was a cross-linked methyl silicone gum, 50m long, 0.2mm diameter and 0.5mm film thickness. Flow rates for the GC were:

hydrogen	=	30 ml/min,
air	=	30 ml/min,
carrier gas	=	1 ml/min,
make- up gas	=	30 ml/min.

The temperature program had an initial temperature of 15°C for 8 minutes, then it was programmed to 210°C at 4°C/min. Figure 2 shows the schematics of the desorber and the GC analyzer.

External standards were used to calculate the response factor of each compound every week. In this analysis, concentrations of the following eleven compounds were obtained Chloromethane, Dichloromethane, Hexane, Chloroform, 1,1,1-trichloroethane, Trichloroethylene, Benzene, Toluene, Perchloroethylene, p,m-Xylene, and o-Xylene. The process involved analysis of four calibration levels for each compound during a given day. The values of limit of detection (LOD) and limit of quantification (LOQ) were determined from the calibration curve. If the instrumental response was linear over concentration range of compounds (which was the case for the eleven compounds considered) a linear equation in the form $Y = A + B \times$, can be employed. Then limit of detection and limit of quantification were defined as:

$$LOD = A + 3.3\ S \qquad LOQ = A + 10\ S$$

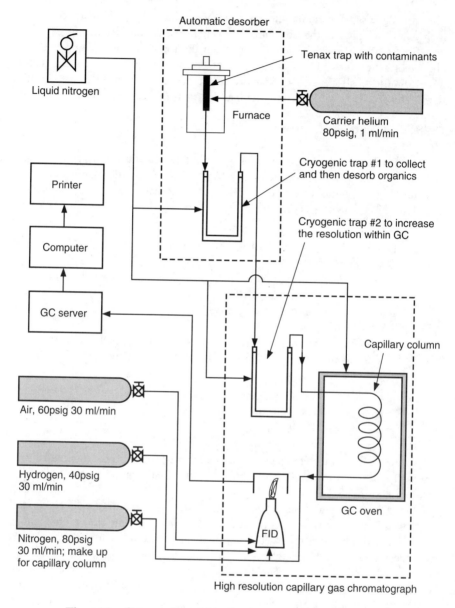

Figure 2 Schematics of Air Quality Analysis Setup.

where A is the intercept of the above equation, and S is the standard deviation of replicate determinations of the lowest concentration level. Table 1 shows the LOD and LOQ values for the eleven compounds.

In this experiment to determine the reproducibility, ten Tenax blank traps were injected with known concentration of standard and then desorbed them into the analytical system. The reproducibility can then be expressed by the coefficients of variation (CV) of the target compounds, which is the ratio of standard deviation to mean of repeated measurements. Table 1 also shows the CV values for ten compounds.

Breakthrough tests were also conducted. In break through tests backup cartridges (two cartridges in series) were used to measure breakthrough percentages of each compound. All backup cartridges contained less than 10 % of the amount of compounds in the front cartridges, except for benzene which had a value of 10.8 %.

Table 1 LOD, LOQ and CV Values for the Compounds Tested.

Compound	LOD	LOQ	CV
Chloromethane	0.540 ppb	1.80 ppb	20.5%
Dichloromethane	0.130 ppb	0.45 ppb	9.20%
Hexane	0.030 ppb	0.95 ppb	11.8%
Chloroform	0.440 ppb	1.50 ppb	9.80%
1,1,1-trichlorethane	0.220 ppb	0.73 ppb	10.5%
Trichloroethylene	0.048 ppb	0.16 ppb	21.4%
Benzene	0.069 ppb	0.23 ppb	N/A
Toluene	0.059 ppb	0.19 ppb	6.70%
Perchloroethylene	0.026 ppb	0.09 ppb	8.70%
p,m-Xylene	0.045 ppb	0.15 ppb	7.50%
o-Xylene	0.056 ppb	0.81 ppb	9.5%

The standard gas for calibration was prepared from a mixture of target compounds. The compounds were injected into an evacuated and clean 13 liter stainless steel cylinder with zero grade helium. The analysis of the standard was performed by Alphagaz, Edison, NJ.

The above method was also used to estimate the concentration of total volatile organic compounds. Since concentration of VOCs is the sum of all the organic compounds separated and detected in GC, the following procedure was used to estimate the concentration of non-methane VOCs. Since the temperature of the cryotrap was not low enough to condense methane, its concentration is not included in the results. After each analysis the total area (summation of all peak areas) from the GC was obtained. Another calibration of GC was performed by direct injection of propane to the FID. Various amounts of propane in helium at 10ppm were obtained from a pressurized cylinder into a gas-tight pressure-lock syringe and injected to GC by passing the capillary column. The volume of injection was recorded against the area from the gas chromatograph. The calibration factor was obtained by multiplying the calibration factor obtained in the propane test by the number of carbon atoms in propane. Hence the non-methane VOC concentration is expressed as parts per million of carbon or parts per million of carbon equivalent.

Field Tests

Air samples were obtained from two asphalt plants. One, a regular batch-mix asphalt concrete plant in New Jersey (Newark Asphalt); and the other a drum-mix plant in New Hampshire (Continental Paving Co. Inc.), which incorporated contaminated soils into hot mix asphalt. Both asphalt plants used heating oil as fuel. In each field test, three environmental samples were taken as controls. Six air samples were taken to determine the VOCs concentrations from the stack of Newark Asphalt plant. The results are shown in Table 2. Four air samples were taken to determine the VOCs concentrations from the stack of Continental Paving Co. plant. This plant uses petroleum contaminated soils in the production of asphalt concrete. The test results shown in Table 2 therefore, includes the contaminants from the burner and from contaminated soils. The time interval for each sample was 15 minutes. In this test Continental Paving Co. added 20.2% contaminated soil with a heating oil concentration of 2000ppm. The contaminated soil left the aggregate dryer with a non detectable heating oil concentration.

Discussion

Emission of volatile organic compounds from stacks of asphalt plants have become a subject of intense interest especially when adding waste products such as contaminated soil into hot mix asphalt concrete. A new method is proposed to determine the concentrations of these compounds. It is accomplished by collecting air samples in Tenax cartridges, then thermal desorption, and GC separation. This method is sensitive and relatively precise. The reproducibility of this method was below $\pm 22\%$ as determined by the coefficient of variation, and hence it is applicable to monitor a wide variety of volatile compounds emitted from asphalt plants.

In regular asphalt plants, some toxic organic compounds are emitted from the stack. For individual compounds, the concentrations did not approach levels that cause severe health problems, however the asphalt plant may be emitting chemical

Table 2 Average Concentrations of Target Chemicals from Asphalt Plants

Compound	Neward Asphalt		Continental Paving Co.	
	Stack Stable (ppbv)	Environmental (ppbv)	Stack Sample (ppbv)	Environmental (ppbv)
Chloromethane	162.83	7.34	1,056.00	17.14
Dichloromethane	109.47	3.21	481.37	2.14
Hexane	149.39	N/D	283.45	N/D
Chloroform	173.66	2.45	1,092.10	N/D
1,1,1-trichloroethane	81.76	1.64	274.72	2.28
Trichloroethylene	133.06	N/D	364.88	364.80
Benzene	131.54	2.10	542.46	N/D
Toluene	24.19	1.48	366.16	4.84
Perchloroethylene	113.06	N/D	292.14	3.96
p,m-Xylene	16.82	1.04	88.23	1.96
o-Xylene	5.87	0.84	99.13	2.46
VOC	26,740.00	180.80	94,750.00	2,650.00

mixtures consisting of several compounds with similar health problems, i.e., carcinogenic and/or mutagenic. Cumulative health effects of these compounds were not assessed. Exposure to complex mixtures probably will be most significant when asphalt is mixed with large amounts of petroleum contaminated soil with high contaminant concentrations.

Processing contaminated soil in hot mix asphalt is a new recycle technique. This method has several advantages: 1. It offers partial incineration for the organic compounds. 2. There are some solidification effects for the organic compounds. 3. Petroleum in the contaminated soil can be use a part of fuel in the processing (Czarnecki, 1988). The most important aspect about this process is how much of the volatile compounds were driven off during the mixing and resulted in VOC emissions. This study showed higher VOC emissions in the hot mix asphalt plant that added contaminated soils. Qualitatively, for individual compounds, the increase was about 1 to 8 times, so the problem of increasing VOCs emission in the processing can not be neglected. It is suggested increasing combustion temperature to totally destroy organic compounds or applying other VOCs control equipment such as after burners to reduce VOC emissions.

REFERENCES

1. Czarnecki, R. C., "Making Use of Contaminated Soil," *Civil Engineering.* (December):, 72, 1988.

2. Chan, C. C., Vainer, L. and Martin, J. W., "Determination of Organic Contaminants in Residential Indoor Air Using an Adsorption-Thermal Desorption Technique," *Journal of Air and Waste Management Association.* 40(1):62, 1990.

3. USEPA, "Emission of Volatile Organic Compounds from Drum Mix Asphalt Plants," EPA-600/2-81-026, U.S. Environmental Protection Agency, Washington, DC, February, 1981.

4. Kostecki, P. T. and Calabrese, E. J., "Soil Contaminated by Motor Fuels: Research Activities and Perspectives of the American Petroleum Institute," *Petroleum Contaminated Soils.* (1):13-19, 1989.

5. Shen, T. T, Nelson,T. P and Schmidt, C. E., "Assessment and Control of VOC Emissions from Waste Disposal," *Critical Reviews in Environmental Control.* 20 (issue 1):43, 1990.

8

CONTROL OF VOLATILE ORGANIC COMPOUND EMISSIONS FROM WASTE TREATMENT AND DISPOSAL FACILITIES

Thomas T. Shen
New York State Department of Environmental Conservation
Albany, NY

INTRODUCTION

Today's rapidly changing industrial technologies, products, and practices frequently carry with them an increasing generation of organic materials in their wastes. Organic materials are of particular interest to environmental professionals in the field of waste management because emission of organic compounds is one of the newly recognized environmental problems that cause public health and ecological implications. Concerns with volatile organic compound (VOC) emissions from waste treatment, storage, and disposal facilities (TSDFs) are increasing because of the emission quantity and toxicity and/or carcinogenicity. VOC emission also has a potential for contributing to ambient ozone formation.

Studies have revealed that even very stable chlorinated hydrocarbons with very low vapor pressure, such as polychlorobiphenols (PCBs) and pesticides, do volatilize (Shen, 1980). Field monitoring data in the 1970s has shown that PCB concentrations near the PCB dump sites and certain contaminated dredge spoil sites were fairly high in the ambient air and in vegetation along the Upper Hudson River of New York State (Shen and Tofflemire, 1979). Thus, VOC emissions from TSDFs are considered significant.

This paper describes the current control strategies to reduce VOC emissions from hazardous waste treatment, storage and disposal facilities (TSDFs). The control strategies include (1) control of TSDF sources by establishment of regulatory requirements and standards such as construction and operating permits, design and operating standards, manifest system, tax credits encouraging the use of waste reduction methods; (2) installation of waste pretreatment and treatment systems; and (3) application of preventive measures that reduce VOCs in waste through physical and procedural means such as waste de-volatilization by sorption or biodegradation, segregation of wastes according to volatility, and also by waste reduction, recovery and reuse at the process plant.

REGULATORY CONTROL

The legal structures to control hazardous pollutants such as VOCs from waste management facilities usually represent a compromise between the public health and welfare on the one hand, and technical, economic, and political factors on the other. The manner in which this compromise is achieved largely depends on the situation existing in each state or regional jurisdiction.

Under Subtitle C of the Resource and Conservation and Recovery Act (RCRA) as amended, U.S. Environmental Protection Agency (EPA) is required to issue regulations setting forth a complete "cradle-to-grave" system for the management of hazardous waste. On May 19, 1990, EPA promulgated several regulations: a regulation identifying hazardous waste (Part 261); regulations governing the issuance of permits and the authorization of States to implement a hazardous waste program (Parts 122-124); interim status standards applicable to owners and operators of "existing" hazardous waste treatment, storage and disposal facilities (Part 265); and administrative, non-technical standards that are to be used in issuing permits to owners and operators of treatment, storage and disposal facilities (Part 264).

EPA implements RCRA Section 3004(n) using a phased approach for TSDF emission sources. In the first phase, EPA regulates air emissions by developing standards for certain hazardous waste treatment processes. The first phase has been completed with standards to reduce organic emissions vented from treating hazardous wastes by distillation, fractionation, thin-film evaporation, solvent extraction, steam stripping, as well as air used for hazardous waste management processes. The second phase involves regulating organic emissions from TSDF tanks, surface impoundments, containers, and certain miscellaneous units. In both the first and second phases, standards are developed that control organic emissions as a class rather than pollutant-by-pollutant. With the third phase, EPA may issue regulations to address the risk remaining after promulgation of the first two phases. If regulations are necessary in the third phase, EPA will likely pose controls on individual toxic pollutants.

Standards proposed by EPA on July 22, 1991, require that specific organic emission controls be installed and operated on tanks, surface impoundments, and containers into which is placed hazardous waste having a volatile organic concentration equal to or greater than 500 ppm by weight (ppmw). The EPA encourages owners and operators to reduce the volatile organic concentration for a specific waste to a level less than 500 ppmw through pollution prevention adjustments and other engineering techniques.

The EPA has already developed RCRA standards to control VOC emissions from hazardous waste incinerators, TSDF tanks, surface impoundments and containers, treatment unit process vents and equipment leaks (Subparts AA and BB in 40 CFR 264 and 265). The implementation schedules for existing facilities with permits are listed in Federal Regulation, July 22, 1991.

The Clean Air Act (CAA) Amendments of 1990 require that each state with one or more areas of ozone nonattainment impose control requirements on VOCs. These control requirements must be sufficient to allow the nonattainment areas to comply with the ambient standard on an aggressive schedule. Many TSDF sites located in ozone nonattainment areas that previously were considered minor sources of VOCs will now be subject to new controls (U.S. EPA, 1991).

The CAA Amendments of 1990 introduce an operating program modelled after a similar program under the federal National Elimination Discharge System (NPDES) law. The purpose of the operating permit program is to ensure compliance with all applicable requirements of the CAA and enhance EPA's ability to enforce the Act. This program, in many ways the most important procedural reform contained in the new law, will greatly strengthen enforcement of the CAA.

Sources are required to obtain permits by the effective date of the permit program. The permit must include a compliance plan, enforceable emission limits and standards, and semi-annual reports on the results of required monitoring. In addition, permits must specify inspection, compliance certification and reporting requirements.

WASTE TREATMENT SYSTEMS

Waste treatment processes may be divided into pretreatment and treatment processes. Pretreatment removes or destroys organics in the waste and reduces organic emissions from treatment, storage, and disposal units handling waste without the need to use add-on emission controls for each of these units. For example, if a waste is pretreated by steam stripping to remove organics, the quantity of organic emissions from all activities that subsequently manage the waste will be reduced relative to the quantity of emissions that would have occurred without pretreatment because of the reduction in the volatile organic content of the waste. Similarly, if a waste is incinerated, then there are no additional waste handling steps, and thus there are no subsequent waste management units that are sources of organic emissions.

Hazardous waste treatment systems are installed to collect, remove or destroy pollutants, including VOCs in the waste. These waste collection and treatment units consist of covers and enclosures, as well as control devices.

Covers and Enclosures

Covers or enclosures reduce organic emissions by suppressing the generation and loss of vapors containing the organics. Appropriate types of covers include fixed roofs, internal floating roofs, and external floating roofs for tanks; covers for containers; and floating synthetic membranes for surface impoundments. Enclosures are structures erected over the entire waste management unit such as an air-supported structure over a surface impoundment or an enclosed building over a drum handling and storage area. However, enclosures are not suitable for organic emissions control without being vented through a control device.

Control Devices

A variety of control devices are available that are capable of achieving high organic emission control efficiencies. Applicability of a specific type of control device to control VOC emissions from TSDFs depends on the size of the facility and the characteristics of the organic vapor stream vented from the facility.

Control devices may be divided into two types: removal and destruction. Organic removal devices extract the organics from the gas stream and recover the organics for potential recycling or reuse. Organic destruction devices destroy the organics in the gas stream by oxidation of the organic compounds, primarily to carbon dioxide and water vapor. A closed vent system to convey the organic vapors from the covered or enclosed waste management unit to a control device is required. The vent system consists of piping, connections and, in some cases, a flow-inducing device to transport the vapor stream.

Organic Removal Devices

Adsorption, condensation, or absorption processes can be used to extract the organics from a gas stream. Considering organic vapor stream characteristics, the organic removal devices most likely to be used for TSDF waste management units are carbon adsorbers and condensers.

Carbon adsorption is the process by which organic molecules in a gas stream are retained on the surface of carbon particles. The gas stream is passed through a bed of carbon particles that have been processed or "activated" to have a very porous structure. However, activated carbon has a finite capacity for adsorbing the organics. When the carbon becomes saturated, there is no further organic emission control because all of the organic vapors pass through the carbon bed. At this point, the spent carbon must be regenerated or replaced with fresh carbon before organic emission control can resume.

Two types of carbon adsorption systems most frequently used for organic emission control are fixed-bed carbon adsorbers and carbon canisters. Fixed-bed carbon adsorbers are used for controlling organic vapor streams with flow rates ranging from 30 to over 3,000 cubic meters per minute.

In contrast to a fixed-bed carbon adsorber, a carbon canister is a very simple device consisting of a drum filled with activated carbon and fitted with inlet and outlet pipes. Use of carbon canisters is limited to controlling organic emissions from TSDF waste management units venting vapor streams with intermittent or low continuous flow rates such as storage tanks or quiescent treatment tanks.

The design of a carbon adsorption system depends on the inlet gas stream characteristics including organic composition and concentrations, flow rate, and temperature. Good carbon adsorber performance requires that: (1) The adsorber is charged with an adequate quantity of high-quality activated carbon; (2) the gas stream receives appropriate preconditioning (e.g., cooling, filtering) before entering the carbon bed; and (3) the carbon beds are regenerated before saturation occurs.

Condensers convert organic gases or vapors to liquid form by either lowering the temperature or increasing the pressure. For TSDF organic emission control applications, surface condensers are most likely to be used. Surface condensers most often consist of a shell-and-tube type heat exchanger. The organic vapor stream flows into a cylindrical shell and condenses on the outer surface of tubes that are chilled by a coolant flowing inside the tubes. The coolant need depends on the saturation temperature or dewpoint of the particular organic compounds in the gas stream. The condensed organic liquids are pumped to a tank for recycling or reuse.

The performance of a condenser is dependent upon the organic composition and concentration in the gas stream as well as the condenser operating temperature. Condensation can be an effective control device for gas streams having high concentrations of organic compounds with high-boiling points. However, condensation is not effective for gas streams containing low organic concentrations or composed primarily of low-boiling point organics because the organics cannot be readily condensed at normal condenser operating temperatures.

Organic Destruction Devices

Organic destruction devices include thermal vapor incinerators, catalytic incinerators, flares, boilers, or process heaters. Because of restrictions in applicability, a particular type of combustion device may not be suitable for controlling certain organic vapor streams vented from covered or enclosed TSDF waste facilities.

Thermal vapor incineration is a controlled oxidation process that occurs in an enclosed chamber. The organic destruction efficiency for a thermal vapor incinerator is primarily a function of combustion zone temperature, the period of time the organics remain in the combustion zone, and the degree of turbulent mixing in the combustion zone. When designed and operated to achieve the proper mix of combustion zone temperature, residence time, and turbulence, thermal vapor incinerators can achieve organic destruction efficiencies of 98 percent and higher for all types of organic vapor streams.

The performance of a thermal vapor incinerator is affected by the heating value of the organic vapor stream to be controlled. Concentrated organic vapor streams normally have sufficient heating value to sustain combustion. However, diluted organic vapor streams can be vented from TSDFs to an incinerator. Consequently, the continuous addition of a supplemental fuel to boost the heating value of these vapor streams is required in order to maintain combustion temperatures in the range necessary for 98 percent organic destruction efficiency. Thus, use of thermal vapor incinerators to control dilute or variable organic vapor streams, may require substantial fuel consumption.

Catalytic vapor incineration is essentially a flameless combustion process that can be used to control certain types of organic vapor streams. The organic vapor stream is passed through a metal or alloy-based catalyst bed that promotes organic oxidation reactions at temperatures in the range of 320 to 380 degrees C. Organic destruction efficiencies of 98 percent or more can be obtained by using catalytic incinerators with the appropriate catalyst bed volume to gas flow rate for certain organic vapor streams.

The applicability of catalytic incineration to controlling organic vapor streams is restricted to fewer organic vapor stream compositions and concentrations than can be controlled by incineration, because catalysts are very susceptible to rapid deactiviation by halogens or sulfur. Thus, catalytic vapor incineration is not suitable for organic vapor streams containing halogen or sulfur compounds.

Unlike vapor incinerators, flares are open combustion devices. The ambient air surrounding the flare provides the oxygen needed for combustion. A natural-gas-fired pilot burner ignites the organic vapor stream. Steam- or air-assisted flares can achieve an organic destruction efficiency of at least 98 percent on organic vapor streams having a heat content greater than 11 megajoules per cubic meter (300 Btu/cu ft). Flares are not suitable for use on organic vapor streams that contain halogens or sulfur compounds because the acid gases formed from these compounds during combustion cause severe corrosion and excess wear of the flare tips.

An existing industrial boiler or process heater can also be used for organic vapor destruction. The organic vapor stream is either premixed with a gaseous fuel and fired using the existing burner configuration, or fired separately through a special burner or burners that are retrofitted to the combustion unit. Industrial boilers and process heaters can destroy organic vapors with 98 percent efficiency, but they are suitable for controlling only organic vapor streams that do not impair the combustion device's performance or reliability.

PREVENTIVE MEASURE

Ample quantities of literature have been published about the benefits of waste reduction at the source, with the logic that, if less VOCs in the waste are generated, less VOCs need to be collected, treated, stored, and destroyed. VOC emissions can be significantly reduced if organic compounds in wastes can be recovered and reused, such as the types and amounts of solvents used in various industrial processes. Significant opportunities also exist for industry to make commercially efficient investment in processes that reduce VOCs at the source by:

- Improvement in housekeeping practices
- On-site recycling and reuse
- Equipment or technology modification
- Process or procedure modification
- Substitution of raw materials
- Improvement in education and training

Organic-contaminated wastes can be reduced effectively by in-plant practices. A typical example is minimizing waste-solvents. Solvents are essential to industry and business, both large and small; they perform important roles in surface coatings, dry cleaning, carriers for other materials in chemical manufacturing processes, and diluting or thinning paints and other products. The types of solvents usually found in wastewater are:

1. Chlorinated solvents: Methylene chloride, trichloroethylene, 1,1,1,-trichloroethane, tetrachloroethylene, chloroform, and carbon tetrachloride.
2. Oxygenated solvents: Acetone, methyl ethyl ketone, butyl-cellosolve and various alcohols, ethers, and amines.
3. Aromatic hydrocarbon solvents: Benzene, toluene, ethyl benzene, and xylene.
4. Fluorocarbons: Freon and chloroflouorocarbons.

Waste solvents can be successfully recovered from almost any industrial process where they are neither destroyed nor incorporated into the final product. However, the recovery of waste solvents during industrial processes is normally based on economic justification. Economic analysis of pollution costs in the past has not included liability costs of socio-economic and ecological damages. Recent events that have increased waste treatment and disposal costs, regulatory requirements, and concern over legal liabilities have also altered the waste generator's attitude towards the effort that should be expended to minimize waste generation.

In many instances, organic liquids used as solvents, lubricants, and cleaning agents can be replaced with aqueous (water-based) solutions. Aqueous alternatives have been developed for metal-working fluids that serve the dual function of lubrication and heat removal during machine cutting. Aqueous alternatives are also available for degreasing operations to replace organic process fluids. Such material substitutions are often motivated by the environmental and economic consequences of the use of the product as well as the management of the waste.

The regulatory system is an important factor in shaping industrial decisions to control waste before generation or after generation. The system should encourage industries to undertake waste prevention as an opportunity and challenge for pollution control. VOC emissions from TSDFs may be reduced through physical and procedural means before they are handled or treated. Sludges and organic wastes which readily generate gases and leachate are known to contribute a significant amount of VOCs. Thus, liquids containing organic compounds in waste landfills should be controlled due to their potential to contaminate not only ambient air but also groundwater. Wastes within the TSDFs can also be segregated to allow effective reductions in VOCs.

SUMMARY AND CONCLUSIONS

The release of volatile organic compounds from waste treatment, storage, and disposal facilities has a number of public health implications, many of which are unknown. VOC control strategies have installed traditionally add-on devices such as an adsorption or condensation unit to the venting systems of the TSDF's covers and enclosures. Recently, VOC controls have focused on the use of manifest systems, waste reduction reporting, toxic release inventories, and permit systems to provide steps and procedures designed at controlling potential toxic emissions including VOCs. Ultimately, the use of preventive measures will be the most cost-effective control strategy to minimize VOC emissions from TSDFs. Preventive measures commonly used include reducing and recycling VOC-bearing wastes at the sources of generation.

The three VOC control strategies are: (1) establishing of regulatory requirements; (2) installing waste treatment systems; and (3) implementing preventive measures. Each strategy has its application and they are mutually complementary. VOC emissions from TSDFs cannot be completely prevented; therefore, VOC treatment systems are needed to remove or destroy VOCs from waste streams that have not been prevented. The regulatory requirements ensure the proper design, operation of the waste treatment, storage, and disposal facilities. Regulatory requirements can also promote waste preventive measures such as waste reduction; VOC recycling, recovery and reuse through the use of various financial incentive approaches.

REFERENCES

1. Federal Register, 1991. Proposed Rules. Vol. 56, No. 140, pp.33505-33514, Monday, July 22, 1991.

2. Shen, T.T. and Tofflemire, J. 1979. "Volatilization of PCBs from Sediment and Water: Experimental and Field Data," Proceedings of the 11th Mid-Atlantic Industrial Waste Conference, University Park, Pennsylvania, July 16-17, 1989.

3. Shen, T.T., 1980. "Fugitive Gaseous Emissions from Land Disposal of Toxic Organic Wastes—Air Quality Impact and Their Control," Proceedings of the 5th International Clean Air Congress, Buenos Aires, Argentina, October 19-25, 1980.

4. Shen, T.T., 1982. "Estimation of Organic Compound Emissions from Waste Lagoons," J. APCA, Vol.32, pp.79-82, January 1982.

5. Shen, T.T., 1984. "Air Pollution Assessment: Toxic Emission from Hazardous Waste Lagoons and Landfills," Proceedings of the International Seminar on Environmental Impact Assessment at University of Aberdeen, Scotland, UK, July 8-21, 1984.

6. US EPA, Implementation Strategy for the Clean Air Act Amendments of 1990. Office of Air and Radiation, Washington, D.C., pp.13-40.

9

EMISSIONS DATA FROM A PYROXIDIZER[®] OPERATING ON REGULATED MEDICAL WASTE

Paul H. Kydd
Envimed Inc.
Rocky Hill, NJ 08553

H-M Chiang
Department of Chemistry and Chemical Engineering
New Jersey Institute of Technology
Newark, NJ

INTRODUCTION

The Model B-20 PYROXIDIZER with a capacity of 22 gallons has been operated over a three month period at the Atlantic City Medical Center - Mainland Division in Pomona, New Jersey. The operation demonstrated highly acceptable operability on actual medical waste. Measurements of air emissions indicated that it should be possible to operate within the permitted limits.

Tests were performed on hospital room waste, operating room waste, pathology lab waste and particularly on "sharps" from general hospital rooms and from the clinical laboratory operations of the hospital. Power consumption levels were measured and are approaching the objective of 1 kilowatt an hour per pound of waste processed.

ACKNOWLEDGEMENTS

The hospitality and cooperation of the management and staff of Atlantic City Medical Center are hereby gratefully acknowledged.

Atlantic Electric Co. provided financial support at an early stage of the development of the field test program.

Pan Stamus, Envimed's control specialist, assisted with the bulk of the runs during which his software performed virtually without modification. Bob Waszazak, Envimed's design and procurement specialist, assisted with the later runs, and supervised the maintenance and repair activities.

PYROXIDIZER

The PYROXIDIZER is a small self-contained device for the disposal of regulated medical waste. It functions by first pyrolyzing the charge of waste in the absence of air and oxidizing the resulting vapors completely to CO_2 and water vapor. The pyrolysis is followed by an oxidation phase in which the char residue is exposed to hot air and is completely combusted to a sterile ash. The device is shown in Figure 1. It is designed to pass through a standard 36-inch door and can be installed anywhere there is 220 volt power and a 4-inch diameter vent. The unit is completely self-contained with an on-board HCL scrubbing system and a microprocessor based control system which permits completely automatic operation with no operator intervention after the start.

The objectives of the field test, which is the subject of this report, were to operate the unit on actual medical waste, which can only be done at a health care facility; to determine the adequacy of the equipment and the automatic control system software; and to measure the emissions of various air pollutants and the characteristics of the ash residue.

MEASURING TRACE ORGANICS

The techniques which have been used for measuring trace organic materials in the exhaust were developed using a one-gallon capacity technology demonstrator in an Innovation Partnership program which was conducted at the Hazardous Substance Management Research Center at New Jersey Institute of Technology (NJIT) in the second half of 1991.

This program was funded by the Center, by Envimed and by the Electric Power Research Institute.

An air emissions permit for the field test was obtained from the New Jersey Department of Environmental Protection on August 19, 1991 and a contract covering the demonstration was signed with Atlantic City Medical Center on October 31, 1991. The prototype unit shown in Figure 1 was installed in the Atlantic City Medical Center - Mainland Division - boiler room on November 8, 1991, and the first run took place on November 20, 1991.

OPERATIONS SUMMARY

The first phase program included a total of 34 runs on the prototype unit in eight campaigns. The feed stocks for each of these campaigns are summarized below:

Tune-up on simulated waste
General patient room waste
Sharps - General
Sharps - Laboratory
Pathology Lab Waste (Body parts)
Sharps boxes in separable containers
Loose sharps in separable containers

Figure 1 Twenty-gallon PYROXIDIZER. Power input 10 KW, 208 V. Capacity 220 cu. ft. of waste per month (3 shifts, 90% utilization).

The results are summarized in Table 1 describing approximately how much material was charged in each campaign and providing the weight of the residue and the percent reduction in mass achieved.

CAMPAIGNS CONDUCTED

The first campaign was conducted on simulated waste to ensure that the unit was running properly following the move and installation. Three runs were made, the second of which was a continuation of a previous run on a large load of wet catalogs. The overall charge for the three runs was 37 pounds. The residue, which was accumulated in the pyrolyzer and not separately weighed for each run, totalled 4.6 pounds, and the mass reduction was 87.7%, largely due to the residue from run number two in which the paper products were pulverized but not completely oxidized. These runs nonetheless showed satisfactory operation, and the unit was ready for charging with actual medical waste.

Campaign number two consisted of runs four, five and six and was conducted on hospital room waste. Bags of waste coming through the normal collection system in the hospital were charged to the unit, selecting relatively lightly loaded bags for the initial tests and ending up with the most heavily loaded bag that could be pushed into the unit. These runs provided highly satisfactory results in terms of operating time and emissions, and the overall mass reduction was 91.7%. They gave a preliminary indication of the wide variety of waste to be encountered in seven tin cans apparently originally containing juice were found in the residue.

Campaign number three addressed the consumption of sharps boxes, plastic containers in which hypodermic syringes are discarded, along with other items which either have sharp edges contaminated with infectious material or can be broken exposing such edges, such as glassware. This material is truly hazardous, and its disposal is closely regulated. It is prescribed that sharps must be disposed of in rigid, puncture-proof containers, and these typically are polyethylene boxes with latching lids which come in all sizes from one quart up to 17 gallons in capacity.

Runs seven through eleven were performed with increasing loads of sharps boxes, which were collected in the normal course of the medical waste disposal routine in the hospital and consisted mostly of boxes from patient rooms, containing disposable hypodermic syringes, and also containing a number of metal objects such as tweezers, forceps, scissors, etc. Performance on these boxes was very satisfactory in terms of the time of disposal and the mass reduction achieved. The total mass for these five runs was 28 pounds, and the mass reduction was 85.6%, despite the fairly large amounts of metal and small amounts of glass.

In an effort to achieve higher loads in the PYROXIDIZER, and therefore better economics, the samples in campaign number four were focused on sharps boxes from the clinical lab, which tend to have a substantially higher density due to the presence of blood vials and culture dishes. Larger sharps boxes which are used to dispose of laboratory waste containing culture dishes were obtained from the

Atlantic City Medical Center - City Division, where most of the biological work is done. The maximum amount that could be charged to the unit in any one run was 14 pounds, and the results generally were very satisfactory, although still higher loads were desired.

Campaign number five was focused on pathology lab waste, and specifically body parts. A 25 pound charge was obtained from the Mainland Division laboratory in a 7 gallon plastic pail. This charge provided the only serious operating problem encountered during the field test program. The run exhibited a very long period during which moisture was being driven off from the sample, followed by an extremely rapid evolution of combustible vapor causing a shut down.

The run was completed on the following day on approximately 15% of the original charge with a significant but controllable exotherm. The residue consisted of 1.2 pounds of material for a 95% overall reduction in mass. Run number 22 was performed on the same type of 7 gallon pail with approximately three quarts of water in it, giving a negligible residue and a fully controllable exotherm. This proved that the residue and the exotherm observed in run 20 were primarily due to the body parts, which were later found to be almost entirely fat. Successful operation on large charges of this type of material will require blowers of higher capacity then were available in the prototype unit during this run.

Campaign number six was the first introduction of a liner which can be filled directly without the interposition of sharps boxes or other packaging.

During campaign number six, which was run with sharps boxes to compare the results using the liner to campaign number four without it, the operation of the PYROXIDIZER with the liner was essentially the same as without it, with minor adjustments in temperature setpoints for the various phases of the operation. The maximum load which could be accommodated was still limited to 18 pounds of sharps contained in a standard 8-gallon sharps box.

Campaign number seven was the first campaign during which loose sharps were loaded directly into the liner from the clinical laboratory. As expected far higher loads were obtainable with this method, and satisfactory operation was achieved with loads of approximately 30 pounds. Somewhat larger loads could be accommodated in the Model B-100 PYROXIDIZER, which has a 28-gallon capacity pyrolyzer, although, as demonstrated in run 31, some modifications to the operating software are required to accomplish the pyrolysis in a shorter time by increasing the average power and preventing premature shifting into oxidation mode with these very high mass but low heating value loads.

Campaign number eight was devoted to operating room waste in the liner. Run 30 was on actual waste from the operating room loaded into the container by operating room personnel, and the charge totalled 10 pounds, which is approximately the maximum that can be accommodated. The mass reduction was very satisfactory, as was the operation of the run, but the limited charge provides rather poor economics. Again, the larger volumes of the B-100 PYROXIDIZER would increase the charge by approximately 33%, and the faster heat up and cool down cycle of that unit will contribute to a more economic operation.

Considerable research was done at Atlantic City on methods of compacting waste, and a compaction technique was developed which avoids the problem of aerialization of pathogens, it was possible to achieve a density of approximately 8 pounds per cubic foot (one pound per gallon) versus the normal density of uncompacted hospital room and operating room waste of 4 pounds per cubic foot. In run 32 the 22 gallon PYROXIDIZER was filled to approximately 3/4 of its volumetric capacity with compacted simulated waste. The run was satisfactory, though rather slow.

Summarizing the results of the eight campaigns of the field test program, operability generally was excellent, except for large loads of body parts and very large loads of clinical lab waste containing massive amounts of glass. The availability of the unit during the Phase I Program was 98%. Only two days were required for maintenance and repair during the 90 day test period.

AIR EMISSIONS

One of the major objectives of the field test program was to develop data on air emissions with actual medical waste charges. The requirements of our air permit are shown in Appendix B. The primary requirements were: less than 50 parts per million HCL; less than 50 parts per million hourly average of carbon monoxide when oxygen is greater than 14%, which it always is in the PYROXIDIZER; particulates less than a 0.1% yield on charge or a concentration in the exhaust of 0.015 grains per dry standard cubic foot; and trace organics limited to less than 1.5×10^{-6} lb. per hour averaged over the 8 hour anticipated run duration, which works back to approximately 8 parts per billion in the exhaust.

The unit was instrumented to measure HCL by bubbling a sample of the exhaust gas through water with an indicator, allowing the HCL to be titrated. Particulates were measured by weighing a full flow filter which was a requirement of our air permit following each run, and then vacuuming the particulates off the filter and reweighing. Carbon monoxide was analyzed with an on-board electrochemical sensor, and oxygen was measured with a miniature fuel cell sensor. CO_2 was determined with a nondispersive IR instrument. Trace organics were determined by absorbing the sample of the gas on Tenax and analyzing by cryofocusing and gas chromatography at NJIT, using techniques which had been developed earlier.

The HCL scrubber was operational for runs 1 through 28, and during this entire period, despite the fact that charges which were known to contain polyvinylchloride were run, the indicator never changed color. The conclusion from this is that the HCL scrubber is highly effective in removing HCL from the gas.

The particulate results are shown in Table 2. It can be seen from the columns titled Yield and Loading that for small loads the results are somewhat marginal, but for the larger loads of the later runs the particulates are comfortably below the permitted level.

Carbon monoxide was recorded continuously during the course of each run, and selected results are shown in Appendix A. Carbon monoxide content in the exhaust is plotted in parts per million and time is in seconds. The permitted level is shown as a horizontal line based on calibrations with span gas containing 113 ppm CO, which are indicated by "Cal." Many of the runs show spikes in the carbon monoxide concentration which are encountered when the air flow changes in response to the combustible vapor flow. Since the permitted level is 50 parts per million as an hourly average, these spikes do not constitute a problem.

The only run which was shut down due to noncompliance with the permit, was run 31 with the largest load of glass blood vials during which the unit shifted into oxidation mode prematurely and CO emissions were clearly above the permit level for a significant period of time. This run was rerun as run 31a, which indicated that the pyrolysis had been completed at the end of run 31.

Trace organics are shown in Table 33 in which the major compound and the next two largest compounds are shown. The original data is shown in Appendix C.

Chloromethane was found in all of these runs in apparently large amounts, whether chlorine was present or not. This is regarded as an artifact and has been ignored. The other compounds which were calibrated were: dichloromethane, hexane, chloroform, trichloroethane, benzene, carbon tetrachloride, trichloroethylene, toluene, perchloroethylene, p- and m-xylenes and o-xylene. The duplicate analyses performed in runs 7, 33 and 34 give an indication of the precision of the method.

The results are very similar to those obtained earlier on the 1 gallon unit in that the most common major component is toluene with benzene and chlorinated species usually being present in substantially lower quantities. A few of these runs, particularly run 7, show high levels of trace organic materials, but as with the particulates, later runs typically showed trace organic concentrations which were at the permitted 8 ppb and below.

OBJECTIVES FOR PHASE II

The primary task and objective of Phase II is to conduct a compliance test of the air emissions from the Model B PYROXIDIZER operating on actual medical waste using EPA approved methods. This test will be done by an independent testing company for maximum credibility.

At the same time Envimed hopes to achieve better integration of the PYROXIDIZER into the day to day operations of the hospital. The initial field test has served to qualify the technology as operable, safe and environmentally acceptable on at least part of the hospital waste stream. It has pointed the way toward a highly economic application of the PYROXIDIZER in the disposal of the most hazardous fraction of the waste, namely sharps, chemotherapy waste and pathology waste. The Phase II operations will be concentrated on providing service to the hospital which achieves their objectives in an optimum way and identifies the waste handling techniques and any auxiliary hardware that is required for the PYROXIDIZER system to provide maximum value to the health care provider.

Table 1. Charges and Residues
Atlantic City Medical Center Demonstration

Campaign	Run No.	Waste Type	Charge (lb)	Residue (lb)	Reduction (%)
1	1–3	Sim. waste	37	4.6	87.7
2	4–6	Hospital room	18	1.7	91.7
3	7–11	Sharps boxes	28	4.0	85.6
4	12–14	Sharps boxes from lab	31	8.8	71.7
	15		12	2.6	78.3
	16		14	3.8	72
	17		13	2.2	82
	18		10	2.2	79
	19		14	3.8	74
5	20, 21	Pathology lab	25	1.2	95
	22	Pail and water	8	0.1	98
6	23	Sharps boxes in liner	4	1.0	75
	24		4	1.0	75
	25		6	1.75	71
	26		10	2.25	77
	27		15	5.75	61
	28		18	12.0	33
7	29	Loose sharps in liner	24.5	15.6	36
	31, 31a		49	28.0	42
	33		27.5	13.9	49
	34		30	15.0	50
8	30	O.R. waste	10	0.7	93
	32	compacted O.R.	14	1.2	91

Table 2. Particulates

Run No.	Filter (grams)	Charge (lb-oz)	Charge (kg)	Yield	Loading (gr/DSCF*)
8	3.5	6-3	2.81	0.0012	0.0081
9	2.12	6-11	3.03	0.0007	0.0048
10	4.0	5-5	2.41	0.0016	0.0092
11	4.26	6-6	2.89	0.0015	0.0098
12	0.3	7-8	3.40	0.00009	0.0007
13	2.75	10-1	4.565	0.0006	0.0063
14	2.52	13-2	5.965	0.00042	0.0058
15	1.86	12-2	5.50	0.00034	0.002
16	5.15	13-13	6.26	0.00082	0.012
17	—	12-11	5.76	—	—
18	1.33	10-1	4.56	0.00029	0.003**
19	0.09	14-12	6.69	0.00001	0.0002
20	4.8	23-12	10.77	0.00045	0.011
21	1.04	residue+ 1-8	11.46	0.00051	0.002***
22	1.15	8-0	3.63	0.00032	0.003
23	1.72	4-0	1.81	0.00095	0.004
24	—	4-0	1.81	—	—
25	0.49	6-0	2.72	0.00018	0.001
26	0.69	10-0	4.53	0.00015	0.002
27	0.49	14-12	6.69	0.00007	0.001
28	0.55	17-13	8.08	0.00007	0.001
29	0.35	24-8	11.12	0.00003	0.001
30	1.01	10-0	4.53	0.00022	0.0023
31	0.10	49-0	22.2	0.000026	0.001
31a	0.48	residue of 30			
32	0.66	14-0	6.36	0.0001	0.0015
33	3.15	27	12.3	0.00025	0.0073
34	0.71	30	13.6	0.0001	0.0015

* Assumes all particulates released in 1-hr pyrolysis
** weighed dry from here on
*** total for runs 20 and 21

Table 3. Major Trace Organics by Tenax Trap and GC Analysis

Run No.	Compound	Amount (Picomoles)	Concentration (ppb*)
3	Toluene	1,850	16
	Perchloroethylene	605	5
	Dichloromethane	253	2
4	Toluene	904	8
	Perchloroethylene	299	3
	Trichloroethylene	192	2
5	Toluene	2,175	19
	Perchloroethylene	566	5
	p, m-xylene	406	4
6	Toluene	1,454	13
	Perchloroethylene	389	3
	Trichloroethylene	256	2
7-1	Toluene	12,814	111
	Perchloroethylene	1,320	11
	Trichloroethylene	93	1
7-2	Toluene	5,813	50
	Trichloroethylene	131	1
	Benzene	72	1
8-2	Toluene	2,013	17
	o-xylene	306	3
	Chloroform	165	1
9	Toluene	1,393	12
	Perchlorothylene	343	3
	Trichloroethylene	161	1
10	Toluene	2,347	20
	o-xylene	1,401	12
	Benzene	391	3
11	Toluene	97	1
	Benzene	16	—
	—	—	—
12	Toluene	6,464	56
	p, m-xylene	212	2
	Perchloroethylene	199	2

*Concentrations were calculated as for particulates by assuming that the trace organics were released during the active pyrolysis process lasting approximately one hour at the maximum measured air flow of 6,500 cfh. On this basis, parts per billion equals picomoles times 0.0866.

Table 3. Major Trace Organics by Tenax Trap and GC Analysis (Continued)

Run No.	Compound	Amount (Picomoles)	Concentration (ppb*)
14	Toluene	3,519	30
	p, m-xylene	119	1
	dichloromethane	115	1
16	Toluene	204	2
	Benzene	122	1
	Trichloroethane	120	1
17	Toluene	406	4
	Perchloroethylene	251	2
	Benzene	226	2
18	Toluene	168	1
	Benzene	139	1
	Trichloroethane	136	1
19	Toluene	152	1
	Benzene	47	—
	p, m-xylene	39	—
20	Toluene	738	6
	Benzene	731	6
	Perchloroethylene	252	2
21	Benzene	681	6
	Trichloroethane	189	2
	Toluene	123	1
22	Trichloroethane	62	1
	Benzene	58	1
	Toluene	31	—
			23
23	Trichloroethane	75	1
	Benzene	74	1
	Toluene	53	—
24	Benzene	909	8
	Toluene	114	1
	Trichloroethane	77	1

* Concentrations were calculated as for particulates by assuming that the trace organics were released during the active pyrolysis process lasting approximately one hour at the maximum measured air flow of 6,500 cfh. On this basis, parts per billion equals picomoles times 0.0866.

Table 3. Major Trace Organics by Tenax Trap and GC Analysis (Continued)

Run No.	Compound	Amount (Picomoles)	Concentration (ppb*)
25	Benzene	1,026	9
	Dichloromethane	385	3
	Hexane	18	—
26	Trichloroethane	119	1
	Toluene	105	1
	Benzene	74	1
27	Toluene	166	1
	Dichloromethane	164	1
	Trichloroethane	118	1
28	Toluene	194	2
	Trichloroethane	168	1
	Benzene	162	1
29	Toluene	246	2
	Dichloromethane	236	2
	Trichloroethane	202	2
31A	Trichloroethane	292	3
	Toluene	229	2
	Benzene	170	1
32	Toluene	192	2
	Benzene	160	1
	Dichloromethane	154	1
33-1	Dichloromethane	937	8
	Toluene	873	8
	Trichloroethane	319	3
33-2	Toluene	517	4
	Benzene	186	2
	p, m-xylene	172	1
34-1	Toluene	1,480	13
	Perchloroethylene	762	7
	p, m-xylene	592	5
34-2	Toluene	949	8
	Benzene	247	2
	Trichloroethane	243	2

* Concentrations were calculated as for particulates by assuming that the trace organics were released during the active pyrolysis process lasting approximately one hour at the maximum measured air flow of 6,500 cfh. On this basis, parts per billion equals picomoles times 0.0866.

APPENDIX A

Selected Temperature-Time Plots

and

Carbon Monoxide Plots

Pyrolyzer and Oxidizer Temperatures in Celsius

Carbon Monoxide in Parts-Per-Million

Time in Seconds

104 Emissions Data

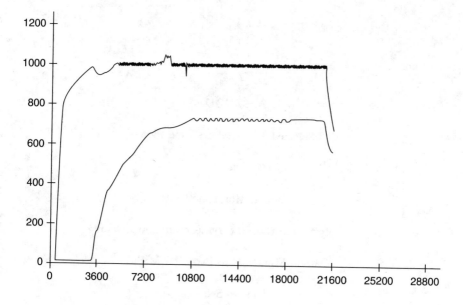

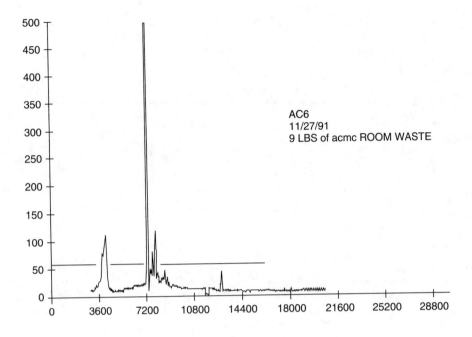

106 Emissions Data

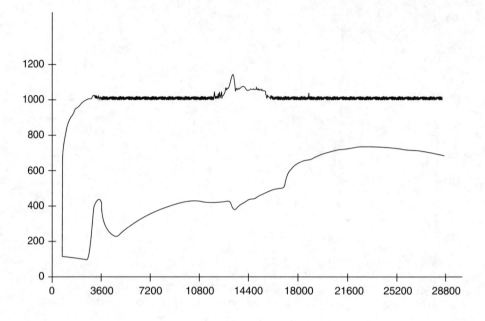

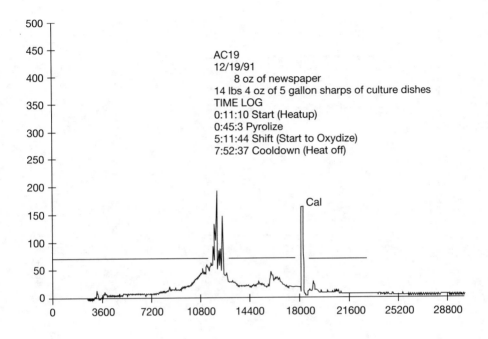

108 Emissions Data

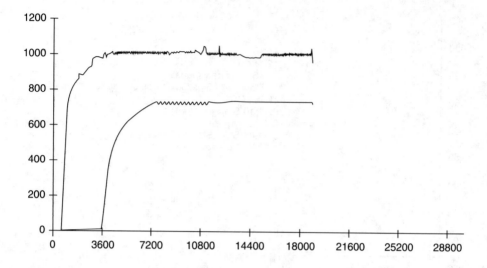

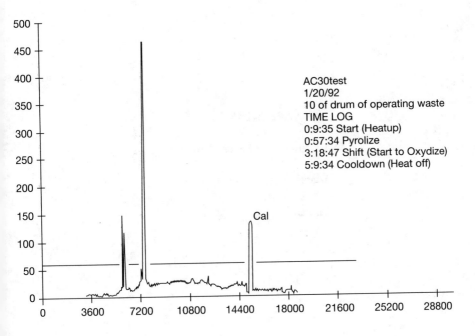

APPENDIX B

Air Emissions Permit Limits

Atlantic City Medical Center

Mainland Division-Installation

Table 1. Air Contaminant Emission Limits[1]

Contaminant	Maximum Rate (lbs/hr)	Maximum Concentration (PPMVD @ 7% O_2)[2]
Particulate (PM)	0.0025	0.015 grains/dscf
Hydrogen Chloride (HCL)	0.0024	50
Sulfur Oxides (as SO2)	0.005	50
Carbon Monoxide (CO)	0.018	100
Non-Methane Hydrocarbons as CH4	1.5 E-6	50
Toxic Volatile Organics	1.1 E-6	—
Nitrogen Oxides (NO_x) as NO2	0.04	—
Lead	0.26 E-6	—
Arsenic	0.20 E-6	—
Cadmium	0.04 E-6	—
Chromium[3]	0.20 E-6	—
Nickel	0.2 E-6	—
Mercury	0.6 E-8	—
2, 3, 7, 8 TCDD	2.5 E-9	—

Notes:
1 Limits are averaged over any consecutive 60 minute period.
2 PPMVD @ 7% O_2 – parts per million by volume, dry basis, corrected to 7% oxygen.
3 All chromium emissions are considered hexavalent Chromium, as worst case.
4 No radioactive waste will be allowed to be incinerated.

B. Opacity
The applicant shall not use the equipment in a manner which will cause visible emissions, exclusive of visible condensed water vapor. Compliance with this provision shall be verified visually by the use of New Jersey Test Method 2 (N.J. A. C. 7:27B-2), or approved equivalent, or by opacity monitoring.

APPENDIX C

Gas Chromatographic Data

Runs AC3 to AC34

	AC3 ss308 Conc (picomole)	AC4 ss605 Conc (picomole)	AC5 ss508 Conc (picomole)	AC6 ss502 Conc (picomole)	AC7-1 ss806 Conc (picomole)	AC7-2 ss908 Conc (picomole)
Chloromethane	1884.116	994.08	1802.112	0	0	0
Dichloromethane	253.561	99.858	283.198	0	0	0
Hexane	0	0	0	35.3152	0	0
Chloroform	0	0	0	0	0	0
Trichloroethane	0	0	0	0	0	0
Benzene	13.14144	18.45504	20.20896	117.7545	39.7008	71.99712
Carbontetrachloride	0	0	0	85.176	0	0
Trichloroethylene	152.801	191.864	311.315	255.751	93.119	131.399
Toulene	1850.309	904.7412	2175.102	1453.599	12814.53	5813.681
Perchloroethylene	605.616	299.0632	565.8368	388.6408	1320.823	0
p, m-Xylene	197.6942	105.2848	405.9086	144.5366	0	57.3942
o-Xylene	175.9188	76.9854	158.72	114.8116	42.0546	22.3448

	AC8-2 ss801 Conc (picomole)	AC9 ss701 Conc (picomole)	AC10 ss705 Conc (picomole)	AC11 ss603 Conc (picomole)	AC12 ss502 Conc (picomole)	AC14 ss808 Conc (picomole)
Chloromethane	729.068	782.116	1599.04	182.628	1240.852	787.208
Dichloromethane	85.351	0	194.109	0	183.785	114.988
Hexane	20.288	58.496	29.696	0	22.1696	0
Chloroform	165.546	0	0	0	0	0
Trichloroethane	0	53.8076	85.8396	0	121.7832	105.7056
Benzene	33.696	92.02464	390.744	15.79392	67.14144	38.73312
Carbontetrachloride	0	179.2	124.992	0	0	0
Trichloroethylene	44.254	161.153	150.568	0	82.882	0
Toulene	2012.678	1393.158	2347.236	97.3152	6464.039	3518.748
Perchloroethylene	75.0048	343.3064	214.2472	0	199.0696	62.6944
p, m-Xylene	116.1914	130.4008	126.3022	0	212.3452	119.0434
o-Xylene	305.6768	111.0544	1401.913	0	92.0266	35.8546

Emissions Data 115

	AC16 ss608 Conc (picomole)	AC17 ss502 Conc (picomole)	AC18 ss508 Conc (picomole)	AC19 ss804 Conc (picomole)	AC20 ss602 Conc (picomole)	AC21 ss101 Conc (picomole)
Chloromethane	1529.652	1872.032	1387.076	1291.772	1517.568	1820.504
Dichloromethane	0	0	0	0	0	0
Hexane	19.3664	61.5168	24.0768	14.272	99.776	28.7104
Chloroform	0	0	0	0	0	0
Trichloroethane	120.2432	161.5152	135.982	121.968	99.0528	189.4816
Benzene	122.5843	226.2211	139.7606	47.52864	731.4019	681.3158
Carbontetrachloride	0	0	0	85.176	0	0
Trichloroethylene	0	0	0	0	0	0
Toulene	203.9118	406.5681	168.1347	152.2224	738.8106	123.2622
Perchloroethylene	63.5624	251.9928	42.5816	0	252.712	39.2584
p, m-Xylene	55.407	151.1468	44.3624	39.008	63.9124	48.3506
o-Xylene	24.0312	75.0882	18.8232	16.2564	52.855	14.5514

	AC22 ss308 Conc (picomole)	A23 ss605 Conc (picomole)	AC24 ss706 Conc (picomole)	AC25 ss607 Conc (picomole)	AC26 ss108 Conc (picomole)	AC27 ss701 Conc (picomole)
Chloromethane	970.52	1003.428	913.292	1105.344	961.628	996.892
Dichloromethane	0	0	0	385.459	0	163.938
Hexane	17.5744	0	18.4192	18.1248	42.4704	37.5168
Chloroform	0	0	0	0	0	0
Trichloroethane	62.1852	74.9672	77.4312	0	119.5656	145.0988
Benzene	58.2336	73.50912	909.7056	1026.103	74.54592	117.7027
Carbontetrachloride	0	0	0	0	0	0
Trichloroethylene	0	0	0	0	0	0
Toulene	30.9411	53.0751	113.9157	0	105.369	166.0422
Perchloroethylene	0	0	0	0	0	0
p, m-Xylene	0	0	44.7488	0	27.554	45.2272
o-Xylene	0	0	13.1936	0	14.9234	25.2092

	AC28 ss705 Conc (picomole)	AC29 ss802 Conc (picomole)	AC31A ss309 Conc (picomole)	AC32 ss808 Conc (picomole)
Chloromethane	1110.208	1224.132	1314.8	1277.636
Dichloromethane	134.568	235.672	212.087	154.504
Hexane	40.3968	43.3152	57.6768	45.6192
Chloroform	0	0	0	0
Trichloroethane	168.1064	202.202	292.2612	242.704
Benzene	162.2419	160.6608	170.8214	160.3843
Carbontetrachloride	0	0	235.48	152.992
Trichloroethylene	58.87	72.442	0	88.537
Toluene	193.5516	26.2082	228.8823	192.4542
Perchloroethylene	37.82	51.956	119.8088	61.2312
p, m-Xylene	55.8762	61.2628	70.2834	57.6794
o-Xylene	30.2126	35.5012	36.735	31.0682

118 Emissions Data

	AC33 ss405 Conc (picomole)	AC33-2 ss706 Conc (picomole)	AC34 ss605 Conc (picomole)	AC34-2 ss702 Conc (picomole)
Chloromethane	3317.476	2559.604	2500.324	2229.004
Dichloromethane	937.348	118.726	258.456	132.61
Hexane	100.8128	53.5296	83.7376	76.5696
Chloroform	0	0	0	0
Trichloroethane	319.0572	155.386	354.2924	242.8888
Benzene	369.576	186.0192	385.7500	246.5856
Carbontetrachloride	0	0	0	0
Trichloroethylene	167.707	72.326	162.893	80.794
Toulene	873.5025	517.2753	1480.215	948.9906
Perchloroethylene	103.7632	76.8056	762.6992	231.4088
p, m-Xylene	256.105	172.2424	592.7606	328.831
o-Xylene	134.447	80.0916	424.6008	208.7478

10

MSW MASS BURN AND RDF INCINERATOR ASH LEACHATE CONSTITUENTS

A. J. Perna
Department of Chemical Engineering
New Jersey Institute of Technology
Newark, NJ

and

D. W. Sundstrom, H. E. Klei, B. A. Weir
Department of Chemical Engineering
The University of Connecticut
Storrs, CT

INTRODUCTION

As landfill space and sites for Municipal Solid Waste (MSW) disposal become less available, methods for decreasing the amount to be disposed of are examined more closely. The State of Connecticut has adopted a solid waste management plan emphasizing recycling and resource recovery facilities that burn municipal solid waste to produce energy. The two major incinerator technologies are mass burn and refuse derived fuel. In a mass burn process, most of the solid waste is fed directly to a grate within the combustion unit. A refuse derived fuel (RDF) facility removes selected noncombustibles and shreds the remaining trash prior to burning. These technologies produce ash with differing physical and chemical characteristics.

The State Department of Environmental Protection has supported a study of the properties and uses of ash from mass burn and RDF incinerators. The research program investigated physical and chemical characteristics of the ash, leachability of chemical components from the ash, and uses of the ash in concrete, asphalt and non-structural fill. A critical aspect of the ash study was the leaching behavior of the raw ash and products containing ash.

Numerous laboratory leaching tests have been developed either to classify wastes or to predict leachate quality (1). The standard EP Toxicity and TCLP tests are used to classify solid wastes as hazardous or nonhazardous based on the concentrations of selected components leached from the waste. Batch and column tests with water or other leaching fluids are used to estimate leachate quality from raw or processed ash (2-9). Batch tests are easier and less expensive to perform, and are more reproducible than column tests. A common batch test is sequential extraction, in which a sample of ash is contacted several times with fresh quantities of water. The leachate from each extraction is analyzed to estimate the effect of liquid quantity on leaching behavior.

Column tests are more representative of field conditions than batch tests and yield information on rates of leaching. In column tests, the raw or processed ash is placed in a column and leaching fluid is fed to the column continuously at a controlled flow rate. Liquid percolates slowly through the ash and mimics landfill conditions more closely than the aggressive agitation of a batch test. The variation of leachate composition with time permits evaluation of rates of leaching of individual ash components.

The major purpose of this study was to compare the leachability of metals in ashes from typical mass burn and RDF plants. Bottom, fly and mixed ashes were subjected to water leaching in batch and column tests. Since proposed uses subject the ashes to different environments, the effect of pH on solubility of metals was also examined.

SAMPLING PROCEDURES

RDF ash was obtained from the Mid-Connecticut facility during the Environment Canada/EPA test program undertaken in February, 1989 (10). Bottom ash was collected continuously in hoppers by diverting it from the conveyor belt which leads from the quench tank. Samples were taken from 14 hoppers during a 4 hour run. Fabric filter (fly) ash was collected from the conveyor belt leading from the baghouse by vacuuming it into a collection truck. Representative samples of fly ash were collected from the back of the truck at the end of a run.

Mass burn ash was obtained from the Bristol plant during May, 1989. Mixed ash samples were collected from a conveyor belt which transports ash to a storage shed. Samples were taken every 20 minutes for a 4.5 hour period. Fly ash was sampled from a screw conveyor every 30 minutes for 4 hours. Both the Mid-Connecticut and Bristol facilities employ lime scrubbers for the treatment of the effluent gases.

ASH PROCESSING

All of the RDF bottom ash was dried at $105^\circ C$ and crushed to pass a 3/8 inch sieve. Material that could not be crushed represented about 15% of the total bottom ash. The mass burn mixed ash for column tests was not ground. Half of the mixed ash was dried so that column tests could be done to assess the effect of drying of ash on metals leachability. The mass burn mixed ash for batch tests was dried at $105^\circ C$ and crushed to pass a 3/8 inch sieve. Since some large pieces in the mass burn ash could not be easily crushed, the ground mixed ash was about 78% of the total sample. Fly ashes from both facilities required no processing before the batch or column tests. Each type of ash was blended separately prior to testing.

BATCH TESTS

Sequential batch extraction tests using deionized water as the solvent were performed on the four ashes. A 50 gram sample of ash was extracted for 18 hours with 1000 ml of water with agitation provided by a rotary shaker at 150 rpm. The ash and liquid were separated by pressure filtration. The spent ash was recontacted with fresh water and the extraction procedure was repeated through four cycles. Liquid to solid ratios (L/S) of 5 and 10 liters per kilogram were also examined for some ashes. All batch extractions were performed in triplicate.

Two types of pH controlled extractions were performed on the ashes at an L/S ratio of 20/1. One type of extraction involved titrating the leachate with acid or base during the course of the extraction to maintain the pH within 0.5 units of the target value. After titrating for 8 hours, agitation was continued for another 16 hours without further adjustment of pH. In the other type of extraction, all of the acid or base was added initially and the suspensions were agitated for 24 hours.

COLUMN TESTS

RDF ashes were tested in 5.5 inch inside diameter Plexiglas columns packed with 18 inches of ash. Column tests were performed on bottom ash, fly ash, and two mixtures containing 15 and 35 weight % fly ash. The mass burn ashes were studied in 9 inch diameter Pyrex columns packed with 30 inches of ash. Wet mixed ash and dried mixed ash were placed in two columns to examine the effect of drying the ash.

All columns were operated in a downflow mode at a flow rate of about 0.3 inches per hour. Deionized water was fed to the columns through distribution systems made of disks of filter paper supported on porous plates. The filter papers provided random distribution of water over the top of the column. Layers of gravel and sand supported the ash within the columns.

Leachate samples were collected periodically until a liquid-to-solid (L/S) ratio of 20/1 was reached. Samples were analyzed by ICP for all metals except lead, which was analyzed by Zeeman AA to achieve greater sensitivity.

Near the end of each column run a tracer test was made in order to characterize the flow patterns. A pulse of potassium bromide solution was injected onto the top of the column. The concentration of bromide ion in the column leachate was monitored continuously with a bromide ion electrode.

SEQUENTIAL EXTRACTIONS

Each type of ash was extracted four times with deionized water at a liquid to solid ratio (L/S) of 20 to 1. The pH values and lead concentrations in the leachates are shown in Figures 1 and 2. The pH of all leachates were above 10 through the four extractions. The mass burn fly ash was the most basic and leached the highest concentrations of lead. The solubility of the lead appeared to be related to the pH of the extract.

Figures 3 and 4 show cumulative plots for sequential extraction of lead from RDF bottom and fly ash at liquid to solid ratios of 5, 10, and 20 to 1. Thus, the bar for the fourth extraction represents the total mass of lead removed in the four extractions. The amount of lead extracted from the fly ash was 5 to 30 times larger than the lead dissolved from the bottom ash. For the bottom ash, each extraction dissolved about the same amount of lead and the L/S ratio had a small effect on the mass of lead released. For the case of the fly ash, most of the water soluble lead was leached out in the first extraction with subsequent extractions contributing a relatively small amount to the total mass of leached lead. Also, the mass of lead extracted from the fly ash increased as the L/S ratio was decreased. For most metals, the L/S ratio had a small effect on the total mass of metal extracted.

EFFECT OF PH

In the study of pH effects on metal solubilities, acid or base was added to the ash by titration or by single addition. For pH levels below 6, single addition of acid generally gave higher concentrations of dissolved metals than titration of acid. When acid is added all at once, the pH initially falls below the target value. As the pH then rises to the final value, some of the metal dissolved at a lower pH may not reprecipitate. Since the method of acid addition did not have a large effect on most metal solubilities, it was neglected in preparing the figures.

Figure 5 shows the acid neutralization curves for the four types of ash. The mass burn fly ash had the largest acid neutralization capacity, requiring about 6 meq of acid per gram to achieve a leachate pH of 7. The high alkalinity results from the lime added in the gas scrubbing system. The acid neutralization capacity of the RDF fly ash was much less than the mass burn fly ash.

The effect of pH on lead concentrations in the leachates is shown in Figures 6 and 7. The solubility behavior of all four ashes was remarkably similar, with minimum lead concentrations (20-100 ppb) between pH 8 and 10. Lead concentrations, however, generally exceeded 10 ppm at pH levels below 4 and above 12. The shape of these curves is characteristic of amphoteric metals that form complexes with hydroxide ions.

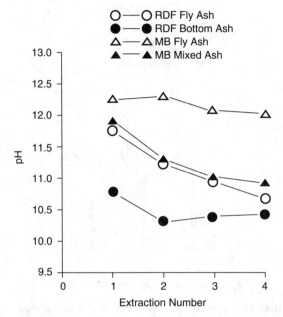

Figure 1 Effect of sequential extractions with water on pH of ash leachates. Liquid to solid ratio was 20 to 1 for each extraction.

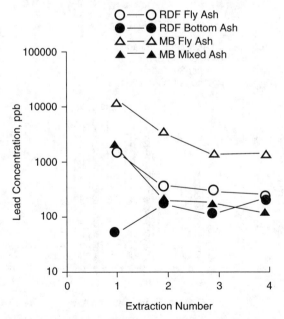

Figure 2 Effect of sequential extractions with water on lead concentrations in ash leachates. Liquid to solid ratio was 20 to 1 for each extraction.

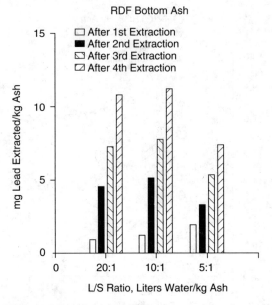

Figure 3 Effect of liquid to solid ratio on total mass of lead removed in sequential extractions of RDF bottom ash.

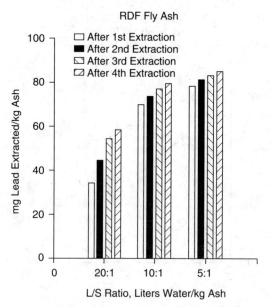

Figure 4 Effect of liquid to solid ratio on total mass of lead removed in sequential extractions of RDF fly ash.

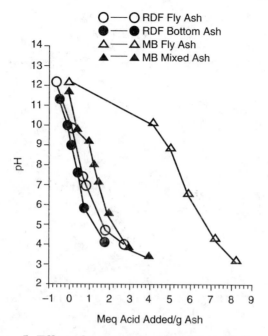

Figure 5 Effect of amount of acid or base added on pH of leachate from batch extractions.

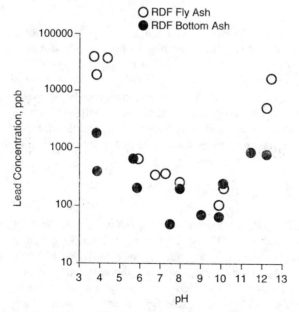

Figure 6 Effect of pH on lead concentrations in leachates from batch extractions of RDF bottom and fly ashes.

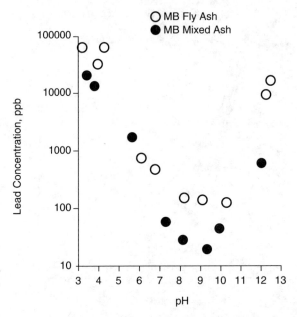

Figure 7 Effect of pH on lead concentrations in leachates from batch extractions of mass burn mixed and fly ashes.

The solubilities of 16 metals as a function of pH was determined using ICP analysis of the leachates. Zinc exhibited amphoteric behavior similar to lead with a minimum solubility below 100 ppb in the pH range of 8 to 10. Barium solubilities of each type of ash were relatively constant over the pH range of 4 to 12. Molybdenum concentrations in the leachates actually increased with increasing pH. Most metals, including cadmium, nickel, cobalt, and manganese, decreased in solubility as pH was increased. Cadmium solubilities for all four ashes, for example, decreased from more than 1000 ppb at low pH to less than 10 ppb at high pH.

COLUMN TESTS RESULTS

Column tests were conducted on RDF bottom and fly ash, mass burn mixed and fly ash, and blends of RDF ash containing 15 and 35 weight % fly ash. The volumes of leachate leaving the columns were converted to liquid to solid ratios to facilitate comparisons between columns and with batch results.

The pH of leachates from RDF and mass burn ash columns are shown in Figures 8 and 9. Leachate pH values from the four columns increased rapidly at the start and then declined to relatively constant levels. The undried mass burn ash retained a high pH level for a longer time than the initially dry mass burn ash

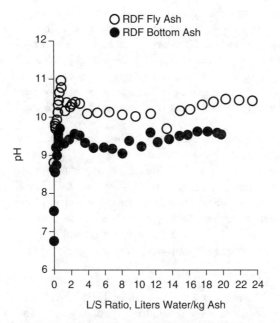

Figure 8 Effect of liquid to solid ratio on pH of leachates from columns containing RDF bottom and fly ashes.

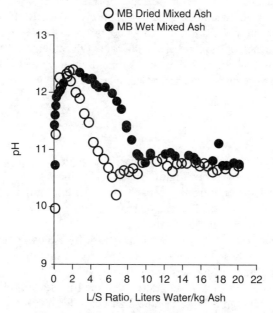

Figure 9 Effect of liquid to solid ratio on pH of leachates from columns containing dried and undried mass burn mixed ashes.

(Figure 9). The steady pH values at L/S ratios above 8 were about 10.8 for mass burn mixed ash, 10.3 for RDF fly ash and 9.3 for RDF bottom ash.

Lead leaching profiles for RDF and mass burn ash columns are presented in Figures 10 and 11. Lead concentrations in the initial leachates from the RDF fly ash and mass burn mixed ash exceeded 10 ppm. Concentrations decreased rapidly to less than 1 ppm at L/S ratios near 2 and reached minima near 10 ppb between L/S ratios of 6 and 12. Lead levels then increased again to the 50-500 ppb range at L/S of 20.

Concentrations of many of the metals in the leachates declined 1 to 2 orders of magnitude at low L/S ratios. As can be seen in Figure 12, the cadmium concentration in leachates from RDF fly ash decreased from about 2000 ppb at the start to less than 10 ppb at an L/S of 1. Although the detection limit for cadmium in leachates is 10 ppb, points below this level are plotted to show that concentrations remained low. Nickel, chromium and copper also exhibited rapid washout to low levels at L/S ratios below 2. One metal that did not washout was aluminum, which leached concentrations above 1 ppm at both low and high L/S ratios for all ashes.

Leaching profiles were similar for most metals in both the initially dry and wet mass burn ashes. The main exceptions were lead and zinc (Figure 13) at L/S ratios below 8 where concentrations in the undried ash exceeded those in the initially dry ash. The probable cause of this solubility difference is the higher pH in the wet ash leachates at L/S ratios below 8 (Figure 9).

Tracer tests using bromide ion were performed on each column near the conclusion of the run. Residence time distributions for the RDF bottom and fly ash columns are shown in Figure 14. The ordinate is a dimensionless age distribution frequency of the fluid elements leaving the column (12). The abscissa is a dimensionless time expressed as actual time divided by mean residence time. The residence time distributions permit comparison of flow patterns in different columns.

BATCH TESTS

All four ashes studied had pH levels above 10 through four water extractions at liquid to solid ratios of 20/1 (Figure 1). The alkaline nature of the ashes are evidenced by the acid neutralization capacities shown in Figure 5. Assuming acid rain of pH 4.5 at a rate of 120 cm/year, over 100 years would be needed to reduce pH to 7 in a 5 cm layer of ash.

Acid neutralization capacity (ANC) of mass burn fly ash was much greater than those for the other ashes (Figure 5). The alkalinity of this mass burn fly ash is fairly typical of ash from lime scrubbers. The ANC RDF fly ash, however, was low (about 0.8 meq/g at pH 7), suggesting that the lime level in this sample was below normal. Although a large sample was taken from a truck, it apparently contained less than the stoichiometric lime needed for acid gas removal.

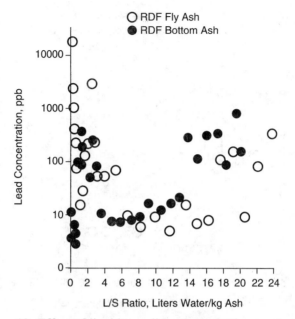

Figure 10 Effect of liquid to solid ratio on lead concentrations in leachates from columns containing RDF bottom and fly ashes.

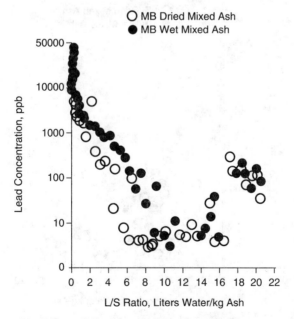

Figure 11 Effect of liquid to solid ratio on lead concentrations in leachates from columns containing dried and undried mass burn mixed ashes.

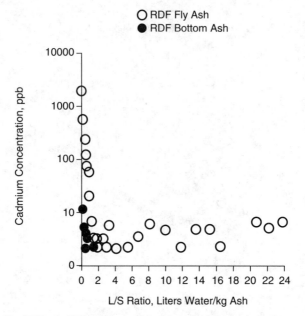

Figure 12 Effect of liquid to solid ratio on cadmium concentrations in leachates from columns containing RDF bottom and fly ashes.

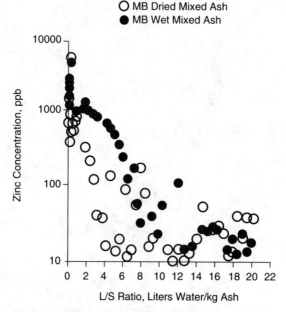

Figure 13 Effect of liquid to solid ratio on zinc concentrations in leachates from columns containing dried and undried mass burn mixed ashes.

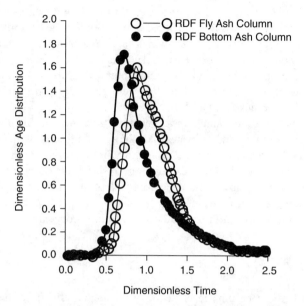

Figure 14 Dimensionless residence time distributions derived from tracer tests on columns containing RDF bottom and fly ashes.

With the exception of RDF bottom ash, lead concentrations in the water leachates decreased with successive extractions (Figure 2). The lead concentrations in leachates from mass burn fly ash were nearly an order of magnitude greater than those in the other ashes. As shown in Figures 6 and 7, lead solubilities increased rapidly at pH levels about 10. Thus, the high pH's in the water extracts from the mass burn fly ash (Figure 1) were a major cause of the high lead concentrations.

Some investigators have suggested the segregation of fly ashes in monofills because of their hazardous nature in regulatory tests. The results of this study indicate that leachate from a monofill containing highly alkaline fly ash might exceed regulatory limits for lead. In this case, the fly ash would require treatment prior to disposal in a landfill.

The effect of L/S ratio on metal solubility varied with type of metal and ash. In general, the total amount of a specific metal extracted from a given ash was comparable at L/S ratios of 5/1, 10/1 and 20/1. Typical behavior is shown in Figure 3 for RDF bottom ash where nearly the same amount of lead was dissolved at each L/S ratio. The main exception was lead in RDF fly ash where most of the soluble lead was removed in the first extraction and the total amount of extracted lead actually increased with decreasing L/S (Figure 4). Although differences in pH and ionic strength between extractions may be partially responsible, the cause of this L/S effect is unknown.

The effect of pH on metal solubility was more dependent on the nature of the metal than the type of ash. For a given metal, the solubility curves had similar shapes for the four types of ash but differed in magnitude. The solubilities of most metals increased with decreasing pH, which is characteristic behavior for many forms of metal compounds. In some cases, such as barium, pH had a minor effect on solubility, and, in one case, molybdenum, solubility decreased with decreasing pH. The effects of pH on metal solubilities were generally consistent with those observed in other studies (8,11).

Lead and zinc exhibited amphoteric behavior with minima in the solubility curves between 8 and 10 for all ashes. As shown in Figures 6 and 7, the lead solubilities usually exceeded the TCLP regulatory limit (5 ppm) at pH levels below 4 and above 12. The solubilities of cadmium in the ashes also exceeded the TCLP regulatory limit at pH below 4 but decreased to less than 10 ppb at pH above 9. These results suggest that ashes yielding leachates with pH levels between 8 and 11 may be desirable to minimize lead and cadmium concentrations in landfill leachates.

COLUMN TESTS

The low initial pH of column leachates may result partially from the release of acidic components from the surface of the ash. The pH then increases rapidly as soluble basic materials leach from the ash. The high pH at low L/S, especially with mass burn ash columns, leads to increased solubility of amphoteric metals such as lead and zinc. For each type of ash the pH of leachate from the column at L/S of 20 was 1 to 2 pH units lower than the pH from the batch test at 20 to 1. The leachate pH's from columns at high L/S were in the range of 9 to 11 where solubility of lead and cadmium is low.

For all columns except RDF bottom ash, initial concentrations of lead exceeded the TCLP regulatory limit but declined rapidly to low levels. Most other metals also exhibited a rapid decrease in leachate concentrations. These results demonstrate the importance of obtaining samples from columns at low L/S ratios. Since liquid to solid ratios in most ash landfills are low and change slowly with time, data at low L/S ratios are needed to relate laboratory studies to field experience.

At liquid to solid ratios between 1 and 9, undried mass burn ash gave higher pH levels than dried mass burn ash. The process of drying the ash prior to testing apparently reduced the rate of dissolution of basic ash components. The higher pH from undried ash was associated with increased concentrations of lead and zinc (Figures 11, 13) in the leachates. Thus, drying of ash before testing may affect the solubility behavior of some metals.

Comparison of RDF fly and bottom ash columns showed that lead, cadmium and barium concentrations were higher in the fly ash leachates at L/S ratios below 1. Bottom ash leachates were higher in nickel up to L/S of 2 and higher in copper for all L/S ratios. The relative concentrations of metals in fly and bottom ash leachates depend upon total concentration in the ash, form of the metal compound and proximity to the surface.

Dry RDF fly and bottom ash were mixed physically to form blends containing 15 and 35 weight % fly ash. The pH and concentration levels from the column with 35% fly ash were generally between those from the separate fly and bottom ash columns. Leaching profiles from columns with this RDF blend and dried mass burn mixed ash were similar in both shape and magnitude for most metals.

The column test with the 15% RDF fly ash blend gave pH and concentration values that were lower than expected. The residence time distribution curve from the tracer test showed significant channeling of the liquid. Tracer tests should probably be included as part of most column leaching studies with ash.

Batch tests leached much more metal from the ashes than column tests. Using lead as an example, the total mass extracted from fly and mixed ash columns up to L/S of 20 was about 5 mg per kg of ash. In contrast, batch tests on the same ashes at L/S of 20 extracted about 40 mg per kg of ash. The higher metal removal in batch tests is caused by the more intimate contact of liquid and solid provided by agitation.

ACKNOWLEDGEMENT

Financial support was provided by the State of Connecticut Department of Environmental Protection and the Environmental Research Institute of the University of Connecticut. Metal analyses were performed by Don Hobro of the Environmental Research Institute.

REFERENCES

1. Wastewater Technology Center, Environment Canada, Compendium of Waste Leaching Tests, Report EPS 3/HA/7, May 1990.

2. Cundari, K. L., and Lauria, J. M., "The Laboratory Evaluation of Expected Leachate Quality from a Resource Recovery Landfill," Malcolm Pirnie, 1986.

3. Francis, C. W., and White, G. H., "Leaching Toxic Metals from Incinerator Ash," *Journal Water Pollution Control Federation* 59, 979 (1987).

4. Hasselriis, F., "How Control of Combustion, Emissions and Ash Residues from Municipal Solid Waste Can Minimize Environmental Risk," AIChE Symp. Ser. 84 (265), 154 (1988).

5. Hjelmar, O., "Characterization of Leachate from Landfilled MSW Ash," Proc. International Conf. on Municipal Waste Combustion, 3B-1, Hollywood, FL, April 1989.

6. Sawell, S. E., Bridle, T. R. and Constable, T. W., "Heavy Metal Leachability from Solid Waste Incinerator Ashes," *Waste Management & Research* 6, 227 (1988).

7. Theis, T. L. and Gardner, K. H., "Environmental Assessment of Ash Disposal," *Critical Reviews in Environmental Control* 20, 21 (1990).

8. Van der Sloot, H. A., "Leaching Behavior of Waste and Stabilized Waste Materials; Characterization for Environmental Assessment Purposes," *Waste Management & Research* 8, 215 (1990).

9. Wadge, A. and Hutton, M., "The Leachability and Chemical Speciation of Selected Trace Elements in Fly Ash from Coal Combustion and Refuse Incineration," *Environmental Pollution* 48, 85 (1987).

10. Kilgroe, J. D. and Finkelstein, A., "Combustion Characterization of RDF Incinerator Technology: A Joint Environment Canada-USEPA Project," Proc. International Conf. on Municipal Waste Combustion, 5A-65, Hollywood, FL, April, 1989.

11. DeGroot, G. J. et al., "Leaching Characteristics of Selected Elements from Coal Fly Ash as a Function of the Acidity of the Contact Solution and the Liquid/Solid Ratio," ASTM STP 1033, 170 (1989).

12. Himmelblau, D. M., and Bischoff, K. B., "Process Analysis and Simulation," Ch. 4, J. Wiley & Sons, 1968.

11

INITIAL SCREENING OF THERMAL DESORPTION FOR SOIL REMEDIATION

James J. Yezzi, Jr. and Anthony N. Tafuri
U.S. Environmental Protection Agency
Edison, NJ

Seymour Rosenthal
Foster Wheeler Enviresponse, Inc.
Edison, NJ

William L. Troxler
Focus Environmental, Inc.
Knoxville, TN

INTRODUCTION

Petroleum-contaminated soils—caused by spills, leaks, and accidental discharges--exist at many sites throughout the United States. Thermal desorption technologies which are increasingly being employed to treat these soils, have met soil cleanup criteria for a variety of petroleum products.

Currently the United States Environmental Protection Agency is finalizing a technical report entitled *Use of Thermal Desorption for Treating Petroleum-Contaminated Soils* to assist remedial project managers, site owners, remediation contractors, and equipment vendors in evaluating the use of thermal desorption technologies for petroleum-contaminated soil applications. The completed report will be available from the Center for Environmental Research Information (CERI) by June 1992.

CONTENTS OF TECHNICAL REPORT

The report will discuss the following areas:

- Thermal desorption theory.
- The relationship of thermal desorption applicability, operations, and efficiency to site, contaminant and soil characteristics, as well as the effects of regulatory requirements.
- Commercial thermal desorption systems.
- Operating costs for thermal desorption systems.

Comprehensive appendices to the report serve as an encyclopedic source with detailed discussions on related topics; for example:

- Thermal desorption theory.
- Site, contaminant, and soil characteristics, and their impact on thermal desorption applicability.
- Regulatory issues affecting the permitting and operation of thermal desorption systems.
- Commercially available thermal desorption systems.
- Project task lists for use of mobile and fixed-based systems.
- Estimation of costs for using mobile or fixed-based thermal desorption systems.
- Comparison of thermal desorbers to incinerators.

THREE-LEVEL SCREENING METHOD

The report will also present a three-level screening method to help a reader predict the success of applying thermal desorption at a specific site. This method utilizes a series of worksheets that will assist the reader in accomplishing the following activities:

- Performing an initial assessment, based on limited data, to determine the applicability of thermal desorption for a given application.
- Identifying thermal desorption and off-gas treatment system requirements.
- Developing an overall cost estimate for treating a site using thermal desorption.

The objective of screening level one is to determine the likelihood of success in a specific application of thermal desorption. It will take into account procedures for collecting and evaluating data on site characteristics, contaminant characteristics, soil characteristics, and regulatory requirements. This level will establish whether or not thermal desorption should be evaluated further for site remediation, whether treatment should occur on-site or off-site, and if on-site is a viable option, what system size will be most cost-effective.

Screening level two will evaluate alternative thermal desorption technologies and factors such as the type of unit operations and operating conditions that are required to achieve specific cleanup criteria. It will also identify the most viable equipment alternatives.

Screening level three will guide in the preparation of an economic evaluation of the treatment alternatives selected in the first two levels. It identifies project tasks that must be conducted and provides typical cost factors for treating petroleum contaminated soils by thermal desorption technologies.

The scope of this paper addresses only screening level one which provides a preliminary assessment of the applicability of thermal desorption to a particular site. This topic encompasses worksheets that are an integral part of the "user friendly" screening process. Level one screening provides a foundation for the subsequent two levels which follow a similar "user friendly" worksheet approach to evaluating thermal desorption technologies and establishing costs for thermal desorption in an overall remediation project.

Figure 1 illustrates the three-level screening method presented in the report.

The screening level one worksheets are developed to simplify the evaluation of thermal desorption effectiveness and are based on the collection of limited data. The worksheets do not constitute a design manual, nor a final basis for choosing thermal desorption as a remedy. They provide a pre-selection screening method to determine if the utilization of thermal desorption to a particular site warrants further consideration.

LEVEL ONE SCREENING

The first level of screening describes six steps for collecting and evaluating key data that will affect the application of thermal desorption at a specific site. These data are defined as "critical success factors."

The worksheets in the report guide the reader through the six steps:

1. Data collection.
2. Waste classification.
3. On-site versus off-site treatment selection.
4. Critical success factor evaluation.
5. Contingency planning.
6. Treatment system selection.

The initial screening accomplished by these six steps limits the number of alternatives that will be subjected to further screening levels.

STEP 1: DATA COLLECTION

This first step in screening level one involves the collection of data in four major categories:

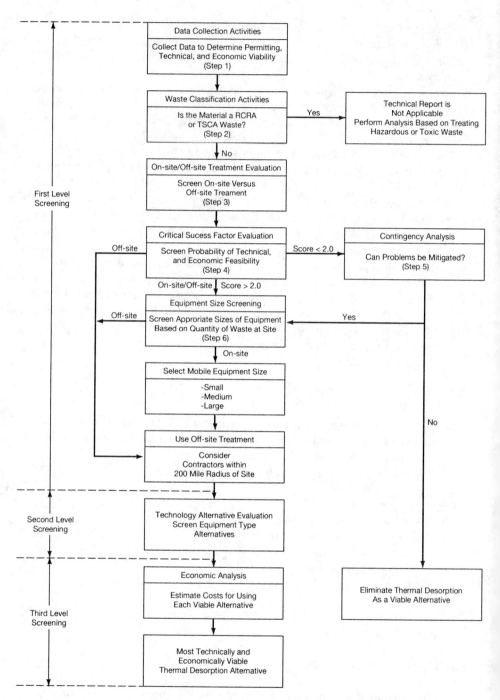

Figure 1 Thermal desorption evaluation decision diagram.

- Contaminant characteristics
- Site characteristics
- Regulatory requirements
- Soil characteristics

Table 1 details these types of data. The method limits the number of parameters in the screening analysis in order to minimize the time and cost of the evaluation. The source of these data, required to complete the critical success factor evaluation, generally include limited field investigations, standard analytical tests, or published sources. The report appendices contain detailed information regarding the potential impact of each item presented.

STEP 2: WASTE CLASSIFICATION

This second step in level one screening uses the data collected in the first step to confirm that the site is appropriate for further evaluation. The report focuses on the cleanup of petroleum-contaminated soil as a non-RCRA and non-TSCA waste. The flow chart in Figure 2 provides a decision tree for classifying the contaminated soil and confirming the applicability of the report. The appendices in the report contain detailed explanations of each element in the flow chart.

STEP 3: ON-SITE VERSUS OFF-SITE TREATMENT SELECTION

Figure 3 presents a decision diagram to compare the economic effectiveness of on-site or off-site treatment. This figure is only a screening tool; it is not a substitute for a detailed economic analysis of alternatives. The report discusses economic analysis in depth as a separate topic.

STEP 4: CRITICAL SUCCESS FACTOR EVALUATION

In this fourth step, worksheets address each critical success factor. Completing these worksheets employs simple qualitative and/or quantitative methods for rating each factor according to the probability for the successful application of thermal desorption. The form ranks each factor as having a least, average, or highest probability for successful use of thermal desorption.

Example - Calculation of the Probability of Success

Table 2 contains an example of a completed critical success factor screening evaluation for an on-site application. The remedial manager first defined the critical contaminant as well as the site, regulatory, and soil characteristics. The manager assumed that an on-site cleanup of 800 tons of soil contaminated with No. 6 fuel oil will occur at a 1.25 acre commercial retail facility in a state having the assumed criteria presented in the example. The contamination at this site resulted from a leaking underground storage tank. The TPH concentration is 12,000 mg/kg and metal concentrations do not exceed state or local criteria.

Table 1 Thermal Desorption Data Requirements Critical Success Factors

Characteristic	Data Collection		
	Rationale	Source	Method[a,b]
Contaminant Characteristics			
Petroleum product type	Selection of required soil treatment temperature	Owner's knowledge of tank usage	Site owner interview
Concentration of TPH in contaminated soil	Determination of treatment and disposal requirements under state and local regulations, selection of required soil treatment temperature and residence time, potential to exceed lower explosive concentration limits in thermal desorption device.	Analytical data from soil boring samples	EPA 418.1 is most common method, state and local requirements may vary
TCLP extract concentration of metals or organics (lead from leaded gasoline is most likely contaminant)	Material may be classified as a RCRA hazardous waste if TCLP extract concentrations exceed values listed in 40 CFR 261. Exclusions apply for wastes from underground storage tanks that are subject to the RCRA Corrective Action requirements in 40 CFR 280. See flow chart in Figure 2. If material is a hazardous waste, this Technical Report is not applicable.	Analytical data from soil boring samples	EPA 1311 (extraction) EPA 6010 (metals) EPA 8260 (volatile organics) EPA 8080 (semivolatile organics)
Concentration of PCBs in contaminated soil	If PCBs are present at a concentration of greater than 50 ppm, the waste is subject to TSCA regulations and this Technical Report is not applicable.	Analytical data from soil boring samples	EPA 8080

Table 1 Thermal Desorption Data Requirements Critical Success Factors (Continued)

Characteristic	Data Collection		
	Rationale	Source	Method[a,b]
Data Requirements Critical Success Factors			
Total metals concentration (As, Ba, Cd, Cr, Pb, Hg, Se, Ag)	State and local regulatory requirements for treatment or disposal of contaminated soil.	Analytical data from soil boring samples	EPA 3050 (acid digestion) EPA 6010 (metals), meet with regulatory agencies
Site Characteristics			
Contaminant source	Exemptions apply for wastes that exhibit the RCRA characteristic of toxicity codes D018-D043 if the waste is from a leaking underground storage tank that is subject to the Corrective Action Requirements in 40 CFR 280. See flow chart in Figure 2.	Identification of contaminant source	Site review
Contaminated soil quantity (tons)	Selection of on-site versus off-site treatment.	Soil borings, concentration of contaminants, soil cleanup criteria	Use approved analytical methods from SW-846
Site usage	Project cost estimate should include revenue loss from normal site activities.	Revenue loss each day that site is out of service	Site owner's cost estimate
Operational area available	Must be sufficient to set up and operate process equipment and maintain feed and treated soils stockpile (on-site treatment only).	Plot plan drawing of area available for operations	Site survey

Table 1 Thermal Desorption Data Requirements Critical Success Factors (Continued)

Characteristic	Data Collection		
	Rationale	Source	Method[a,b]
Surrounding land use	Adjoining land uses such as schools, parks, health care facilities, or dense urban development may preclude on-site treatment	Map showing surrounding land uses	Site survey
Distance to stationary thermal desorption facility (off-site treatment only)	Potential cost of soil transportation, evaluation of on-site versus off-site treatment options.	Location of stationary thermal desorption facilities in geographical area	Contact state regulatory agency
Ambient temperature	Frozen soil is difficult to excavate, pretreat, and process in thermal treatment devices.	Average temperature at time of treatment	Weather of U.S. Cities, Vol 1 & 2, Gale Research, Detroit, Michigan, 1985.
Regulatory Requirements			
No. of permits required	Total permitting cost (on-site treatment only).	Review of state and local requirements	Meet with regulatory agencies
Site specific performance testing requirements	Testing costs and project schedule impacts, including analytical turnaround (on-site treatment only).	Review of state and local requirements	Meet with regulatory agencies
TPH target residual level	Soil treatment time and temperature requirements, soil disposal alternatives.	Review of state and local requirements	Meet with regulatory agencies

Table 1 Thermal Desorption Data Requirements Critical Success Factors (Continued)

Characteristic	Data Collection		
	Rationale	Source	Method[a,b]
BTEX target residual level	Soil treatment time and temperature requirements, soil disposal alternatives.	Review of state and local requirements	Meet with regulatory agencies
Transportation restrictions	Some states may restrict off-site transportation of petroleum contaminated soils.	Review of state and local requirements	Meet with regulatory agencies
Soil Characteristics			
Moisture content	Materials handling properties, drying duty of thermal desorption process	Analytical data from soil boring samples	ASTM D-2216
Soil classification (coarse grained soils)	Material size reduction requirements.	Analytical data from soil boring samples	USCS
Soil classification (fine grained soils)	Material carryover from TD device, material plasticity characteristics.	Analytical data from soil boring samples	USCS
Soil classification (organic soils)	Potential for TPH analysis interferences because of naturally occurring organic matter	Analytical data from soil boring samples	USCS

(a) SW-846 - "Test Methods for Evaluating Solid Wastes, Physical/Chemical Methods". U.S. EPA, SW-846, Third Edition, November, 1988. Methods 6010, 418.1, and 1311 are analytical methods described in SW-846. Methods ASTM D-2216 is an analytical method described in American Society for Testing and Materials (ASTM), latest approved method.
(b) USCS - Unified Soil Classification System.

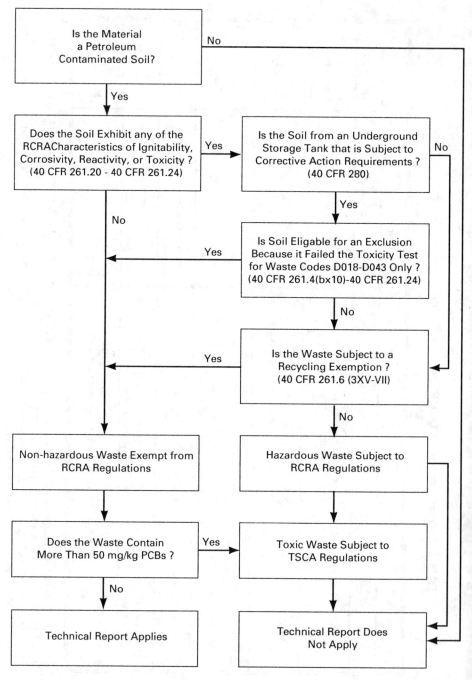

Figure 2 Waste classification decision diagram.

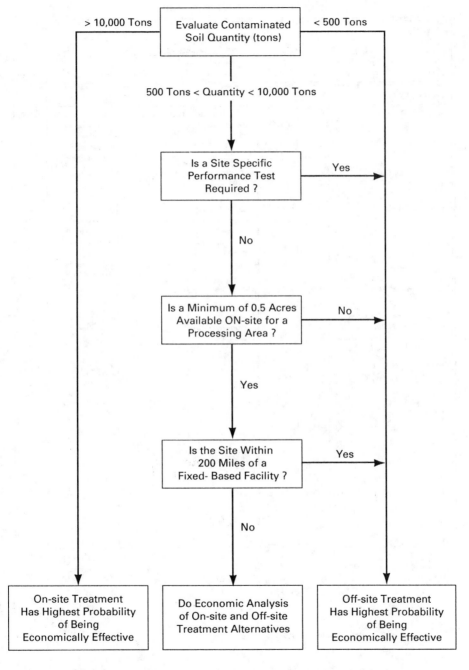

Figure 3 On-site versus off-site treatment decision diagram.

Table 2. Example of Thermal Desorption Critical Success Factor Evaluation

Evaluation Type: On-Site X Off-Site Factors	Units	Condition or Value at Site	Critical Success Factor Category			Score
			Least (1)	Average (2)	Highest (3)	
Contaminant Characteristics						
Petroleum Product Type	NA	No. 6 Fuel Oil	No. 6 fuel oil, lube oil, used motor oil, crude oil	No. 1 fuel oil (kerosene), No. 2 fuel oil (diesel fuel), No. 3 fuel oil, No. 4 fuel oil	Naptha (light or heavy), aviation gasoline, automobile gasoline, jet fuel A and B	1
Concentration of TPH in Contaminated Soil	mg/kg	12,000	>30,000	5,000 – 30,000	<5,000	2
Contaminant Source	NA	UST	Other (Waste may exhibit hazardous characteristics and be subject to RCRA)	NA	Underground storage tank subject to Corrective Action Requirements in 40 CFR 280	3
Concentration of PCBs in Contaminated Soil	mg/kg	<2	>50	NA	<50	3
Total metals concentration (State specific disposal criteria. Rank as least likely to succeed if criteria is exceeded for any metal).			At least one parameter greater than criteria	NA	All parameters less than criteria	3
State or local criteria:						
Antimony NA	mg/kg	5				
Arsenic 55	mg/kg	7				
Barium 2,750	mg/kg	1,450				
Cadmium 55	mg/kg	11				
Chromium 275	mg/kg	34				
Lead 77	mg/kg	45				
Mercury 17	mg/kg	<1				
Selenium 165	mg/kg	<2				
Silver 165	mg/kg	<5				
Thallium NA	mg/kg	<2				
Other _____ _____	mg/kg	_____				

Table 2. Example of Thermal Desorption Critical Success Factor Evaluation (Continued)

Evaluation Type: On-Site X Off-Site Factors		Units	Condition or Value at Site	Critical Success Factor Category			Score
				Least (1)	Average (2)	Highest (3)	
TCLP extract concentrations (RCRA criteria listed below, rank as least likely to succeed if criteria is exceeded for any parameter and no exclusions apply as shown in Figure 2.				At least one parameter greater than criteria	NA	All parameters less than criteria or exclusion applies	3
D004 Arsenic	5.0	mg/kg	<1.0				
D005 Barium	100.0	mg/kg	<5.0				
D006 Cadmium	1.0	mg/kg	<0.5				
D007 Chromium	5.0	mg/kg	<1.0				
D008 Lead	5.0	mg/kg	<4.0				
D009 Mercury	0.2	mg/kg	<0.1				
D010 Selenium	1.0	mg/kg	<0.5				
D011 Silver	5.0	mg/kg	<2.0				
D012 Endrin	0.02	mg/kg	ND				
D013 Lindane	0.4	mg/kg	ND				
D014 Methoxychlor	10.0	mg/kg	ND				
D015 Toxaphene	0.5	mg/kg	ND				
D016 2,4-D	10.0	mg/kg	<2.0				
D017 2,4,5-TP (Silves)	1.0	mg/kg	<1.0				
D018 Benzene	0.5	mg/kg	6.3	NOTE: Benzene exclusion applies because waste is from an UST remediation.			
D019 Carbon tetrachloride	0.5	mg/kg	ND				
D020 Chlordane	0.03	mg/kg	ND				
D021 Chlorobenzene	100.0	mg/kg	ND				
D022 Chloroform	6.0	mg/kg	ND				
D023 o-Cresol	200.0	mg/kg	ND				
D024 o-Cresol	200.0	mg/kg	ND				
D025 p-Cresol	200.0	mg/kg	ND				
D026 Cresol	200.0	mg/kg	ND				
D027 1,4-Dichlorobenzene	7.5	mg/kg	ND				
D028 Dichloroethane	0.5	mg/kg	ND				
D029 1,1-Dichloroethylene	0.7	mg/kg	ND				

148 Thermal Desorption

Table 2. Example of Thermal Desorption Critical Success Factor Evaluation (Continued)

Evaluation Type: On-Site X Off-Site Factors		Units	Condition or Value at Site	Critical Success Factor Category			Score
				Least (1)	Average (2)	Highest (3)	
D030 2,4-Dinitrotoluene	0.13	mg/kg	ND				
D031 Heptachlor	0.008	mg/kg	ND				
D032 Hexachlorobenzene	0.13	mg/kg	ND				
D033 Hexachlorobutadiene	0.5	mg/kg	ND				
D034 Hexachloroethane	3.0	mg/kg	ND				
D035 Methylethylketone	200.0	mg/kg	ND				
D036 Nitrobenzene	2.0	mg/kg	ND				
037 Pentachlorophenol	100.0	mg/kg	ND				
D038 Pyridine	5.0	mg/kg	ND				
D039 Tetrachloroethylene	0.7	mg/kg	ND				
D040 Trichloroethylene	0.5	mg/kg	ND				
D041 2,4,5-Trichlorophenol	400.0	mg/kg	ND	ND = Nondetected. Detection limits were below TCLP criteria for all parameters.			
D042 2,4,6-Trichlorophenol	2.0	mg/kg	ND				
D043 Vinyl chloride	0.2	mg/kg	ND				
Site Characteristics							
Contaminated soil quantity (evaluate for on-site treatment only)		tons	800	<500	500 – 2,000	>2,000	2
Site Usage		NA	Retail	Retail	NA	Other	1
Operational area available (evaluate for on-site treatment only)		acres	1.25	<0.5	0.5 – 2.0	>2.0	2
Surrounding land use (evaluate for on-site treatment only)		NA	Commercial	Public use areas such as schools, health care facilities, or parks; dense urban development; heavy residential)	Commercial, light residential	Industrial	2
Distance to stationary thermal desorption facility (evaluate for off-site treatment only)		miles	NA	>200	100 – 200	100	—
Average ambient temperature at time of treatment		°F	75	<32	NA	>32	3

Table 2. Example of Thermal Desorption Critical Success Factor Evaluation (Continued)

Evaluation Type: On-Site X Off-Site Factors	Units	Condition or Value at Site	Critical Success Factor Category			Score
			Least (1)	Average (2)	Highest (3)	
Regulatory Requirements						
Number of permits required (evaluate for on-site treatment)	NA	3	>4	3 – 4	0 – 2	2
Site specific performance testing required (evaluate for on-site treatment)	NA	No	Yes	NA	No	3
TPH target residual level	mg/kg	50	<1	<10	<100	3
BTEX target residual level	mg/kg	<2	<1	<5	<10	2
Soil Characteristics						
Moisture Content	%	22	>25	10 – 25	<10	2
USCS Soil Classification: Coarse Grained Soils (rate either coarse, fine, or organic soils category, not all three)	NA	NA	GW, GP, GC, Cobbles, Boulders	GM, SP	SW, SP, SM	—
USCS Soil Classification: Fine Grained Soils (rate either coarse, fine, or organic soils category, not all three)	NA	CH	CL, CH	MH, OH	ML, OL	1
USCS Soil Classification: Organic Soils (rate either coarse, fine, or organic soils category, not all three)	NA	NA	OH, Pt	NA	NA	—
Evaluation Summary						
A. Total Sum of Scores in All Categories						38
B. Total Number of Parameters Rated						17
C. Average Composite Score (A/B)						2.23
The data in this worksheet compiled a total score of 38 from 17 rated parameters, with an average score of 2.23. This score warrants further consideration of thermal desorption. The manager should continue through the two additional screening levels.						

150 Thermal Desorption

Benzene concentrations in the leachate exceed the TCLP standard. Moderate regulatory considerations require three permits, little or no performance testing, and residual target levels of 50 mg/kg TPH and <2 mg/kg BTEX. The soil is fine-grained inorganic clay with a moisture content of 22%.

Using the site values recorded on the example worksheet, the reader calculates the appropriate score for each critical success factor. A score of 3 has a "highest" probability of success; 2 indicates "average"; and 1 is the "least likely to succeed." In some instances a particular success factor may not be applicable to an alternative, or data may not be available. Duplicate evaluations must consider on-site and off-site treatment separately, since several data factors apply to only one of these alternatives.

An evaluation summary appears at the bottom of the worksheet. By calculating the total score for all categories and dividing by the number of factors that were rated, the reader can compile an overall composite score. This score indicates the probability for success in this application of thermal desorption.

The composite score is a relative indicator of technical difficulty and treatment cost. Sites that receive a composite score greater than 2.0 are the most technically and economically viable candidates. Treatment costs for these applications will generally range from $35 to $65 per ton. A score below 2.0 indicates lower viability and higher costs ($65 to $125 per ton).

The data in this worksheet compiled a total score of 38 from 17 rated parameters, with an average score of 2.23. This score warrants further consideration of thermal desorption. The manager should continue through the two additional screening levels.

STEP 5: CONTINGENCY PLANNING

The reader can use Table 3 to prepare contingency plans for any critical success factors with a "least" probability for success. In many cases, engineering or administrative procedures can mitigate the possible effects of a parameter with a "least" probability rating.

STEP 6: TREATMENT SYSTEM SELECTION

Figure 4 contains a diagram for determining the most cost-effective size of thermal desorption equipment as a function of contaminated soil volume at a site. A vertical line drawn from the site size value on the x axis will intersect with one or more horizontal operating range bars that represent various sizes of treatment equipment. The systems identified (by the intersection of the line with bars representing them) should continue on to second and third level screening.

Table 3 Critical Success Factor Contingency Analysis.

Characteristic	Reason for impact	Contingency plan
Contaminant Characteristics		
Petroleum product type	Petroleum product requires high treatment temperature.	Selection of thermal desorption devices with appropriate operating temperature range.
Concentration of TPH in contaminated soil	High (>2-3%) concentration of TPH in contaminated soil may cause concentration of organics in thermal desorption offgas to be above lower explosive limit for directly heated thermal desorption devices.	Blend highly contaminated soil with lower TPH concentration soils to reduce overall average concentration or use indirectly heated thermal desorption device.
TCLP extract concentration of metals or organics	Concentration of parameter in TCLP extract exceeds criteria.	Material must be handled as a RCRA hazardous waste. Technical Report does not apply.
Total metals concentration (As, Ba, Cd, Pb, Hg, Se, Ag)	Exceeds state regulatory criteria for preferred treated soil disposal alternative.	Use alternative treated soil disposal option or stabilize treated material.
Concentration of PCBs in contaminated soil	PCB concentration greater than 50 ppm.	Material must be handled as a TSCA toxic waste. Technical Report does not apply.

Table 3 Critical Success Factor Contingency Analysis (Continued)

Characteristic	Reason for impact	Contingency plan
Site Characteristics		
Contaminanted soil quantity	Small quantity of soil (< 500 tons).	Use off-site treatment.
	Large quantity of soil (> 10,000 tons).	Use on-site treatment.
Site usage	Revenue lost from site's normal commercial operations because site is out of service.	Use off-site treatment.
Operational area available	Insufficient operational area available for on-site treatment (Note: area required depends on capacity of mobile thermal treatment system).	Use off-site treatment.
Surrounding land use	Adjoining land uses such as schools, parks, health care facilities, or dense urban development.	Use off-site treatment.
Distance to stationary thermal desorption facility	Transportation cost to ship soils.	Use on-site treatment.
Ambient temperature	Low ambient temperature may cause soil to freeze and be difficult to screen and difficult to thaw in thermal desorber.	Perform project during warmer weather. Crush material before processing in thermal desorption device.

Table 3 Critical Success Factor Contingency Analysis (Continued)

Characteristic	Reason for impact	Contingency plan
Regulatory Requirements		
No. of permits required	Permitting cost.	Use off-site treatment. Performance testing cost. Use off-site treatment or use stack testing data from similar application if appropriate.
TPH target residual level	Capability of meeting performance criteria.	Select technology with appropriate soil treatment temperature and residence time.
BTEX target residual level	Capability of meeting performance criteria.	Select technology with appropriate soil treatment temperature and residence time.
Soil Characteristics		
Moisture content	Soil moisture content too high to feed and process soil properly.	Air dry soil if sufficient area is available, weather is appropriate, and project schedule allows time for drying (may need to consider control of fugitive emissions).
USCS Soil Classification	Soils are classified as group GW, GP, GC, cobbles, or boulders (coarse grained soils).	Screen soil to remove oversize material. Wash rocks or crush rocks to a size that can be processed in thermal desorption system (typically <2.0 inches diameter).

Table 3 Critical Success Factor Contingency Analysis (Continued)

Characteristic	Reason for impact	Contingency plan
USCS Soil Classification	Soils are classified as group CL or CH (fine grained soils).	Reduce soil feed rate and burner firing rate (if applicable) to reduce carryover. Air dry material or blend with lime, kiln dust, or dry soil so that it is below the plastic limit.
USCS Soil Classification	Soil is classified as group OH or Pt (organic soils).	Use alternative analytical technique which is not subject to interferences from humic materials. Correct TPH analytical results on treated soils for apparent background levels in thermally treated soils which have no known petroleum contamination.

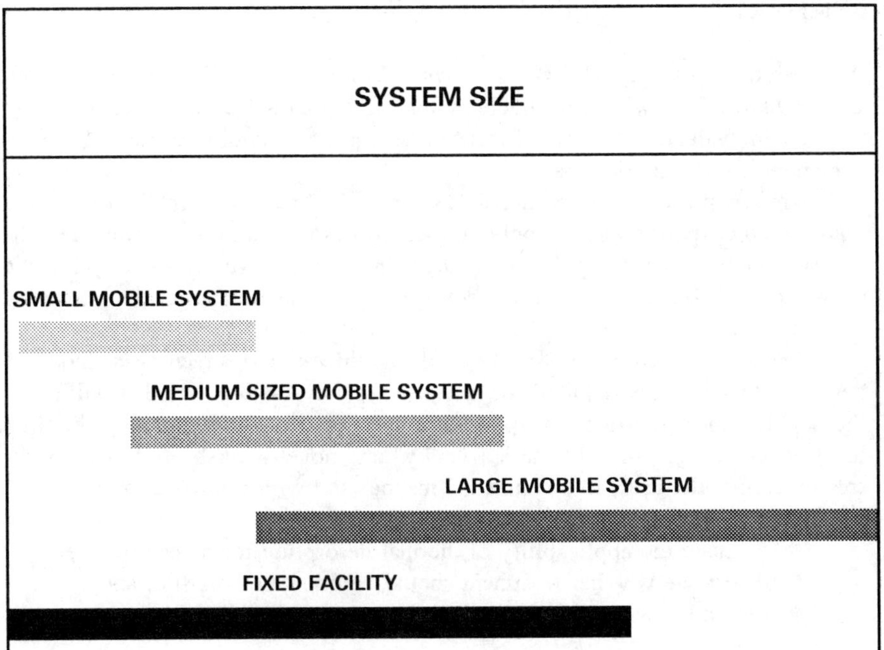

SYSTEM SIZE

SMALL MOBILE SYSTEM

MEDIUM SIZED MOBILE SYSTEM

LARGE MOBILE SYSTEM

FIXED FACILITY

System Characteristics	System Type			
	Small Mobile	Medium Mobile	Large Mobile	Fixed Facility
Number of Trailers	1-2	3-6	7-10	25-150
Primary Burner Capacity (MM Btu/hr)	5-15	15-30	25-50	25-150*
Secondary Burner Capacity (MM Btu/hr)	5-15	15-30	25-50	25-150
Soil Processing Capacity (tons/hour)	5-15	15-30	25-50	25-150

*Some fixed facilities do not include afterburners

Site Size (tons): 0, 2,000, 4,000, 6,000, 8,000, 10,000, 12,000-

Figure 4 Therm desorber size versis site size.

SUMMARY

A research report entitled *Use of Thermal Desorption for Treating Petroleum-Contaminated Soils* will be available by June 1992 that will assist government and industry in evaluating the use of thermal desorption technologies for petroleum-contaminated soil applications.

The report presents a three-tiered "user-friendly" screening method to help a reader predict the success of applying thermal desorption at a specific site. The report also includes comprehensive appendices that serve as an encyclopedic source on all aspects of thermal desorption as they relate to the screening methodology.

This report focuses on the first level of the three-tiered screening method and describes the six steps required for collecting and evaluating data that will affect the application of thermal desorption at a specific site. By completing this first level of screening, using the "user friendly" and novel worksheets and decision-tree diagrams, the user, in a simplified manner and with limited data can:

- evaluate the applicability of thermal desorption for a specific site,
- determine whether treatment should be on-site or off-site, and
- determine what system size will be most cost-effective.

12

THERMAL DESORPTION OF ORGANIC CONTAMINANTS FROM SAND USING A CONTINUOUS FEED ROTARY KILN

Samuel H. T. Chern, Anthony LaRosa and Joseph W. Bozzelli
Department of Chemical Engineering, Chemistry and
Environmental Science
New Jersey Institute of Technology
Newark, NJ

INTRODUCTION

A continuous feed rotary kiln was designed and constructed to study the thermal desorption behavior of organic contaminants from sand with respect to temperature and residence time. Sand was uniformly contaminated with a known amount of target organic compound and fed into a rotary kiln operating at a pre-determined temperature and average residence time with a constant purge gas flow. The analysis of the contaminated sand and the treated sand (thermal treatment with purge in rotary kiln) were both quantitatively analyzed for the target species levels by Gas Chromatography (GC), with flame ionization detection (FID). Tests were run at temperatures between 50 to 250°C, and residence times on 6, 12 and 20 minutes on 1-dodecene and 1-hexadecene, compounds representative of non aromatic petroleum hydrocarbons. The results show that the rate of desorption from sand with increases in temperature and residence time, for both compounds. The results also indicate a removal efficiency of over 97.4 percent.

Thermal desorption is a physical separation process that inputs moderate temperatures or energy to the contaminated soil relative to complete incineration of the soil matrix. An inexpensive carrier gas helps transport the vapor concentrations out of the soil matrix and this limits readsorption as well as increases the degree of desorption. Subsequent incineration or concentration (collection) of this vapor phase can then be accomplished more easily and economically, because it does not include a large mass of soil. This technology is viable, it uses less energy than complete incineration of the soil mass, allows recycle of the soil, and reduces the volume of the contaminants.

Various types of thermal technology have been tested and shown useful for site decontamination. These thermal processes include : rotary kiln, infrared conveyor furnace, fluidized bed, and hybrid thermal treatment.[1]

A rotary kiln design may comprise one of the most versatile and universal waste disposal systems available.[2] Of the approximately 340 incinerators that have come into service in the U.S. since 1969, 12.3 percent are of the rotary kiln design.[3] Solid material can be fed through entrance chutes, concurrently and parallel with the gas flow. The kiln rotation continuously exposes fresh surfaces for desorption/purge transport and also provides for constant solids removal. Kiln angle and rotation speed determine the solids residence time. Thus, a rotary kiln accomplishes the task of providing turbulent mixing of the contaminant species with excess purge gas at high temperatures and at residence times sufficient to remove the organic contaminant to within acceptable levels.

Rotary kiln incinerators have been shown to be generally safe and effective under steadystate continuous feed operation.[4] The information available in the literature on the thermal desorption of contaminants from soils is, however, quite limited.

The objective of this research is to collect data on thermal desorption of organic contaminants from sand matrices by using a rotary kiln thermal desorber so that some working thermal desorption rules of operation based on fundamental parameters of the target contaminants can be developed. This will enable us to establish conditions of temperature, residence time and purge flow to achieve a known final concentration of the contaminant remaining in the sand after the thermal desorption.

The soil material studied in this initial rotary kiln desorption process was sand. This allows simplification of the operating parameters and desorption model since sand has no pore structure compared to soils. The compounds studied in this experiment were 1-Dodecene and 1-Hexadecene.

A continuous feed rotary kiln was designed and constructed in this research. Several parametric experiments regarding reactor temperature and residence time were performed. In thermal desorption experiments, sand was uniformly contaminated with a known amount of organic compound and fed into the rotary kiln operating at a predetermined operation temperature and residence time with a constant purge gas flow. The contaminated sand and the thermal desorption treated sand were both quantitatively analyzed for the target species levels by a Gas Chromatography (GC) with Flame Ionization Detector (FID).

Rotary Kiln Thermal Desorber

A bench scale rotary kiln thermal desorption system was constructed and tested using contaminated sand as the initial evaluation matrices. The apparatus is shown schematically in Figure 1. This is a continuous feed system; that is, the

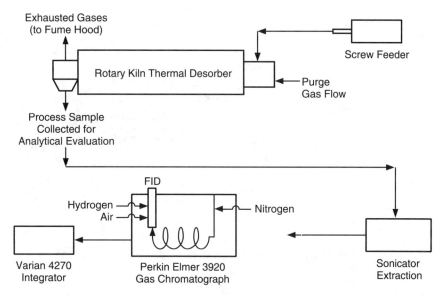

Figure 1 Process schematic rotary kiln thermal desorption system.

contaminated sand is continuously loaded into the rotary kiln by a pulsating screw feeder and desorbed for a predetermined residence time. The bench scale rotary kiln, shown in Figure 2, is 16 inches in length and 4.5 inches in outside diameter. The screw feeder equipped with remote controller for grove rotation and bed pulsation rate is shown in Figure 3.

Two cartridge heaters, Model MWF * 302751 KF, are purchased from Ogden Manufacturing Co., 48 Seegers Road, Arlington Heights, Illinois. The cartridge heaters are used to control the temperature. These heaters are the primary heat source for the system and are 1/2 inch in diameter by 18 inches in length. They operate with a maximum power of 1,500 watts.

The temperature of the reactor is monitored by two type K thermocouples positioned at the center and at the end of the reactor. These thermocouples are placed within a 5 mm ID quartz tube for protection along the reactor. The temperature of the reactor (center thermocouple) is controlled by an Omega Model temperature controller.

An inert air flow is used to purge the reactor and to help remove the organic compounds from the sand. The air flowrate is held at a linear velocity (1.7-1.8 cm/min), with the flow monitored by a rotameter calibrated with a soap bubble meter. The air is supplied from the building's air compressor and prefiltered with activated charcoal and a molecular sieve both at 4060 mesh.

Analysis of the desorption results is performed with a Gas Chromatograph, Perkin Elmer 3920, equipped with a Flame Ionization Detector. The GC column was made of 316 stainless steel, 1/8 inch diameter by 2 meter length, and packed with 15% OV 101 on Chromasorb WHP 80/100. The GC oven temperature is set

160 Thermal Desorption

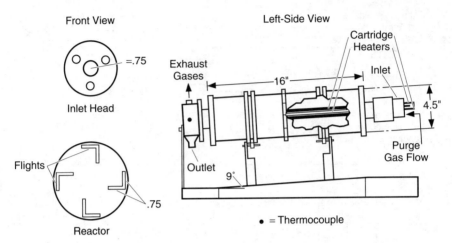

Figure 2 Rotary kiln thermal desorber: left-side view and front view.

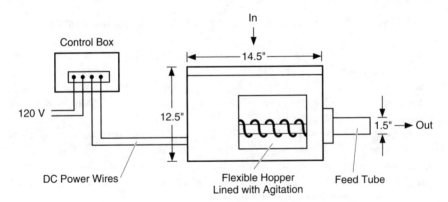

Figure 3 Screw feeder with remote controller.

between 120°C and 190°C in order to separate the target compound from the solvent and the internal standard. The FID output signal is connected to a Varian 4270 reporting integrator.

Preparation and Contamination of Sand

The sand used in the experimental process is placed into an oven (120°C) the night prior to contamination. This removes water and any other volatile species. The mesh size of the sand used in this study was set between 30-50 (particle size 300-600 μm), by use of a sieving apparatus.

The sand is contaminated by taking a known volume of target compound in acetone (solvent) and adding a known amount of sand. The solution and sand are mixed together in a 5-gal drum with a roller tumbler for one hour in order to obtain identical 3 kg batches. This mixture is allowed to stand overnight for thorough absorption of the contaminant into the sand. Afterward, the solvent is evaporated from a flat tray bed 15.5" * 12" width in a fume hood at room temperature for 24 hours until the sand appears dry.

Several different ratios of target desorbent compound in acetone volume to sand were tried. The 5% (w/w) ratio was identified as giving the most uniform concentration of contaminated sand. At a contamination level feasible for our experiments, the level of contamination determined by analysis was consistently observed to be less than that calculated by mass balance. This illustrates that some evaporation of our two target contaminants occurs in this preparation process. Extensive repetition of these contamination procedures consistently yielded the same quantity of mass loss (94.7 percent of mass loss for 1-Dodecene; 70 percent of mass loss for 1-Hexadecene).

Solvent Extractions

Extractions are performed on the uncontaminated sand to ensure that the sand is clean prior to contamination. They are also performed on the desorbed and decontaminated sand to accurately determine target compound concentrations over these sand matrices with no contaminant peaks observed in these studies on clean or cleaned materials. Acetone was selected as the solvent. The concentration of the target organics in the sand is also determined prior to each experiment.

The extraction performed to determine the contaminant concentration in the process sample is described below. A process sample of 7.9 g is placed in a clean 30 ml screw cap vial equipped with a teflon cap liner. Then, 10 ml of solvent (acetone) are added to each of these vials. Each vial is labeled, tightly capped, and placed on a sonicator at room temperature for 30 minutes of extraction. Each vial is allowed to stand overnight, so that the fine particles could settle. These extracts are analyzed on a Perkin Elmer GC 3920 that is equipped with dual columns and a FID. All GC settings for these injections are the same as the parameters for the standard curves. A one microliter liquid injection was performed in duplicate.

Experimental Parameters

The primary parameters affecting the desorption rate of contaminants in this experiment are residence time and kiln temperature. The solid feed rate, in combination with kiln slope and rotation speed, normally determines the solids residence time in a practical rotary kiln. The kiln slope is always maintained at 9 degrees during these experiments. A series of experiments were performed to obtain and verify the residence time distributions on different soil feed rates and kiln rotary speeds. Approximately thirty grains of rice particles were used as tracer materials. The soil feed rate and kiln rotation speed are maintained at a predetermined value prior to each run. Tracer injection is initiated after the operating system reaches steady state. Each residence time data point is recorded thereafter as the time required for a tracer passing through the kiln reactor.

Kiln temperature is another key parameter in this experiment. The temperature employed in the desorption experiments ranges from 50 to 250 degrees Celsius. The representative kiln temperature used as a control parameter is measured by a K type thermocouple at the kiln center.

Experimental Procedure

Before the start of each test, the thermal desorber system is brought up to operating temperature via the cartridge heaters in the kiln for about one hour. The heaters are adjusted by the temperature controller to maintain a constant operating temperature throughout each run. The inert air flow is maintained at 150 ml/min at all times (The linear velocity ranged between 1.7-1.8 cm/min). For this experiment, the rotary kiln is positioned for cocurrent operation; the sand travels in the same direction as the inert air flow.

The process of feeding sand to the kiln is started only after the system is operating satisfactorily at a temperature steady state condition. The feed rate to the kiln and the kiln rotation speed are maintained at the value required to reach the operating residence time. Upon maintaining constant sampling temperature, residence time, and air flow rate the treated sample was collected after one residence time plus 3.3 minutes for analysis.

In analyzing the treated samples that were collected during each run, two steps are performed. First, the samples are extracted by solvent (acetone) in a sonicator. Second, the extracted solutions are analyzed by a gas chromatograph (GC) for the concentration of remaining target compound.

Residence Time Distributions (RTD)

The results of residence time distributions at a kiln rotary speed of 1, 2 and 4.5 rpm are shown in Figures 4-6. As illustrated, the tracer output profile is a rapid sharp curve for a fast feed rate, whereas it turns to be a slow increase and decrease for a slow feed rate, owing to well mixing with the sand particles.

The curve that describes the concentration time relationship of the tracer in the exit stream of a vessel in response to an idealized instantaneous pulse injection is defined as the "C curve" (Levenspiel, 1972). The mean residence time is given by the equation as shown below:

$$\bar{t} = t\,Ce\,/\,Ce \qquad (1)$$

Where Ce is the tracer concentration in the effluent.

The mean residence times of Figures 4-6 are calculated by equation (1) and reported in Table 1. This data is plotted as soil feed rate versus mean residence time in Figure 7. The results demonstrate that as soil feedrate or kiln rotation

Table 1. Summary of Calculated Mean Residence Times (min)

Soil Feed Rate (g/min)	Kiln Rotary Speed		
	1 rpm	2 rpm	4.5 rpm
99	—	9.5	6.2
87	14.8	—	6.8
72	—	10.4	—
56	17.9	12.2	8.4
43	19.9	13.5	10.2
37	21.6	—	—
28	24.2	16.6	11.7

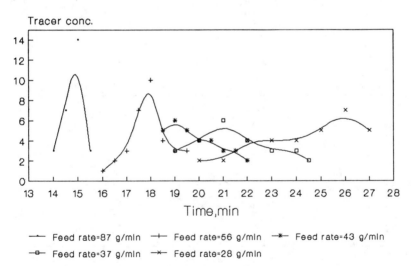

Figure 4 Mean residence time distributions (Kiln rotary speed = 1 rpm)

164 Thermal Desorption

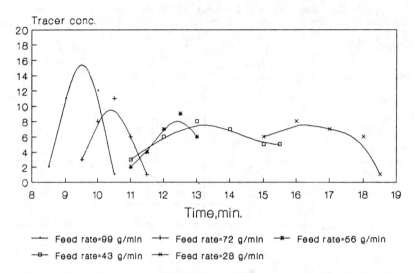

Figure 5 Residence time distributions (Kiln rotary speed = 2.0 rpm)

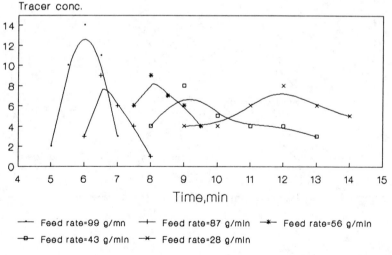

Figure 6 Residence time distributions (Kiln rotary speed = 4.5 rpm)

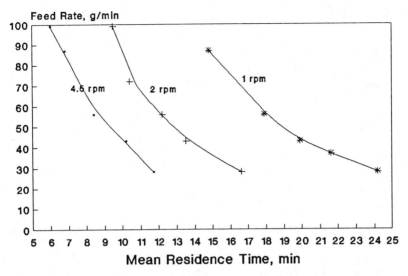

Figure 7 Effect of soil feed rate and kiln rotation speed on mean residence time

speed increases, the residence time decreases. The soil feed rate ranges from 28g/min to 99 g/min in the tests. The kiln rotation speeds used were 1, 2, and 4.5 rpm. The residence times employed in the desorption experiments were of 6, 12, and 20 minutes.

One might suspect whether this is actually representative of the tracer material, because the rice particle has a different size than the sand which is used for the desorption process (mesh #3050). However, another series of residence time distribution experiments were performed using the same size sand (mesh size #30-50) colored by ink. The results confirmed the results described above.

Temperature Profiles

Figures 8 and 9 present two temperature profiles which were controlled at 100°C and 250°C, each measured at high (20 min), medium (12 min), and low (6 min) residence times respectively. These results show that lower temperatures exist in the front part of the kiln due to the continuous feed of cold sand and purge gas. However, the continuous mixing of sand causes uniformity of temperature in the center and rear part of the kiln. The results also illustrate that the higher kiln temperature and shorter residence time give less uniformity in the temperature profiles.

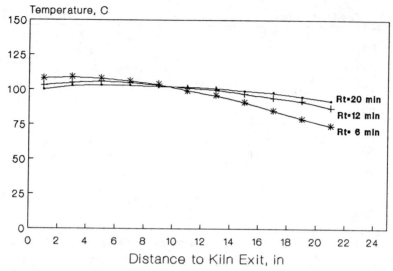

Rt = Residence Time

Figure 8 Temperature profile—100 °C controlled condition

Thermal Desorption Results

Tables 2 and 3 summarize the desorption results of 1-Dodecene and 1-Hexadecene contaminated sand. The percent removal versus operation temperature on high (20 min), medium (12 min), and low (6 min) residence times is shown as Figures 10 and 11, respectively. The results show that the thermal desorber system is highly effective in removing organic contaminants from sand. As

**Table 2. Desorption Results for 1-Dodecene
(Concentrations Remaining, ppm)**

Temperature (Degrees C)	Kiln Rotary Speed		
	6 min	12 min	20 min
50	93	85	51
70	59	38	32
85	42	24	20
100	20	16	ND
125	13	ND	ND

ND = None Detected (Detection limit = 7 ppm)
1-Dodecene initial concentration = 267 ppm

Table 3. Desorption Results for 1-Hexadecene (Concentrations Remaining, ppm)

Temperature (Degrees C)	Kiln Rotary Speed		
	6 min	12 min	20 min
50	1,523	1,334	1,384
	1,514	1,368	1,357
100	1,080	878	788
	1,060	882	779
150	72	53	22
	65	53	23
175	—	—	7
	—	—	8
200	8	3	ND
	8	3	ND
250	2	ND	ND
	3	ND	ND

ND = None Detected (Detection limit = 2 ppm
1-Hexadecene initial concentration = 1,559 ppm

expected, the temperature and residence time are two primary parameters affecting the desorption results. Higher temperatures and longer residence times result in higher removal efficiency.

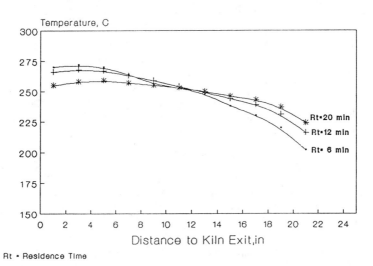

Rt = Residence Time

Figure 9 Temperature profile—250 °C controlled condition

Table 4 lists the removal efficiency of our desorption results. The reduction of the target compounds is greater than 97.4 to 99 percent at temperatures between 100 and 250°C. It should be noted that although the boiling points of 1-Dodecene and 1-Hexadecene are as high as 213°C and 284°C respectively, a high removal efficiency could still be obtained at lower temperatures.

Table 4. Removal Efficiency of Desorption Results, (%)

Analyte	Initial Concentration	Residence Time		
		6 min	12 min	20 min
1-Dodecene	267 ppm	>97.4% (150 deg C)	>97.4% (150 deg C)	>97.4% (100 deg C)
1-Hexadecene	1,559 ppm	>98% (250 deg C)	>99% (250 deg C)	>99% (250 deg C)

CONCLUSIONS

Low temperature thermal desorption was tested with a bench scale rotary kiln on 1-Dodecene and 1-Hexadecene contaminated sand. Tests were run at temperatures from 50 to 250°C, and residence times of 6, 12, and 20 minutes. The results indicate that low temperature desorption removes over 97 percent of the target compounds from sand. Bench scale tests of low temperature thermal desorption indicate that this technology is promising for effective treatment of organic contaminated sand.

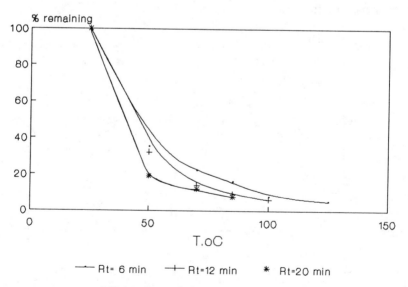

Figure 10 1-dodexene desorption.

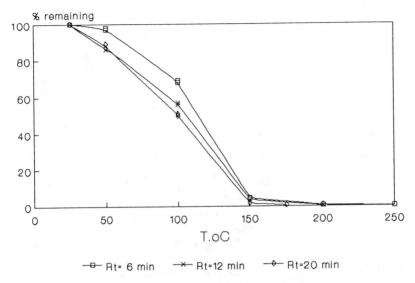

Figure 11 1-hexadecene desorption.

REFERENCES

1. Cudahy, J.J. Decicco, S.G., and Troxler, W.L.," Thermal Treatment Technologies for Site Remediation," IT Corporation, 312 Diractions Drive, Knoxville, TN 37923; Presented at the International Conference on Hazardous Material Management, Catta noga, TN; June 9, 1987.

2. C. Brunner, "Design of Sewage Sludge Incineration Systems" in *Pollution Technology Review Series No.71*, Noyes Press, Ridge Park, NJ,1987.

3. W.P. Linak, J.D. Kilgroe, J.A. McSorley, J.O.L. Wendt, J.E. Dunn, "On the Occurrence of Transient Puffs in a Rotary Kiln Incinerator Simulator: I. Prototype Solid Plastic Wastes," JAPCA 37: 54(1987).

4. S.L. Daniels, D.R. Martin, R.A. Johnson, A.D. Potoff, J.A. Jackson, R.H. Locke,"Experience in Continuous Monitoring of a Large Rotary Kiln Incinerator for CO, CO_2 and O_2" 79th APCA Annual Meeting, June 1985.

13

ENERGY RECOVERY FROM SCRAP TIRES

Hilary Davidson
Duke Power Company
Charlotte, NC

Jy S. Wu and Ravi Vallabhapuram
Department of Civil Engineering UNC
Charlotte, NC

INTRODUCTION

Scrap tires account for about 1% of the nation's solid waste stream. However, tire piles scattered throughout the country not only present an eyesore and potential health and environmental hazards, but also illustrate the lack of utilization of this energy resource. Options appropriate for managing scrap tires are presented, with an emphasis on potential energy recovery and large-scale management.

Managing scrap tires has become an environmental challenge for the 1990's. Stockpiling and landfilling have been widely practiced as two dominant methods of disposal. Stockpiling may pose mosquito nuisance and increase the risk of malaria. Scrap tires in whole occupy a relatively large landfill space, tend to rise to the surface, and are not decomposed in landfills. Tires can ignite between 550 and 650°F and yield a heat of combustion of about 15,000 Btu per pound. The largest tire fire occurred in February 1990 consumed most of the 14 million tires at a tire recycling facility in Ontario.[1] The next-to-largest tire fire in Virginia burned for 9 months and consumed about 7 million tires.[2] A stream of pyrolysis oil was generated at a rate of 50 gallons per minute, requiring over one million dollars from the Superfund to respond to this tire fire.[3]

The United States has stockpiled between 750 million and 2 billion used tires. Of the 280 million tires scrapped each year, about 70% are either landfilled or stockpiled. Eleven percent are being used for energy recovery and the remaining go into a variety of uses including agricultural use for low speed vehicles, reclaim and export as fuel, and reuse in reefs, barriers and breakwaters etc. A small percentage is for rubber modified asphalt paving and pyrolysis decompositions.[4]

This paper examines the current regulatory approaches, and presents the various management options which would be economically feasible and environmentally acceptable for large-scale management of scrap tires.

Legislative Status

Since the introduction of scrap tire management bills by the Congress in 1990, more than 50% of the states have developed regulations to prevent uncontrolled disposal of scrap tires and incentive programs to enhance the scrap tire management. As of January 1991, more than 23 states have passed scrap tire laws governing the storage, processing and hauling of scrap tires. Funding revenues for scrap tire management vary among states including charging $0.5-2 for tire retail, 1-2% additional sales tax, $0.5-2 per vehicle registration or $0.25-1 per tire disposal.

For instance, the state of South Carolina has proposed a comprehensive tracking procedure for waste tire haulers; collection, processing, and disposal facilities; and closure and post closure. A waste tire processing facility cannot accept tires for processing if the storage limit has reached 30 times the daily through-put of the processing equipment. At least 75% of both the waste tires and processed tires must be removed for disposal or recycle from the facility during the year. At closure, tires should be completely removed, and on-site land grading, cleaning and seeding with native vegetation are required.

The State law of North Carolina mandates each county to develop its scrap tire disposal procedures as part of the county's solid waste management plan. Acceptable disposal options include recycling; and monofill or co-disposal for split, ground, chopped, sliced or shredded tires at a permitted landfill or recycling facility. A ban of co-disposal is being considered. A collection facility must clearly indicate an identified, legitimate end use (monofill, co-disposal or recycling) in order to obtain a permit. The number of tires on site must be kept to an absolute minimum. This requirement appears to be less specific in comparison to South Carolina's on-site storage allowance. In contrast, collection/dump sites in Wisconsin are required to pay an escrow of $75 per ton for storage of used tires. The escrow will be completely returned to the depositor once tires are removed from the site, or a partial return of $50 per ton when tires on site are shredded.

Management Options

Tires are made through a process called vulcanization. Unlike glass, newspaper or aluminum, scrap tires cannot be recycled to make new tires. In general, management alternatives for scrap tires can be divided into two categories: non-tire-derived fuel (non-TDF) and tire-derived fuel (TDF) options.[5]

Non-TDF Options

The non-TDF options include use of scrap tires in retreading, rubberized asphalt concrete, pyrolysis units, gasifiers, artificial reefs, and rubber products.

Retreading is the reuse of a tire casing by applying a new tread to replace the original tread. This extends the useful life of a tire from three to five years and, at the same time, reduces the amount of scrap tires for disposal. Retreading of truck tires appears to be a promising market because of the relatively higher cost for a new truck tire. On the other hand, the passenger retread business has reportedly decreased by 50% in the mid 80's due to the low cost of imported and domestic tires.[6]

Rubberized asphalt concrete has several advantages over regular asphalt concrete. The life expectancy of rubberized asphalt concrete is about twice longer than conventional asphalt concrete. In cold areas, it resists ice cracking common to asphalt concrete roads. The potential leach of metals and polynuclear aromatic hydrocarbons could pose an environmental concern. Also, rubberized asphalt concrete typically costs approximately 30-50% more than conventional asphalt concrete.

Scrap tires can be converted into fuel in pyrolysis units. Carbonaceous materials are thermally decomposed in the absence of oxygen; producing oil, producer gas and carbon black. The quality of tire-derived oil is equivalent to #6 fuel oil. The fluctuations in the oil, gas and crumb rubber markets render this option economically unstable.

Tire chips can be converted in a gasifier into producer gas of low Btu values. Because of the marginal market, the use for gasifier is not an economically feasible alternative for the volume of scrap tires generated.

Artificial reefs can be made by tying tires together, putting concrete on them and dropping them into the ocean. Reefs are advantageous to fisherman and provide a habitat for fish. However, this option has limited potential for the volume of scrape tires produced.

Scrape tires can be processed as reclaimed rubber for tires, tubes, shoes, belts and hoses, and shock absorbing rubber for railroads. Other minor uses include soil erosion control, rubber dock bumpers, door mats and mud flaps. There was an obvious drop of reuse of reclaim rubber due to low cost of raw materials for making rubber.

Landfilling has been a common practice for scrap tires, particularly the use of monofills. For instance, a 87-acre monofill site in North Carolina provides a temporary storage of 300 million tires. When tires are delivered to the site, initial sorting is conducted to recover tires that can be retreaded. The remaining tires are taken to a rim crusher to separate the tire from its rim. Tires are shredded into 2" × 2" chips for landfill. The steel rims can be sold at approximately $60 per ton. Tubes can also be sold for reuse. The operation helps save the county's landfill space and provide a future retrieval of the buried tires for energy recovery.

In brief, all of the above non-TDF options currently have limited potential for solving the scrap tire problem due to either high cost or lack of capacity to handle the volume of scrap tires generated.

TDF Options

In comparison to coal, TDF can be an excellent alternate fuel. Table 1 indicates that TDF has a relatively higher Btu, and a lower content of moisture, ash and sulfur. The major concern of using TDF would be its air emission which is inherent to any combustion processes. As seen from Table 2, no significant changes are noted due to combustion of TDF when compared with coal burning.

Conversion of scrape tires into TDF has experienced significant growth. TDF can be used as a supplemental fuel for various types of combustion units including boilers, incinerators and cement kilns. A number of factors must be considered when evaluating the feasibility of burning TDF including costs of the fuel, tire collection, converting tires into TDF, and equipment modification; and permitting associated with burning TDF.

Table 1 Fuel Analysis of Coal and TDF.

Parameters	Typical Coal	TDF
Btu/lb	13,400	15,500
Moisture, 10%	5.2	0.6
Ash, %	6.2	4.8
Carbon, %	73.9	83.4
Hydrogen, %	4.8	7.1
Nitrogen, %	1.8	0.2
Oxygen, %	6.4	2.2
Sulfur, %	1.6	1.2
Chlorine, %	0.1	0.1

Table 2 Comparison of Emission from Boilers Burning TDF/Coal.

Parameters	TDF Performance Compared to Coal
Excess Air	No Change
Combustion Efficiency	No Change
Particular Emission	No Increase in Cement Kilns
Sulfur Oxides	Decrease
Nitrogen Oxides	Decrease
Organic Compounds	No Significant Change
Heavy Metals	No Significant Change

Industrial Boilers

Most large capacity utilities are fired with suspension burners that require coal to be finely pulverized.[7] Smaller capacity utilities employ stoker, wet bottom, cyclone, and fluidized bed boilers. They are used for power and steam generation in co-generation facilities. Table 3 lists TDF requirements for different boilers.

The utility industry is becoming aware of the energy potential in TDF. Attempts have been made to explore the development of a tire-to-energy plant. This includes the conversion of an existing boiler to burn tires as a supplementary fuel, or to build a dedicated tire-fired boiler. Most utility boilers are fired with suspension burners that require pulverized coal. A utility boiler must modify the existing

Table 3 TDF Requirements for Different Boilers.

Boiler Design	Tire Form Requirement
Suspension Fired	Dry, De-wired and Pulverized
Stoker w/Traveling Grate	Shredded, De-wired
Cyclone	Shredded, De-wired
Wet Bottom	Shredded or Whole
Fluidized Bed	Shredded

suspension fired boilers to include a stoker-fired traveling gate, together with redesign of the ash handling and other control systems. The shredded tires, with steel removed, would travel on the stoker grate to allow complete burning. The pulverized coal would be fired in the upper furnace, providing 50 to 75% of the heat input.

The smaller capacity boilers are often fired with mechanical or air-swept spreader stokers. A number of utilities such as the Ohio Edison Utility, the Illinois Power Company, and the Otter Trail Power Company of South Dakota have performed test burns for TDF. Often they have had tire supply problems as well as operational difficulties. Wet bottom boilers allow the molten slag from burning whole tires; they are usually older and do not have appropriate pollution control equipment to handle the emission. A test burn of TDF in a cyclone has demonstrated the potential of emission reduction in sulfur dioxide and trace metals. Fluidized-bed combustion units are vertical, refractory lined and contain a bed of uniformly shaped granular materials (sand, dolomite or limestone). Scrap tires are fed into a chamber where hot particles of air-blown sand and lime ignite the shredded tires.[8]

Emission data for tires compared to refuse-derived-fuel (RDF) are presented in Table 4. As seen from this table, burning TDF appears not to contribute negative environmental factors. There are other cost factors to be considered. Theoretically, the 200 million tires generated annually are equivalent to about 2 million tons of bituminous coal. The tire pieces of 2" or 3" sizing are $10-15 per ton, while 1" pieces are $15-20 per ton.[9] The economic drawbacks of using TDF are the costs associated with purchase of shredding equipment, modification of the fuel handling and feed equipment, higher cost of operation and maintenance for slag,

Table 4 Emission Comparison for TDF and RDF.

Parameters	TDF, ppm	RDF, ppm
CO	30	20
NOx	222	412
NOx w/Control	46	47
SO_2	630	-
SO_2 w/Lime	60	-
VOCs	None Detected	None Detected
HCl	0	250
CHl w/Lime	0	25

replacement of a new ash handling equipment, and NOx control due to high temperature combustion. An estimate of some of these costs is: delivery ($20-58 per ton), operations ($55-70 per ton), tip fee required ($14-40 per ton). Utilities have little incentive to switch from conventional fuel to TDF.

Incinerators

There are two major facilities in the U.S. that use waste tires as their primary source of fuel: the Oxford Energy Plant located in California and the Firestone Plant located in Illinois. The Oxford facility with a capacity of 800 tires/hr operates like a power plant having two boilers with reciprocating stoker grates at the base.[10] These boilers, included 4 burning zones, are designed to burn tires at temperatures in excess of 2500°F. The steam generated is used to drive a turbine generator which produces 14.4 megawatts of electricity. The steel slag formed is recovered for use in road beds. The pollution control system consists of an ammonia injection unit, followed by a filter baghouse and a lime scrubber.[11] The Oxford facility has experienced some operational problems with its feeding system and the accumulation of slag. The company has also experienced financial loss for the initial several years. Several barriers to building a dedicated tires-to-energy facility can be attributed to the high capital cost (almost 2 to 7 times more than a conventional coal plant), the uncertainty of continuous and steady supply of scrap tires, and the difficulty of siting a new facility.

Cement Kilns

The use of TDF in cement kilns has been practiced in Europe for approximately thirty years. In Japan, a large percentage of scrap tires are being burned in cement kilns. There are several major differences between cement kilns and incinerators. Cement kilns are operated at higher temperatures, longer residence times, and relatively uniform temperature within the kiln. There is practically no ash residue since ash is chemically incorporated into the cement clinker. Due to different designs, some kilns are better suited than others for burning tire chips or whole tires. The maximum tire use as a percent of total fuel is given in Table 5.[12]

Cement kilns provide a means of neutralizing the acid gases (90% of the raw materials is limestone), and complete destruction of toxic organic compounds (combustion at temperatures in excess of 3500°F). Table 6 presents the emission data from the Calmat cement kiln in Rillito, Arizona.[12] It indicates that the plant showed no detrimental change when burning TDF. In addition, some cost saving could be realized. By utilizing whole tires as a supplement, the South Carolina Blue Circle Cement has estimated a saving of $180,000 could be resulted from the first year's operation.

Table 5 Maximum Tire Use As Percent of Total Fuel.

Kiln Type	Tire Chips, %	Whole Tire, %
Long Dry Kiln	5	0
Long Wet Kiln	5	0
Long Wet Kiln w/ Scoops	20	0
Pre-calciner Kiln, Air Through	25	15
Pre-Calciner Kiln, Air Separate	50	10
Grate Preheater Kiln	20	15

Table 6 Test Run Data from a Cement Kiln Plant.

Rubber, %	0	30	31	41	33
Coal, %	58	51	49	27	47
Gas, %	42	19	20	32	20
Hydrocarbon, %	0.002	0.002	0.004	0.004	-
CO, %	-	-	0.041	0.019	-
O2, %	8	8	5	4	-
Mercury, ppb	0.069	0.114	0.150	0.055	0.014
Selenium, ppb	<0.9	<11.6	<7	<7	<7
Arsenic, ppb	<0.3	<0.9	<0.6	<0.6	<0.6
Antimony, ppb	<0.3	<0.1	<0.1	<0.1	<0.1
Lead, ppb	<12	<2.2	<1.4	<1.3	<1.3

Other economic consideration is equipment modification and materials handling costs. The modifications are largely dictated by the cement kiln design which, in turn, determines the type of tire feeding system. An automated whole-tire feed system is approximately $700,000 to $1,000,000. A mechanical system to feed shredded tires are $300,000 to $400,000, or $200,000 for a pneumatic system. Shredding and screening may add costs to the tire-derived-fuel, but savings in freight and handling may off-set this.

Product quality is a major concern by the cement industry. Chlorine should be kept below 0.3% of the tire weight.[12] Zinc and lead levels in the clinker are expected to be less than 0.1% and 0.2%, respectively. In general, no adverse impact has been reported on the quality of the cement produced by incorporating TDF.

The availability of fuel and distance from the fuel source is another major economic consideration. There are approximately 120 cement plants distributed throughout the United States which would keep the hauling cost to a minimum. The future capacity of cement kilns to manage scrap tires is largely untapped, both in the number of kilns that have not entered the waste management market and the types of wastes that can be efficiently and effectively managed by cement kilns.

CONCLUSIONS

Each of the non-TDF options examined in this paper may help manage scrap tires, however, the non-TDF options are not apt to handle the large volume of scrap tires generated in the United States. The TDF option could be an attractive alternative for handling the large volume of scrap tires. Of these, energy recovery of scrap tires using cement kilns appears to be an economically and environmentally feasible solution due to (a) tires are excellent and inexpensive fuel source, (b) cement kilns have existing combustion facilities scattered through the county which help reduce the transportation costs, and (c) whole tires can be fed into most cement kilns eliminating shredding equipment and associated costs.

ACKNOWLEDGEMENT

Most of the work reported in this paper is based on Ms. Hilary Davidson's thesis research at the University of North Carolina at Charlotte. The original report contains more than 180 citations. Additional field survey was performed by Mr. Ravi Vallabhapuram.

REFERENCES

1. "It's Out," *Scrap Tires News*, vol. 4, no. 2, March/April 1990, p.4.

2. "Two States Battle Tire Fires," *Scrap Tire News*, vol. 4, no. 2, July/August 1990, p.5.

3. Tejada, S. "Tire Fire Lights Up a National Problem," Proceedings of the Scrap Tire Processing and Recycling Tour and Seminar, Recycling Research Institute, April 1989, pp. 1-2.

4. Serumgard, J.R. "Special Waste: Scrap Tires," Presented at the Recycling Business and Technology Conference, Rubber Manufacturing Association, February, 1990, pp. 3-4.

5. Davidson, H.S. "Energy Recovery from Scrap Tires," M.S. Thesis, The University of North Carolina at Charlotte, 1991.

6. "Scrap Tires: Deflating a Growing Problem," *Compressed Air Magazine*, December 1988, p.10.

7. "Scrap Tires: A Private Solution to a Public Concern," *Scrap Tire News*, vol. 4, no. 5, September 1990, p.6.

8. Pope, K.M. "Tires to Energy in a Fluidized Bed Combustion System," Proceedings Waste Tires as A Utility Fuel, January 1991, p.6.

9. Schwartz, J.W., Jr. "Engineering for Success in the TDF Market," Proceedings of the Scrap Tire Processing and Recycling Tour and Seminar, Recycling Research Institute, April 1989.

10. Dick, P. "Out to Confound the Skeptics," *Business*, January 7 1990, p.E.

11. "The Modesto Energy Project," Oxford Energy Corporation, California, 1988.

12. Kohl, R. "Use of Scrap Tires in the Cement Industry," Proceedings from the Scrap Tires Processing and Recycling Tour and Seminar, Recycling Research Institute, May 1989, p.1.

14

FATE OF TOXIC METALS AND PHENOL IN THE BOTTOM SEDIMENT OF WET DETENTION/RETENTION PONDS RECEIVING HIGHWAY RUNOFF

L. Y. Lin
Christian Brothers University
Civil Engineering Department
Memphis, TN

Y. A. Yousef
University of Central Florida
Civil and Environmental Engineering
Orlando, FL

J. A. Feuerbacher
University of Central Florida
Civil and Environmental Engineering
Orlando, FL

INTRODUCTION

Sediments collected from the bottom of four highway detention ponds were extracted using EPA's Toxicity Characteristic Leaching Procedure (TCLP). The extracted fluids were analyzed for 6 trace metals, 68 semi-volatile, and 36 volatile organic priority pollutants. The results indicated that only several pollutant compounds were found in the extraction. In accordance with the EPA regulations, all of the concentrations were below the toxic level. Therefore, the sediments cannot be classified as hazardous materials. In addition, the fate of Cu, Pb, Zn, and phenol in sediments was studied through batch experiments. It appeared that the sediment has strong affinity with these pollutants. The Freundlich isotherm and a first order equation can describe the equilibrium. Although heavy metals and phenol compounds have minimal leaching capacity from the sediments, the leaching potential may increase because of a high accumulation rate. Periodical remediation is required in order to secure the groundwater resources.

Highway detention ponds have been demonstrated to be one of the most promising techniques to control runoff. The facility provides a temporary storage basin to store the excess highway runoff and relieve flooding conditions following rainfall events. In the meantime, it also functions as a reactor and settling chamber. The pollutants associated with the particles in the runoff can be directly settled on the bottom of the pond or indirectly removed within the pond by physical, chemical, or biological processes (Yousef et. al, 1990). As a result, runoff quality will be improved before it discharges to the receiving waters

Previous studies had found that heavy metals and phenol were the most detected species in the highway runoff and sediments because of automobile and wear of road surface (Startor and Boyd, 1972). A study of highway runoff in the Washington D.C. area by Wilber and Hunter (1977) found that Cu, Pb, and Zn are the three major metals in runoff, comprising approximately 90-98% of all metals observed. Since metals are easily liganded with hydrophilic ions, the compounds may settle and accumulate in the basins of detention ponds. The accumulation of heavy metals in the bottom sediment of highway detention pond has been widely documented (Wigington, 1983; Harper, 1985; Yousef, 1986, 1990). However, the potential leaching of those contaminants to underground sources has rarely been reported. Phenol may contribute from gasoline spills. According to the Nationwide Urban Runoff Program (NURP) report, three compounds out of eleven organic pollutants frequently detected in highway runoff were phenols. A study conducted by the U.S. Geological Survey (Schiffer, 1989) on a Longwood, Florida retention pond found that high concentrations of phenol compounds, ranging from 1.9 μg/l to 18 μg/l, were presented in the sediment samples. It appeared that the highway and parking lot were the sources contributing to these compounds.

Because most of the pollutants deposited on the bottom sediments have very limited susceptibility to photochemical or biological oxidation that could induce the accumulated sediment to reach relatively toxic levels. In order to protect the water resource underneath of the detention pond, the leaching and the fate characteristics were investigated. The specific objectives of the present research were to (1) measure the potential leaching concentrations of volatile, semi-volatile, and toxic metals including Cr, Cd, Cu, Ni, Pb, Zn in the bottom sediments using the Toxicity Characteristic Leaching Procedure (TCLP); (2) study the fate of Cu, Pb, Zn, and phenol through the batch experiments; (3) provide the general remediation guidelines for maintenance of highway detention ponds and protection of the groundwater resources.

MATERIALS AND METHODS

Four highway detention ponds located in the State of Florida (Figure 1), each constructed and operated by the Florida Department of Transportation, were

1. Silver Star Road, Orlando
2. S.R. 40 and S.R. 25, Ocala
3. S.R. 507 and Crown St., Palm Bay
4. S.R. 580 and U.S. 19, Clearwater

Figure 1 Location of highway detention ponds in Florida.

selected for this research. The selection was based on the age, pond surface area, drainage basin characteristics, surrounding traffic volume, and construction features. The age of each pond was calculated from the beginning of the operational period of the pond until the sampling date. A determination of the pond surface area, drainage basin area, and percent of land use was made by an ALTEK digitizer provided by the U.S. Geological Survey. The average daily traffic volumes (ADT) were provided by the FDOT for each road that contributed stormwater runoff to each of the ponds. The construction was measured by conducting a survey for each pond.

Following the designated procedure (Yousef and Lin, 1990), fifteen to twenty bottom sediment core samples were collected from various locations throughout each pond. Once the core sample was obtained, it was securely placed in a freezer and transferred to the laboratory. The sediment properties were conducted to determine sediment density, moisture content, volatile content, grain size distribution. Also, the frozen core samples were extruded from the sample core and sectioned into six layers: top loose layer, 0-1 cm, 1-3 cm, 3-5 cm, 5-10 cm, and greater than 10 cm of parent soil. Identical sediment sublayers from each core sample were combined and mixed to be a composite sample.

The Toxicity Characteristic Leaching Procedure (TCLP) is designed to determine the mobility of organic and inorganic contaminants present in liquid, solids, and multiphase wastes. The testing procedure was published by the U.S. EPA in the November 7, 1986 Federal Register. The Millipore Hazardous Waste Filtration and Zero Head Extractor were used for the present research. During the preliminary evaluations, each composite sediment was determined for the percent of solids, the particle size, and which of the two extraction fluids will be used for extraction of volatile, semi-volatile, and heavy metals. As the preliminary evaluations were completed, the sediment sample was extracted with the selected solution for volatile, semi-volatile, and heavy metals. A Bechmen Spectra-Span Direct Coupled Plasma Spectrometer was used for metal analysis. Volatile and semi-volatile were examined using GC/MS at the UCF priority Pollutants Laboratory.

Uptake of Cu, Pb, and Zn in the bottom sediment was studied with adsorption experiment. The pond water collected from the field was filtered and spiked with various concentrations of Cu, Pb, and Zn stock solutions. The concentrations varied from 500 $\mu g/l$ to 25000 $\mu g/l$. The mixture was then placed in one liter plastic bottles on a shaker for 24 hours, pH in the solution was adjusted using 1.0 N of NaOH or HNO_3 to the original condition. Five grams of composite oven-dry sediments was added to each bottle and shaken continuously. At 1 hr, 6 hr, 1 day, 3 days, 5 days, 12 days, and 30 days, 20 ml of aliquots were withdrawn centrifuged, filtrated, acidified, and analyzed for Cu, Pb, and Zn.

The fate of phenol in the sediment was to delineate the sorption behavior in sediments as well. Two experiments using oven-dried and top fresh sediments were conducted for each studied pond. Five grams of oven-dry sediments and twenty-five grams of fresh sediments were transferred to clean Erlenmeyer flasks separately. 100 mls of phenol solution prepared with reagent grade phenol (C_6H_5OH) and distilled/deionized water was added to each flask. The concentrations ranged from 1 ppm to 200 ppm. After phenol's solutions were spiked into the samples, a double layer of parafilm was immediately placed on the top of the flask. The flasks were then placed on a shaker operating at 30 rpm. Each flask was removed at 1 day, 2 days, 4 days, 9 days, 11 days, and 15 days. The supernatant was centrifuged and decanted. The resulting solutions were analyzed using the 4-aminoantipyrene method (EPA, 1986).

RESULTS AND DISCUSSIONS

The laboratory analysis of the sediment core showed that the top accumulated sediment had a difference in color from the original parent soil. The sediment had higher moisture content, volatile (organic) content, and porosity than that of the parent firm soils (Table 1). The moisture content in the top layer varies from 52.3% at Ocala to 63.2% at Palm Bay. A rapid decline occurs in the percent moisture content in the sediment. The top layer also showed a high percentage of organic content, varying from 4.3 at the Clearwater pond to 7.1 at the Orlando pond; because the top sediment receiving particulate matters, plants, and animal debris accounts for the higher percentage of moisture content and volatile content. The grain size distribution analysis indicates that more than 90% of the sediment consists of particles less than 0.425 mm in size. The uniformity coefficients of the studied ponds calculated from 1.75 (Orlando) to 2.22 (Ocala). Based on the soil analyses, it can be concluded that the bottom sediments were poorly graded sands, diverse from medium to fine-sand.

Table 1. Sediment Characteristics of Four Highway Detention Ponds

Pond	Porosity		% Moisture		% Volatile		Wet Density	
	Loose	Parent	Loose	Parent	Loose	Parent	Loose	Parent (g/cm^3)
Orlando	0.45	0.72	62.8	34.7	7.1	2.9	1.27	1.68
Ocala	0.45	0.67	52.3	18.5	6.5	1.8	1.38	1.99
Palm Bay	0.46	0.74	62.2	22.3	4.5	0.8	1.28	1.92
Clear Water	0.44	0.78	59.8	24.0	4.3	1.8	1.31	1.87

Metal concentrations extracted from the TCLP solution are listed in Table 2. In general, Pb and Zn were two major species found in the TCLP solution because of high accumulation (Yousef and Lin, 1990). The Orlando pond seems to have the most abundance of Pb and Zn concentrations. High content of organic and fine texture in the sediment may be the reason the metal bound with the sediment (Lin and Yousef, 1990). As the external environment changes, the metals were released from the sediment. Other EPA priority pollutants including 68 semi-volatile organic compounds and 36 volatile organic compounds in the TCLP solution were also analyzed. With the exception of diethyl phthalate and bis(2-ethylhexyl) phthalate, none of the semi-volatile compounds were detected. Also, only four volatile organic compounds, which are methylene chloride, acetone, 4-methyl-2-pentanone, and chloroform, were found in the TCLP solution. The summary of semi-volatile and volatile pollutants in the sediment is listed in Table 3 and Table 4.

All concentrations of volatile, semi-volatile, and heavy metals detected during the course of this study were lower than the EPA toxic guidelines (40 CFR,

1986). Therefore, it can be concluded that the bottom sediment of highway detention ponds are not hazardous materials.

Due to insignificant leaching from the sediment, the potential contamination of groundwater from the existing sediments is minimal.

Table 2. Concentrations of Metals in the TCLP Extraction Solution

Pond	CD (µg/l)	Cr (µg/l)	Cu (µg/l)	Ni (µg/l)	Pb (µg/l)	Zn (µg/l)
Orlando	79	217	27	127	1902	3683
Ocala	9	52	47	336	1432	693
Palm Bay	317	45	42	612	1455	377
Clear Water	23	127	140	717	1664	553

Batch sorption studies of selected metals (Cu, Pb, Zn) and sediments showed that although adsorption and desorption occur simultaneously in the solution, it seems that the rate of adsorption significantly exceeded the rate of desorption. The equilibrium concentration was rapidly achieved within 72 hours as shown in Figure 2. It evidenced that the metals have strong affinity with the bottom sediment. Among the three selected metals, Pb had the highest adsorption rate to the sediment, followed by Cu and then Zn. The Orlando pond appears to have the highest uptake capacity to adsorb metals. A high percent of fine texture and organic contents could be the reason (Weber, 1972). Based on the mass balance in each batch, the ratio of the metal removed from the sediments was calculated. Figure 3 shows that the Freundlich isotherm provides the best fit for sorption. This result agrees on the conclusions made by Zabowski and Zasoski (1987). The distribution coefficients of Pb, Cu, Zn were 4.88-1.79 l/g, 4.08-1.51 l/g, and 0.68-0.11 l/g, respectively.

Experimental results of phenol removal in the oven-dried sediment showed a rapid decline of phenol concentration with increasing contact time for various initial concentrations. However, the removal was dependent on the initial solution concentration used. For initial solution concentrations of 1 ppm, 9.410 ppm, and 97.77 ppm, almost 100 % of the phenol spiked were removed after contact times approximate to 2, 4, and 9 days, respectively. The 196.7 ppm initial concentration reached an equilibrium between 9 and 11 days, removing only 60% of the initial phenol. The maximum phenol uptake by the oven-dried sediment was about 3,000 µg/g dry sediment.

The batch experiment using fresh, untreated sediment to remove phenol seemed to have faster rates than the oven-dried sediment. At 7 days, almost 90 percent of

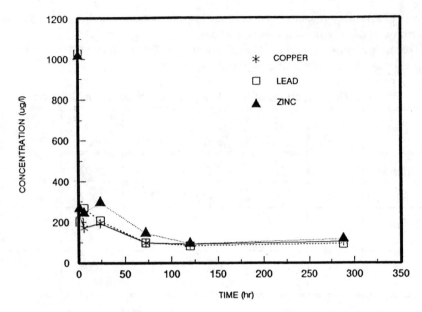

Figure 2 Metal sorption in bottom sediment of Palm Bay Pond.

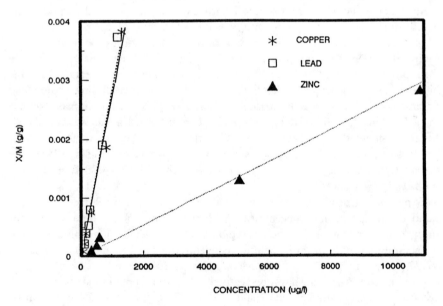

Figure 3 Freundlich isotherm for heavy metals in bottom sediment of Palm Bay pond.

the phenol was removed from the fresh sediment. A comparison of the oven-dried and the fresh sediment is shown in Figure 4. It reveals that the fresh sediment had three to six times higher removal rate than the oven-dried sediment. It could be the biological activity involved. The identical result was reported by Genther (1989). He found that freshwater and estuarine sediments have higher capacity to degrade chlorophenol. Also, Hwang and Hodson (1986) found that microbial degradation is the primary transformation process from phenol.

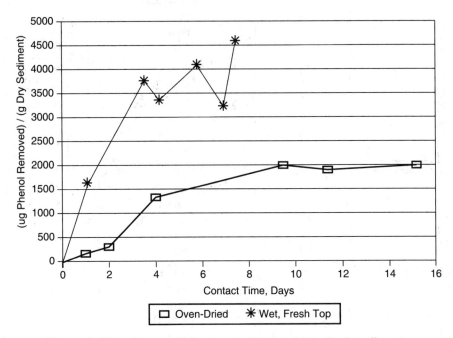

Figure 4 Phenol removal from oven-dried and wet fresh sediment by batch sorption.

In addition, the top layer sediment contains a higher organic content and fine materials. This is the fact influencing the greater rate of phenol adsorbed onto the sediment.

The rate of phenol removal was conducted using the computer program SYSTAT (1990). The removal of phenol from the oven-dried and fresh sediments can be described by the following first-order equation:

$$C_t = C_o \, e^{-kt} \qquad (1)$$

where: C_t is the concentration of phenol in solution at a particular time, ppm; C_o is the initial concentration of phenol in solution, ppm; k is the rate constant; hr^{-1}; and t is the time of contact with sediment, hrs. The rate constants of these studies

Table 3. Semi-Volatile Organics Detected in TCLP Extracts for Detention Pond Sediments

Pond	Layer of Sediment	Semi-volatile Organic Compound	Concentration (ug/l)
Orlando	Top	none	—
	1	none	—
	2	Diethl Phthalate	11
	3	none	—
	4	none	—
	5	none	—
Ocala	Top	none	—
	1	none	—
	2	none	—
	3	none	—
	4	none	—
	5	none	—
Palm Bay	Top	Bis(2-ethylhexyl)phthalate	10
	1	Bis(2-ethylhexyl)phthalate	20
	2	none	—
	3	none	—
	4	none	—
	5	none	—
Clearwater	Top	none	—
	1	none	—
	2	none	—
	3	none	—
	4	none	—
	5	none	—

Table 4. Volatile Organics Detected in TCLP Extracts for Detention Pond Sediments

Pond	Layer of Sediment	Volatile Organic Compound	Concentration (ug/l)
Orlando	Top	Methylene Chloride	6
		Chloroform	7
		4-Methyl-2-Pentanone	21
	1	none	—
	2	Acetone	26
		Chloroform	7
		4-Methyl-2-Pentanone	17
	3	4-Methyl-2-Pentanone	17
	4	4-Methyl-2-Pentanone	21
	5	Chloroform	9
		4-Methyl-2-Pentanone	19
Ocala	Top	none	—
	1	none	—
	2	Methylene Chloride	5
		Acetone	120
	3	Acetone	90
	4	Methylene Chloride	7
	5	Methylene Chloride	7
		Acetone	95
Palm Bay	Top	none	—
	1	none	—
	2	none	—
	3	none	—
	4	none	—
	5	none	—
Clearwater	Top	Methylene Chloride	6
		Acetone	40
		4-Methyl-2-Pentanone	21
	1	4-Methyl-2-Pentanone	18
	2	Acetone	14
		4-Methyl-2-Pentanone	19
	3	4-Methyl-2-Pentanone	17
	4	Methylene Chloride	16
		Acetone	30
		4-Methyl-2-Pentanone	22
	5	Methylene Chloride	7
		Acetone	15
		4-Methyl-2-Pentanone	19

are listed in Table 5. The partition coefficient (K_p) using the Karichhoff equation (1979) was calculated as well. The predicted adsorption using Karckhoff K_p and the maximum possible phenol removal in the sediment were very low compared with that observed in all cases examined. This indicated that some mechanism(s) other than adsorption, such as biological activity, as described by the Karickhoff model are most likely responsible for the major removal of phenol in the sediment.

Table 5. Summary of the Rate Constants for Phenol Adsorption

Sediment	Initial Concentration (mg/l)	Rate Constant (hr^{-1})	R^2
Oven-Dry	1.000	0.500	0.981
	9.410	0.015	0.892
	97.770	0.009	0.939
	196.700	0.005	0.957
Wet	97.770	0.006	0.969

When applying equation (1) to the highway detention pond system, it shows that bottom sediment had high affinity for phenol and large quantities of phenol can be removed from aqueous solutions. Half of the phenol in water is deemed to be removed within less than five days. Since the rate constant in equation (1) was increased exponentially as the initial solution concentration decreased, it implies that any phenol entering a highway detention pond will be removed in a few hours.

CONCLUSIONS

Sediment samples obtained from the bottom of highway detention ponds were investigated to determine if significant quantities of the priority pollutants, including 6 trace metals, 68 semi-volatile and 36 volatile organic compounds may leach from the sediment using the Toxicity Characteristic Leaching Procedure. As a result, most compounds were non-existent, and the remaining were detected under EPA's regulations. Overall, the sediments cannot be considered as hazardous waste. Also, it appeared that there is a minimal threat to groundwater due to leaching from the sediments.

The fate of the selected metals (Cu, Pb, Zn) and phenol in the sediment were studied through the batch experiments. All of the species with sediments can achieve equilibrium within several days. It indicated that selected metals and phenol tend to be less mobile due to their high affinity with fine texture and organic matters in the sediment. The Freundlich isotherm for the selected metals and a first-order equation for phenol were the best-fit models to describe the equilibrium between adsorbent and adsorbates. Wet, fresh sediments appeared to remove phenol from the solution three to six times faster than oven-dried sediments. From Karickhoff's model, the biological activity associated with the sediment may be involved.

Accumulation of metals and phenol on the bottom of highway detention ponds are expected to be increased due to high affinity of those species with the sediments. Results of the leaching potential will be increased instantly. In order to secure the underground water resources, periodical remediation needs to be conducted. It suggests that the removal cycle is approximately 25 years per cycle, therefore, it can prevent the groundwater contamination.

ACKNOWLEDGEMENT

This research was supported by the U.S. Federal Highway Administration and Florida Department of Transportation. The authors wish to acknowledge the financial support and technical assistance of Messers Win Lindeman and Gary Evink.

REFERENCES

1. Genther, B.R., Price, W.A. Price, and Pritchard, P.H., "Anaerobic Degradation of Chloroaromatic Compounds in Aquatic Sediments Under a Variety of Enrichment Conditions," *Applied and Environmental Microbiology*, Vol 55, pp. 1466-1471, 1989.

2. Hwang, H., and Hodson, R.E., "Degradation of Phenol and Chlorophenol by Sunlight and Microbes in Estuarine Water," *Environmental Science and Technology*, Vol 30, pp 1002-1007, 1986.

3. Karickhoff, S.W., Brown, D.S., and Scott, T.A., "Sorption of Hydrophobic Pollutants on Natural Sediment" *Water Research*, Vol 13, pp. 241-248, 1979.

4. Lin, L. Y., and Yousef, Y.A., "Metal Transport Through Bottom Sediments of Wet Detention Ponds Receiving Highway Runoff," *Proceeding of International Conference on Computer Applications in Water Resources*, Vol 2, pp. 730-737, 1991.

5. Schiffer, D.G. "Effects of Three Highway-runoff Detention Methods on Water Quality of Surficial Aquifer System in Central Florida," U.S. Geological Survey Report 88-4170, 1989.

6. "SYSTAT- Computer Program," SYSTAT Inc., Version 5.01, 1990.

7. Startor, J.D., and Boyd, G.B., "Water Pollution Aspects of Street Surface Contaminants," U.S. Environmental Protection Agency, EPA R2-72-081, Nov. 1972.

8. Yousef, Y.A., and Lin, L.Y., "Toxicity of Bottom Sediments in Detention Ponds," *Environmental Engineering Proceedings of ASCE*, p. 809-816, 1990.

9. Yousef, Y.A., Hvitved-Jacobsen, T., Harper, H.H., and Lin, L.Y., "Heavy Metal Accumulation and Transport Through Detention Ponds Receiving Highway Runoff," *The Science of the Total Environmental*, Vol 93, pp. 433-440, 1990.

10. US EPA, "Hazardous Waste Management System: Land Disposal Restrictions," 40 Code Federal Register, 1986.

11. US EPA, "Test Methods for Evaluating Solid Waste," Vol 1B: Laboratory Manual, Physical/Chemical Methods," 1986.

12. Weber, W.J. "Physicochemical Processes for Water Quality Control," John Wiley & Sons, 1972.

13. Winginton, J.P., Randall, C.W., and Grizzard, T. J., "Accumulation of Selected Trace Metals in Soils of Urban Runoff Detention Basins," *Water Resources* Bulletin, Vol 19, pp. 709-718, 1983.

14. Wilber, W.G., and Hunter, J.V., "Aquatic Transport of Heavy Metals in the Urban Environment," *Water Resources Bulletin*, Vol 13, pp. 721-734, 1977.

15. Zabowski, D., and Zasoski, R.J., "Cadmium, Copper, and Zinc Adsorption by a Forest Soil in the Presence of Sludge Leachate," *Water, Air and Soil Pollution*, Vol 36, pp. 103-113, 1987.

15

IN-SITU VOC REMOVAL FROM GROUNDWATER USING AIR STRIPPING WICKS

Walter Konon
Department of Civil and Environmental Engineering
New Jersey Institute of Technology
Newark, NJ

INTRODUCTION

The treatment of contaminated groundwater, particularly where potable water supplies are affected, is a major environmental problem. With the support of the Hazardous Substance Management Research Center at NJIT professor Walter Konon has invented a cleaner wick system that is effective in removing volatile organic compounds (VOCs) from groundwater where dewatering by direct pumping is impractical.

During the last fifteen years, the installation of prefabricated vertical wick drains has replaced the use of sand drains as the most cost effective method of achieving soil consolidation. These plastic geotextile wicks can be installed to depths of over 100 feet and serve as vertical water migration paths in poorly draining soils. This established drain wick technology, used for soil consolidation, was modified by the Principal Investigator so that the air stripping cleaner wicks can be used to achieve in situ removal of VOCs from groundwater. The principle used is the transfer of VOCs from solution in a liquid to solution in a gas. This is accomplished by a diffused aeration system at each wick. An individual air stripping wick consists of a hollow flexible plastic tube installed in the center of a conventional drain wick core (Figure 1). Air under pressure is forced down the tube and exits at the bottom inside the wick core void. The air mixes with the contaminated groundwater (which has flowed inside the core void through the filter fabric) and drives the water and volatiles up through the core (Figure 2).

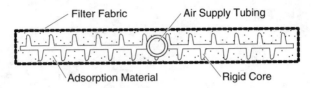

Figure 1 Schematic of a cleaner wick.

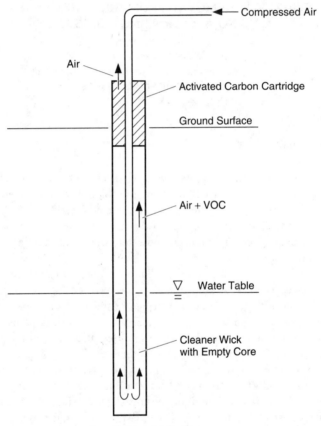

Figure 2 Modified cleaner wick with activated carbon cartridge.

The VOCs are discharged through the wick at the ground surface and can be adsorbed by AC filter cartridges located on top of each wick or can be collected for surface treatment.

Air stripping wicks can be installed to depths of up to 100 ft. by conventional drain wick installation equipment. Typical depths of installation would be 40 feet deep. The lateral spacing of individual cleaner wicks at a particular site would depend on soil permeability and would range from 3 ft. to 10 ft. centers. The cleaner wick system is comprised of multiple wicks installed in a checkerboard pattern over the contaminated groundwater plume area, or as an alternate, the in situ cleaner wicks can be installed as an air stripping wall with wicks at approximately 5 ft. on centers and with the wall several stages deep.

The air stripping cleaner wick system can be used to economically remove the VOCs from contaminated groundwater and reduce the levels of these contaminants to the required MCLGs by in situ treatment without the problems associated with pumping, surface treatment and discharge of treated waters. There are particular advantages for this system in poorly draining soils where well pumping is ineffective or where the water table should not be lowered, such as at the perimeter of landfills and contaminated sites where the water table outside the landfill must be kept at a higher elevation than inside the landfill to induce groundwater flow toward the contaminated area and prevent further off site contamination.

DESIGN MODEL

A design model for water flow in the cleaner wick system and the concurrent air stripping of VOCs (when natural groundwater flow occurs due to a hydraulic gradient), can be subdivided into several distinct processes (Figure 3).

- a) Rectilinear flowing water is affected by the air lift discharge of a cleaner wick and water is drawn toward the wick.
- b) Water flows through the wick filter fabric, enters the core then circulates upwards through the inside of the core due to the upward movement of the compressed air.
- c) The compressed air bubbles cause diffused aeration to occur inside the wick core. This causes air stripping and transfers the VOCs from solution in a liquid to solution in a gas.
- d) The water flowing up the core rises above the top of the groundwater surface, flows through the filter fabric and circulates back into the groundwater.

AIR STRIPPING

In water treatment, aeration is a unit process in which the water and air are brought into contact with each other for the purpose of transferring volatile substances from the water. Removal efficiency depends on the air to water ratio, hydraulic loading rate, and some of the physical properties of the contaminants like solubility, molecular weight, vapor pressure and Henry's constant.

Favorable equilibrium is a prerequisite for air stripping of trace organic compounds. The greater the Henry's constant, the greater the driving force for the transfer of the solute from the aqueous to the gas phase. The Henry's constant of some organic compounds increases by a factor of approximately 1.6 with each 10 degree centigrade rise in the temperature. Thus, temperature is a key parameter affecting the extent of volatile organic compounds in gas-liquid contacting processes.

Organic contaminants can enter an aquifer and be transported great distances because they have little affinity for soils. There is a long list of organic compounds typically found as contaminants in the ground water, some of them are:

1) Trichloroethylene
2) 1,1,1-Trichloroethane
3) Tetrachloroethylene
4) Carbon tetrachloride
5) Cis-1,2-dichloroethylene
6) 1,2-Dichloroethane
7) 1,1-Dichloroehtylene
8) Toluene
9) Benzene

The majority of these compounds are halogenated carbon compounds which are widely used in commerce as solvents. Contamination by organic chemicals is generally associated with situations in which they are made, spilled, used or discarded.

AIR STRIPPING MODEL

To develop a typical performance model for in situ treatment by air stripping using cleaner wicks, the following model criteria were assumed:

1. Hydraulic conductivity of soil
   $K = 500$ gpd/ft^2 K-2.36×10^{-2} cm/sec
2. Hydraulic gradient 1%
3. Air Flow Q=1 ft /min per wick
4. Distance between two wicks (across groundwater flow) = 5 ft
5. Wick depth D = 40 ft
6. Groundwater Temperature (20 to 24 degrees centigrade)
7. Four rows of Wicks each 5' apart (Figure 4).
8. All the water flowing 2.5' left and right of the wick will pass through the wick due to the action of the air lift.

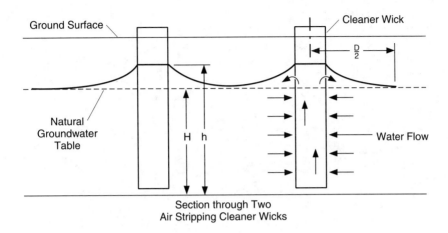
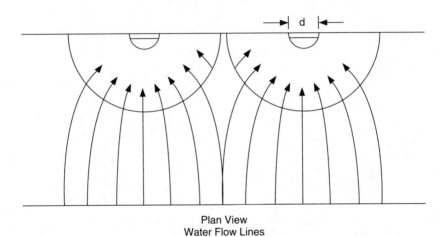

Figure 3 Schematic of water flow at air stripping cleaner wicks

198 VOC Removal

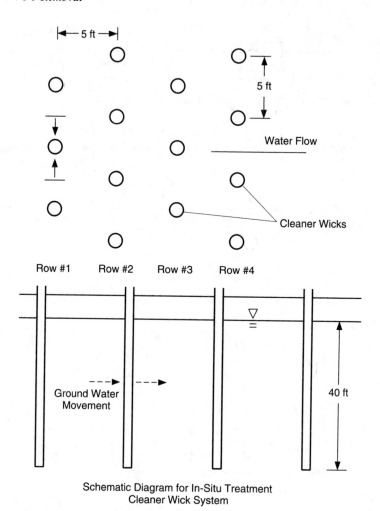

Figure 4 Schematic diagram for in-situ treatment cleaner wick system.

Based on this criteria water passing through each wick per day was calculated to determine actual air-to-water properties of the contaminant, the removal efficiency (for different volatile organic contaminants found commonly in groundwater in the USA) was calculated.

For the stated model criteria for a 4 row system of wicks and groundwater contaminated with Tricholoethylene at 1000 ppm the concentration of contaminant would drop to 126 ppm after the first row, to 16 ppm after the second row, to 2 ppm after the third row and to less than one ppm after the fourth row.

The air stripping model results show that under typical groundwater flow conditions the cleaner wick air stripping barrier system can achieve significant VOCs contaminant removal due to the air to water ratios provided by the system.

GROUNDWATER FLOW MODEL

The results indicated by the air stripping model can only be achieved if all the water flowing through the air stripping barrier is, in fact, captured by the wicks and is air stripped. The flow of rectilinear groundwater is diverted toward the wicks due to the suction produced by the airlift inside the wick. The proper center to center spacing of wicks to achieve a significant percentage of water capture by the wicks is critical to the effective operation of the system. This center to center spacing is affected by specific site conditions such as the permeability of the soil and hydraulic gradient and system operating conditions such as the air pressure supplied to the wicks.

WICK FLOW TESTS

In an effort to obtain data regarding the amount of water flowing toward the wick (which is equivalent to the water flowing up the wick) laboratory tests were conducted on an individual cleaner wick. The first series of tests placed the wick into a water tank and measured the water flow up and out of the wick once a steady state water flow condition had been achieved. The laboratory set up shown in Figures 5, 6, 7 consisted of the following: The bottom of the air tube was 5 ft. below the water surface. The wick filter fabric with a reported $K = 2 \times 10^{-2}$ cm/sec was covered with a plastic cover for 12 inches below and above the water table (exit pipe) level. Air flowed down the wick air tube and exited at the bottom within the wick core. The air flow pressure, air flow volume and resulting water flow volume out of the wick core were measured. The water level in the tank was maintained 1 inch below the level of the exit pipe by continuously adding water to the tank during the test. The results of this test #1 are shown in Table 1. These results show that the cleaner wick can capture and lift a volume of water equal to approximately 25 of the air volume supplied to the wick. This shows that a significant amount of water flows toward the wick.

A second series of tests placed the cleaner wick into a water tank which was filled with sand and again measured the water flow up and out of the wick once a steady state water flow condition had been achieved. The results of a grain size sieve test for the sand which filled the tank showed a SW soil. The D_{10} size of this sand material was 0.26 mm with a coefficient of permeability in the 3×10^{-2} cm/sec range. The laboratory set up shown in Fig 5 was the same as used in the first series of tests except that the sand encased the wick up to a depth of 6" below the water table (exit pipe) level. The results of this test #2 are shown in Table 1.

The results from the second series of tests show that when water flow to the wick core is restricted by soil there is a significant reduction in water flow out of the wick. Under these conditions the cleaner wick captured and lifted a volume of water equivalent to 3% of the air volume supplied which resulted in a water flow of some 1.5 liters of water per minute.

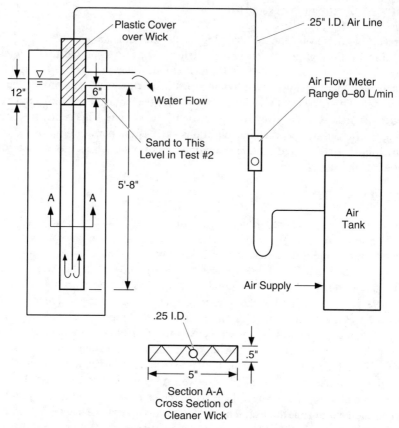

Figure 5 Laboratory test set up for cleaner wick water flow testing.

Figure 6 Cleaner wick with air supply tube attached to air flow meter and compressed air tank.

Figure 7 Cleaner wick operating at 5 psi air pressure with water flowing from exit tube into measuring tank.

Test 1. Wick in Water Only, No Soil – NJIT 6/3/91

Air Flow Pressure PSI	Air Flow Volume		Water Flow Volume		Water to Air Flow Percent	No Water Flow, Static Rise of Water Above Water Table Inches
	Cu. Ft/Min	(L/Min)	Cu. Ft/Min	(L/Min)		
5	1.06	(30)	.36	(10.1)	34	5
10	1.94	(55)	.44	(12.6)	23	8
15	2.30	(65)	.54	(15.2)	23	12

Test 2. Wick in Water with Sand Around It – NJIT 6/4/91

Air Flow Pressure PSI	Air Flow Volume		Water Flow Volume		Water to Air Flow Percent	No Water Flow, Static Rise of Water Above Water Table Inches
	Cu. Ft/Min	(L/Min)	Cu. Ft/Min	(L/Min)		
5	1.06	(30)	.04	(1.1)	44	1
10	1.94	(55)	.05	(1.4)	3	2
15	2.47	(70)	.06	(1.7)	3	3

Based on the water flow results recorded in the laboratory with the cleaner wick placed in the sand (0.05 cu ft/min) and using the Figure 4 air stripping model example (which has a K similar to the K of the lab soil) a water flow of 72 cu ft/day would be expected to enter the wick. This would only be 54% of the water flowing past an individual wick (100% is 134 cu ft/day) in the model example. The air stripping model example, however, extends 40 ft. below the water table, while the experimental wick extends only 5.67 ft. below the water table. It is anticipated that if the laboratory wick test length were increased in length seven fold to 40 ft. there would be a seven fold increase of wick surface and a significant increase in water flow to the wick.

The laboratory results for unrestricted water flow toward the wick (water only condition) are the upper limit on the maximum flow to the wick. The capacity of the cleaner wick in soil to move water up the wick is limited only by the quantity of water entering the wick (up to the maximum limit). Using the above stated argument an expected 300 to 400 cu ft of water per day would flow toward and be circulated up the cleaner wick in the air stripping example shown in Figure 4. Since the daily flow past one wick in this example is 134 cu ft/day the expected wick flow would provide 2 to 3 full circulations of the volume of water passing the wick. This confirms that the assumed 5 ft. center to center spacing of the wicks in the example would insure that most of the groundwater passing the wick would flow up the wick.

RESULTS TO DATE

a) Development of air stripping and groundwater flow model for the system
b) Laboratory testing to determine water flow up the wick cores
c) Peer review of items (a) and (b) by Geraghty and Miller - Groundwater Consultants (Dr. Rao and A. Vernic) with their recommendation that the cleaner wick air stripping system should be field tested
d) Proposals to and meetings with site representatives for possible field testing of the system

CONCLUSIONS

Removal efficiency of the cleaner wick system depends on providing the required air to water ratio. The required air to water ratio can be provided in situ by establishing multiple air stripping rows of cleaner wicks.

The cleaner wick system attracts, lifts, aerates, and circulates significant amounts of surrounding groundwater. The experimental laboratory data on water flow up the wick and theoretical calculations on the diameter of influence of the system indicates that cleaner wick spacing of 3 ft. to 10 ft. on centers would insure that the rectilinear water traveling between wicks will be drawn to the cleaner wick and aerated.

The cleaner wick system can remove VOCs by air stripping and now needs to be field tested at a suitable site. In a contaminated groundwater flow situation this would consist of establishing an air stripping barrier and installing observation wells to test the level of contaminants before and after the barrier. Only such a field test can conclusively demonstrate the level of effectiveness of the cleaner wick system.

REFERENCES

1. W. Konon, "In-situ Cleaner Wick System for Removal of Groundwater Contaminants," Proc. Twelfth Annual Madison Waste Conference, Madison Wisconsin, Sept. 20, 1989.

2. Patents: A U.S. patent (No. 4883589) was issued on Nov. 28, 1989 for "System for Removing Contaminants from Groundwater." The scope of the patent may be expanded.

3. W. Konon, "Testing of Air Stripping Wicks," Hazardous Substance Management Research Center, NJIT, SITE-24 Final Report, Jan. 1992.

16

NEUTRON ACTIVATION ANALYSIS OF HEAVY METAL BINDING BY FUNGAL CELL WALLS

Theodore C. Crusberg and John A. Mayer
Departments of Biology & Biotechnology and
Mechanical Engineering
Worcester Polytechnic Institute
Worcester, MA

INTRODUCTION

Aqueous effluents are produced during nuclear power and nuclear weapons development activities which frequently contain low levels of dissolved radioactive nuclides. Reactor coolant waters, evaporator condensate, fuel-reprocessing waste streams, wastewaters and runoff from uranium ore processing and contaminated ground-waters all pose challenges to radionuclide removal processes. Removing radioactive metal ions from these and other effluents is an important consideration of the growing nuclear industry which is under substantial and growing pressures from both federal agencies and environmental organizations. Inorganic zeolites have proven effective in reducing the release of radionuclides into the environment (Grant and Skriba, 1987; Blanchard et al., 1984).

A number of laboratories are now focusing attention to renewable biological materials to provide traps for low concentrations of dissolved radioactive metal ions in wastewater effluents. Uranium (VI) biosorption studies have been reported using yeast, algae and several other organisms (Bryerley, et al., 1987). A group at Oak Ridge National Laboratories (TN) has reported the use of a bacterium encapsulated in gelatin to remove low levels of Sr and Cs, the only contaminants found in wastewaters at ORNL, from water (Faison, et al., 1990; Watson, et al., 1990). Chitosan and chitin from marine species have also been shown to effectively bind heavy metals (Coughlin, et al., 1990), and marine red algae cell walls are commercially employed in removing heavy metal ions from industrial wastewaters (Darnall and Gabel, 1989). Fungi of different species have been shown to effectively remove dissolved uranium and thorium from aqueous solutions (Tsezos and Volesky, 1982; Kuyucak and Volesky, 1988; Marques, et al., 1990).

All of these materials represent living or non-living cells or even cell components which can in most cases reversibly bind metal ions and remove them from aqueous solution. The term BIOTRAP can be used to describe such materials, and in this laboratory cell wall preparations of the fungus Penicillium ochro-chloron have been employed to demonstrate their capacity and affinity to reversibly bind and remove copper(II), an important component in electroplating, and other industrial wastewaters, from solution (Crusberg, 1991). The purpose of the research reported here was to: (1) investigate the feasibility of three preparations of biomaterials as effective BIOTRAPS for removing metal ions from solution and, (2) to test that BIOTRAP which provided the best overall characteristics for its ability to bind mercury (Hg^{+2}).

Since neutron activation analysis (NAA) was readily available, that method was one of several applied to this problem as a suitable analytical methodology to study heavy metal-to-BIOTRAP interactions. Copper and mercury provide good examples of metals which are capable of undergoing activation by thermal neutrons. In NAA, ^{63}Cu (69.1% natural abundance) is converted to ^{64}Cu which has a half live of 12.7 hr, and ^{202}Hg (29.7% natural abundance) is converted to ^{203}Hg which has a half life of 46.,6 d. Both emit gamma photons which are easily detected, ^{64}Cu at 1.34 MEV and ^{203}Hg at 0.279 MEV. The possible application of a BIOTRAP to the removal of radioactive metal ions from aqueous solution could be inferred from these experiments.

EXPERIMENTAL

Copper and mercury standard solutions and chitin were purchased from Sigma Chemical Co. (St. Louis, MO, U.S.A.). **Porphyra tenera** was purchased locally as the dried product, and the growth of 3-4 mm beads of the fungus **Penicillium ochro-chloron** has been previously described (Crusberg, et al., 1991). All candidate BIOTRAPS were washed extensively in deionized water, then incubated in 90 g/L Na_2CO_3/60 g/L $NaHCO_3$, pH 9.5 for 30 min., washed 3 times with 5 times their settled volume of deionized water, followed by a wash with gentle agitation (100 rpm) in 1 N HCl for 3 hrs. The BIOTRAPS were then washed thoroughly with deionized water until the effluent attained a pH of 4. The BIOTRAPS were then dehydrated with three washes of acetone (the final one decanted only after 12 hrs.), air dried and stored in a desiccator over sulfuric acid. One gram amounts of BIOTRAP were used in each experiment. The BIOTRAP was incubated with 1 L of solution at pH = 4, containing 1000 mg of Cu or the given amount of Hg, for 1 hr. at 30°C, with gentle stirring. The BIOTRAP was collected in a Buchner funnel and washed 6 times with 100 mL aliquots of deionized water adjusted to pH = 4 with 1 N HCl. This volume was demonstrated to successfully wash all unbound metal from the BIOTRAP preparation by assaying the effluents for copper according to a standard spectrophotometric method (American Public Health Association, 1981). The BIOTRAP was allowed to dry overnight in a hood which facilitated placing it into the vials for activation analysis.

Neutron activation analysis (NAA) was conducted in a General Electric Corp. 10 KW open pool reactor with low enriched uranium (20% ^{235}U) fuel, operated at 1 KW. (nominal flux 10^{10} n/cm^{2}sec) for 30 min. Samples were introduced and removed from the core with a manual insertion device. Activated specimens were analyzed after a suitable cooling period to allow decay of radioisotopes with short half lives. The vials were placed (one at a time) in a Coaxial Germanium Detector System linked to a 8K-multichannel analyzer (MCA) interfaced with a PC. Decay rates were extrapolated back to the time of removal from the core. Employing standard methods, gold foils were used to determine the neutron flux density in the vicinity of the BIOTRAPS in the reactor core. Also, known concentrations of both Cu and Hg were simultaneously activated to serve as standards, since count rates are proportional to the total number of atoms of an isotope present in the sample. The concentration of metal ions in solution was determined from the difference between the total amount known to be present in the 1 L volume and the amount found adsorbed to the BIOTRAP by NAA.

Following activation, the various BIOTRAPS were allowed to decay for 4 half lives and then were treated with 1 L of 1 N HCl at room temperature with gentle agitation to desorb the bound metal ions. After a wash in a Buchner funnel with three 100 mL aliquots of pH 4 deionized water the filter cakes were air dried and again subjected to NAA to determine the efficiency of desorption.

RESULTS

Typical energy peak spectra for samples of marine shellfish chitin and beads of **P. ochro-chloron** cell walls incubated for 60 min. with 1000 mg/L Cu^{+2} in deionized water at pH 4, and then washed thoroughly with deionized water and dried are shown in Figures 1 and 2. Copper presents a peak at 1.347 MEV. Spectra were also recorded for the **Porphya tenera**-Cu^{+2} system, and in all cases a cooling off period of 4 hr. was employed to allow the short half life nuclides (such as ^{24}Na) an opportunity to decay. The energy peak spectrum of **P. ochro-choron** beads alone showed only the presence of ^{40}K which might be expected since naturally occurring ^{39}K (which can absorb a neutron and become radioactive ^{40}K) is so ubiquitous in the environment. Table 1 shows the results of the copper-binding studies, with all count rates extrapolated back to the time each specimen was removed from the reactor core. Desorption of Cu^{+2} with 1 N HCl was complete since no ^{64}Cu signal was observed after treatment with diluted acid.

Similar experiments were performed for 1g samples of **P. ochro-chloron** cell walls (in bead form) in the presence of 500, 1000, 2000, and 5000 µg/L Hg^{+2}.

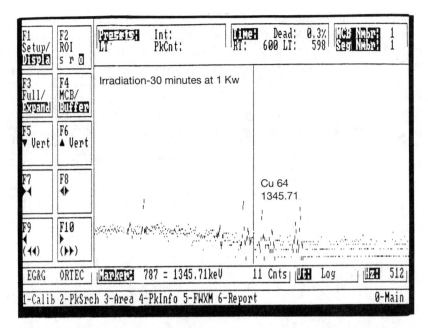

Figure 1 Energy peak spectrum for a sample of chitin with adsorbed Cu.

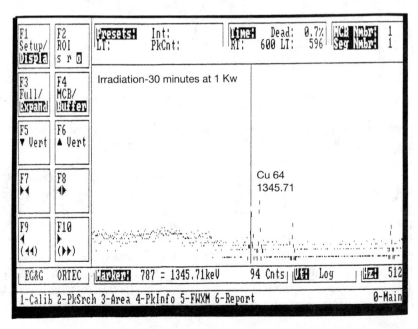

Figure 2 Energy peak spectrum for a sample of p. ochro-chloron cell wall with adsorbed Cu.

cooling off period of 4 half lives was routinely employed in these studies before desorption of metal ions with 1 N HCl was attempted. Figure 3 shows the energy peak spectra for fungus alone, fungus after incubation with 500 µg/L Hg^{+2}, and the background. ^{203}Hg exhibits its characteristic energy peak at 0.279 MEV. The samples were saved for 4 months and then when the radioactivity was at a safe level, they were treated with 1 N HCl in hopes of desorbing the bound mercury. Figure 4 shows the energy peak spectra for the fungus after incubation with 1000 µg/L Hg^{+2} and after washing with HCl. No appreciable desorption occurred.

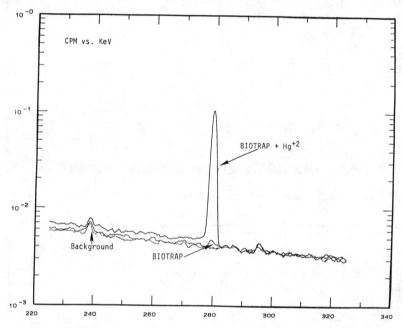

Figure 3 Energy peak spectra of p. ochro-chloron BIOTRAP treated with 1000 mg/L Hg^{+2}, BIOTRAP alone, and background.

Since the aqueous concentrations of Hg^{+2} present in the solutions after adsorption of metal ion by the fungus could be calculated from the difference between total mercury added to the L of water in which the fungal beads were dispersed, and the amount adsorbed to the fungus, a Langmuir-type sorption coefficient could be determined. Figure 5 shows the double reciprocal plot of $1/r$ (r = µg Hg^{+2} bound/g fungus) vs. $1/m$ (m = mercury ion in solution at equilibrium) from which the maximal binding (r_{max}) can be found by extrapolation to the ordinate and the binding constant K_c can be found from the slope = K_c/r_{max}. The data were fitted by linear regression, giving the equation:

$$1/r = 0.00147 + 2.44 \ (1/m) \wedge R^2 = 0.862$$

From this, the maximum binding of Hg^{+2} by the fungus is 680 μg/g and the sorption coefficient is about 1700 ug/L (1.7 mg/L).

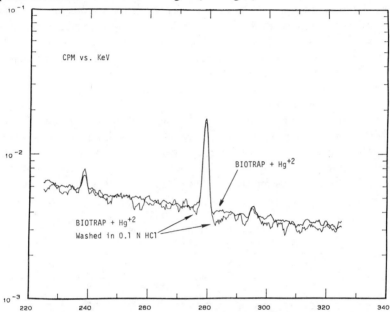

Figure 4 Energy peak spectra of P. ochro-chloron BIOTRAP after treatment with 500 mg/L Hg^{+2} and after further treatment with 0.1 N HCl.

Table 1. Neutron Activation Analysis of Three Biotraps Incubated with Equal Amounts of Cu^{+2}

DISCUSSION

The purpose of the first part of this study was to identify a suitable BIOTRAP for further study. The fungus **P. ochro-chloron** proved to be the best candidate overall. The affinity of the marine alga **Porphyra tenera** for Cu^{+2} was indeed better than was the affinity of fungus for that same metal. However **Porphyra** is very difficult to handle, forming mats which absorbed water and swelled. Filtration of water through the algal mat during the binding studies proved difficult, taking several hours to prepare a single 1 g sample. On the other hand the fungal beads were porous as expected from scanning electron microscope observations (Crusberg, et al., 1991) and were easily handled. Chitin too was porous and easily manipulated but did not prove as good a sorbent for copper ion.

Red marine algae are used in a commercial process for heavy metal removal and recovery from industrial wastewasters, but must be entrapped in a gel matrix (Darnall, 1989).

Figure 4 shows that Hg^{+2} is not removed after adsorbing to the BIOTRAP even after attempting to elute with 1 N HCl. Since the fungus is a renewable BIOTRAP it may be better to accept this apparent irreversibility of Hg^{+2} binding and either try to remove the heavy metal by another physicochemical process or to dispose of it while still sorbed onto the fungus in a secure landfill, since mercury is so very toxic in the environment. Biological detoxification of mercury involves conversion of Hg^{+2} to a volatile form, such as dimethylmercury or elemental mercury, or to precipitate the metal as a sulfide or other insoluble inorganic salt (Aiking, et al., 1985). Removal of bound Cu^{+2} from fungus BIOTRAP by 1 N HCl is indeed complete however.

The demonstration here that **P. ochro-chloron** cell wall BIOTRAP functions with both copper and mercury ions suggests that it could be incorporated into a commercial process for removing these and perhaps other heavy metal ions from waters containing low levels of radionuclides of concern to the nuclear power industries and to the government-owned installations involved in nuclear weapons production.

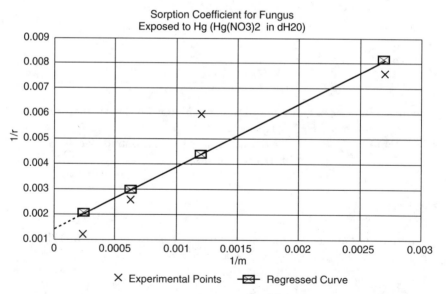

Figure 5 Linear regression analysis of Hg^{+2} binding to penicillium ochro-chloron cell walls.

REFERENCES

1. Aiking, H., H. Govers, and J. van't Riet, "Detoxification of Mercury, Cadmium, and Lead in Kebsiella aerogenes NCTC 418 Growing in Continuous Culture," *Appl. Environ. Microbol.* Vol. 50, pp. 1262-1267, 1985.

2. American Public Health Association, *Standard Methods for the Examination of Waters and Wastewaters*, 15th ed., Procedure 313C, Washington, D.C., pp. 193-194, 1981.

3. Byerley, J.J., and Scharer, J.M., "Uranium(VI) Biosorption from Process Solutions," *Chem. Eng. J.* Vol. 36, B49-B59, 1987.

4. Blanchard, G., Maunaye, M., and Martin, G., "Removal of Heavy Metals from Waters by Means of Natural Zeolites," *Water Res.* Vol. 18, pp. 1501-1507, 1984.

5. Coughlin, R.W., Deshaies, M.R., and Davis, E.M., "Chitosan in Crab Shell Wastes Purifies Electroplating Wastewaters," *Environ. Prog.* Vol. 9, pp. 35-39, 1990.

6. Crusberg, T.C., Weathers, P.J., and Baker, E.F., "Biotrap for Heavy Metal Removal and Recovery from Industrial Wastewaters," pp. 91-96, In H.M. Freeman and P.R. Sferra, eds., *Innovative Hazardous Waste Treatment Technology Series*, Vol 3, Biological Processes, Technomic Publ. Co., Lancaster, Pennsylvania, 1991.

7. Darnall, D.W., and Gabel, A. "A New Biotechnology for Recovering Heavy Metal Ions from Wastewaters," pp. 217-224, Proc. Third Intern. Conf. on New Frontiers for Haz. Waste Management, U.S. Government Printing Office, Washington, D.C., EPA600/9-89/072, 1989.

8. deRome, L., and Gadd, G.M., "Use of Pelleted and Immobilized Yeast and Fungal Biomass for Heavy Metal and Radionuclide Recovery," *J. Indust. Microbiol.* Vol. 7, pp. 97-104, 1991.

9. Faison, B.D., Cancel, C.A., Lewis, S.N., and Adler, H.I., "Binding of Dissolved Strontium by Micrococcus luteus," *Appl. Environ. Microbiol.* Vol. 56, pp. 3649-3656, 1990.

10. Giesy, J.P., and Paine, D., "Uptake of Americum-241 by Algae and Bacteria," *Prog. Water Technol.* Vol. 9, pp. 845-857, 1977.

11. Grant, D.C., and Skriba, M.C., "Removal of Radioactive Contaminants from West Valley Waste Streams Using Natural Zeolites," *Environ. Prog.* Vol. 6, pp. 104-109, 1987.

12. Kuyucak, N., and Volesky, B., "Biosorbents for Recovery of Metals from Industrial Solutions," *Biotechnol. Lett.* Vol. 10, pp. 137-142, 1988.

13. Marques, A.M., Bonet, T.R., Simon-Pujol, M.D., Fuste, M.C., and Congregado, F., "Removal of Uranium by an Exopolysaccharide from **Pseudomonas** sp.," *Appl. Microbiol. Biotechnol.* Vol. 34, pp. 429-431, 1990.

14. Tsezos, M., and Volesky, B., "The Mechanism of Thorium Biosorption by Rhizopus arrhizus," *Biotechnol. Bioeng.* Vol. 24, pp. 955-969, 1982.

15. Watson, J.S. , Scott, C.D., and Faison, B.D., "Adsorption of Sr by Immobilized Microorganisms," *Appl. Biochem. Biotechnol.* Vol. 20/21, pp. 609-709, 1989.

17

STRUCTURAL ANALYSIS OF A FUNGAL BIOTRAP FOR REMOVAL OF HEAVY METALS FROM WASTEWATERS

T. C. Crusberg, G. Gudmundsson, S. C. Moore,
P. J. Weathers, and R. R. Biederman
Worcester Polytechnic Institute
Departments of Biology & Biotechnology,
Mechanical Engineering and Biomedical Engineering
Worcester, MA

INTRODUCTION

With the advent of federal regulations (40 CFR Part 433) industries are no longer permitted to discharge their heavy metal contaminated wastes into waterways. To meet the challenge to make the environment cleaner and safer many physical and chemical technologies were rapidly developed including chemical oxidation and reduction, precipitation, electroplating, evaporation, ion exchange, ultrafiltration and reverse osmosis to reduce heavy metals in industrial wastewaters to below federal discharge limits (Freeman, 1990a,b; Lindsay, 1989). The older processes often generated other forms of hazardous wastes, such as heavy metal hydroxide sludges which also require highly regulated and costly disposal.

An alternate to these often costly and inefficient older methods are the use of so-called 'traps' or materials capable of removing heavy metals from wastewaters that have advantages over the traditional treatments. There are two different kinds of traps, biological and non-biological. An example of the latter are natural zeolites (Blanchard, et al., 1984), and an intermediate type (Dudley, et al., 1991) made up of certain types of soils. The term BIOTRAP has been coined (Crusberg, et al., 1989a) to describe biological materials or biosorbents (intact living cells or even dead cells or cellular components) which are capable of reversibly binding heavy metal ions in solution. BIOTRAPS are renewable resources which offer attractive solutions to many problems associated with the more conventional methods, such as lower costs, and potential for recycling and reuse of the trapped metal ions.

Many microorganisms have been studied as possible biosorbents including bacteria (Giesy & Paine, 1977; Kuhn & Pfister, 1990), yeasts (Norris & Kelly, 1977; deRome and Gadd, 1991), and algae (Darnall & Gabel, 1989; Mahan, et al., 1989). Likewise shellfish chitin and chitosan have been shown to bind heavy metals (Coughlin & Deshaies, 1990; Nishi, et al., 1987). Fungi have been found growing in saturated solutions of copper sulfate in electroplating industries (Fukami, et al, 1983; Stokes and Lindsay, 1979). One operational problem in adapting most of these biosorbents for possible industrial processes in removing and recovering heavy metals from waste streams is that they must be encapsulated in gel form for packaging into a batch or column adsorber system.

This study reports the application of the fungus **Penicillium ochro-chloron** (ATCC 42177) as a BIOTRAP for removing heavy metals from wastewaters. The advantage of this fungus is that it can be grown in the laboratory into a small spherical structure such that a single 3-mm diameter bead possesses as much as 20 cm^2 of surface area. Encapsulation is not required. This fungus has been shown to have properties which make it easier to handle and more efficient as a BIOTRAP than other materials including shellfish chitin and cell walls of the red alga Porphyra (Crusberg, et al., 1991b). To assess the structure of the fungal beads of **P. ochro-chloron** two methods were employed. First, scanning electron microscopy was used to investigate the structure of the bead, the arrangement of hyphae which make up the mycelia and the binding of heavy metal ions. Second, magnetic resonance imaging (MRI) was used to obtain real-time images of the entry and exit of Cu^{+2} ions into and out of the fungal beads during column flow. In MRI, protons are aligned into a strong magnetic field. The presence of paramagnetic ions such as Cu^{+2} affects the orientation of water protons in this strong field and the MRI system is designed to record such differences (Dixon & Ekstrand, 1982). Direct images of 'slices' through the object can be observed and even the concentrations of paramagnetic ions in regions within the image can be quantified.

EXPERIMENTAL

Fungal beads were grown from spores at pH = 4, as previously described (Crusberg, et al., 1989a,b) in polycarbonate flasks in a glucose-minimal salts medium at 30° for 5 days, shaking at 200 rpm in a rotatory incubator.

Scanning electron microscopy (Crusberg, et al., 1991a) was carried out on beads which were prepared by washing 4 times with deionized water, treated with a solution of 60 g/L $NaHCO_3$ and 90 g/L Na_2CO_3, pH 9.5, and then washed with deionized water adjusted to pH 4.0, until the effluent recorded a similar pH value. Fungal beads were treated with various concentrations of $CuSO_4$ in pH = 4 deionized water, and desorption of Cu^{+2} was carried out also in deionized water adjusted to the given pH with 6 N HCl. Thoroughly washed beads were cryogenically cleaved by placing one bead into a test tube filled with hexane equilibrated in a dry ice box. As the bead descended into the tube at -80° C, it rapidly expanded while freezing and was cleaved almost perfectly in half in less than 3 sec. Specimens were freeze dried, coated with carbon in a vacuum evaporator, or coated with Au/Pd 60:40 to 100 A in a sputter coater, and observed in a JEOL 820 SEM, equipped for energy dispersive X-ray (EDX) analysis with a Kevex detector and image analysis system. Micrographs were recorded on Polaroid Type 53 film, and X-ray microprobe microanalysis results were obtained on a conventional plotter.

MRI images were obtained from 4 mm dia. beads placed one on top of the other in a glass column initially filled with pH 4 deionized and degassed water, and centered in the vertical solenoid radio-frequency (rf) coil near the iso-center of a GE 2.0 Tesla/45 cm CSI-II imaging spectrometer equipped with 20 gauss/cm self-shielded gradients, operating at 85.56 MHz for protons. The field of view of the 128 × 128 pixel image was 1 cm × cm, providing a resolution of 76 μm/pixel. The slice thickness was 1 mm. Images were acquired in 1 min. using a standard spin-echo sequence, with an echo time of 50 msec and a repetition time of 200 msec. A solution of 5 mM $CuSO_4$ in a plastic reservoir was perfused under gravity flow through the column and images were obtained periodically for 40 min on Polaroid Type 53 film. The flow rate decreased initially from 1 mL/min when deionized water at pH 4 was perfused through the beads to 0.2 mL/min when the copper containing solution contacted the fungal beads presumably due to volume changes in the individual hyphae making up the mycelia. In 40 min. 8 mL of 5 mM $CuSO_4$ perfused the bead in the center of the MRI image. Mean pixel brightness was found to be related to the concentration of Cu^{+2}. A simple FORTRAN program was written to relate signal strength obtained from MRI images to the relaxation times T1 and T2 of the protons, which correlate with the concentration of Cu^{+2}. Using the digitized data obtained from the MRI measurements the concentrations of Cu^{+2} immediately outside the surface of the bead and in the central and the peripheral regions of the bead were computed to determine the rate of entry and egress of those ions during perfusion.

RESULTS

Figure 1a shows the secondary electron image of the surface features of a fungal bead by SEM, and the internal structure of a cryogenically cleaved bead is shown in Figure 1b. An MRI image of a bead in a 4 mm glass column surrounded by deionized water at pH 4 is shown in Figure 2a. When 5 mM $CuSO_4$ is perfused through the bead the pixel brightness increases due to changes in the relaxation times for the water protons in the vicinity of the Cu^{+2} ions (Figure 2b). The brightness of the peripheral hyphae in the bead shown in Figure 2a is probably due to a yet unidentified paramagnetic impurity which also can alter the proton relaxation times of water in that same region of the bead. A diagram of the beads in a glass MRI column is shown is Figure 3. Regions of the bead at the center are appropriately labeled.

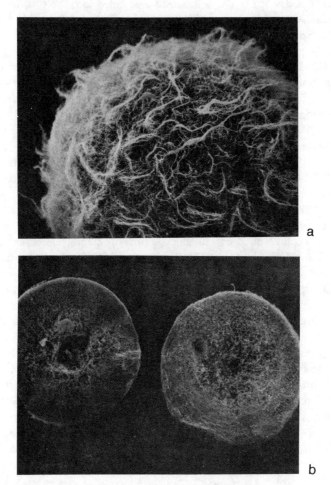

Figure 1 Scanning electron micrographs of P. ochro-chloron BIOTRAP 4 mm beads (a) surface (b) interior.

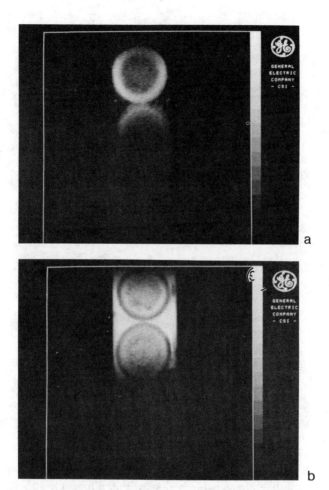

Figure 2 MRI images of 4 mm P. ochro-chloron BIOTRAP beads. (a) in pH 4.0 H_2O (b) in 5 mM $CuSO_4$.

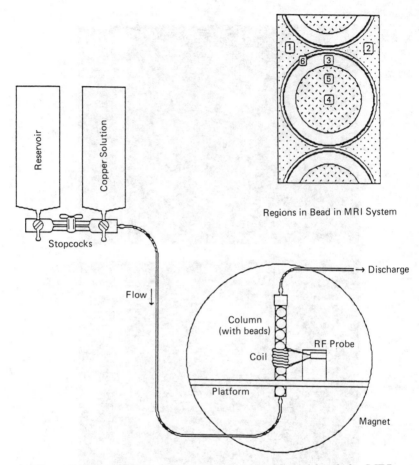

Figure 3 Upward flow system for observing fungal beads by MRI and diagram showing regions within a fungal bead.

EDX analysis of a 0.92 mm² region in the center of beads treated with various concentrations of Cu^{+2} and thoroughly washed with deionized water at pH 4 are shown in Figures 4a-d. Peak area is directly related to bound Cu^{+2}. Figures 5a, b show EDX analysis of beads equilibrated with 1,000 mg/L Cu^{+2} and washed with deionized water adjusted to pH = 3 (Figure 5a) and pH = 2 (Figure 5b).

Approximate Cu^{+2} concentrations (mM) in the solution surrounding the beads (Figure 3, regions 1 and 2) of **P. ochro-chloron** in the glass column during the MRI perfusion experiment were determined from mean pixel brightness. About 40 min. were required for the concentration of Cu^{+2} surrounding the beads to reach that of the feed reservoir (4.72 mM or 300 mg/L) as shown in Figure 6a. The Cu^{+2} concentration in the center of the bead (Figure 3, region 4 and Figure

6b) increased steadily during perfusion but only attained 1.6 mM after 40 min. (8 mL total volume perfused). In region 3 (cf. Figure 3) of the fungal bead the concentration of Cu^{+2} leveled off after about 30 min. (Figure 5c) at about 2.5 mM. When 0.1 M HCl replaced the $CuSO_4$ solution the flow rate increased to 1 mL/min. and the Cu^{+2} in region 3 was progressively removed (Figure 6d). However, even after 40 min. considerable copper remained in the bead.

DISCUSSION

Mycelia beads of **P. ocho-chloron** cultured in minimal glucose medium containing Tween 80 grow into beads with a dense outer periphery and an almost hollow core. These beads adsorb copper ions onto their mycelial walls, and the copper does not enter the cytoplasm of the cells (Fukami, et al., 1983). Sorption of Cu^{+2} by mycelia is optimal at pH 4, but is desorbed efficiently at pH 2 or below, as shown by EDX analysis. MRI experiments provided real-time measurements of Cu^{+2} entry and egress from beads perfused in a small glass column in the core of the magnet. Although Cu^{+2} enters the peripheral regions of beads rapidly, movement into the core is not complete even after 40 min. Furthermore, egress of Cu^{+2} requires considerable time and, even when perfusing with 0.1 M HCl for more than 50 min. more time will be needed to observe complete reduction of that ion throughout the bead. The complex nature of the mycelia making up the bead as seen in the SEM micrographs provides insights to the limitations of this technology for possible commercial development. The use of mycelia beads in adopting a column technology for removing heavy metals from industrial wastewaters may not be appropriate. The beads do however bind and release Cu^{+2} and other heavy metal ions reversibly (Crusberg, et al., 1991b). Sequential batch adsorption studies will be carried out using **P. ochro-chloron** mycleia beads as BIOTRAP for heavy metals. In batch form the beads could be more easily agitated during sorption/desorption phases, and a sequential adsorption system would permit the movement of liquid through a series of tanks in which heavy metal ions would be progressively removed until the concentrations in the wastewater is below federal discharge limits. SEM and MRI provide excellent tools for observing and understanding the dynamics of metal ion sorption/desorption processes which occur within macroscopic biological structures which may have possible future roles in the development of novel industrial wastewater treatment technologies.

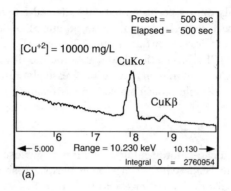

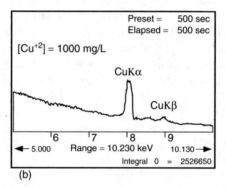

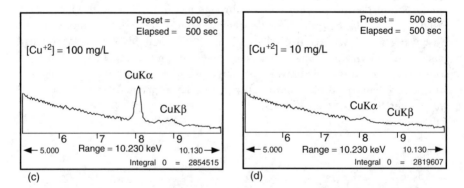

Figure 4 EDX analysis of copper binding with P. ochro-chloron BIOTRAP beads treated as indicated and cryogenically cleaved. Analysis was performed in a region 0.92 mm² in area near the center of each bead.

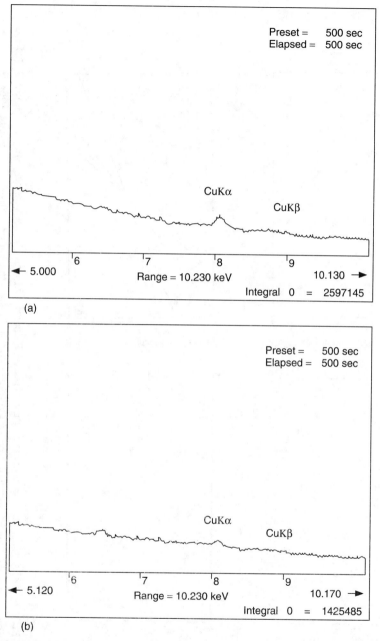

Figure 5 EDX analysis of beads of P. ochro-chloron BIOTRAP treated with 1000 mg/L Cu^{+2} and washed in: (a) 0.001 N HCl (b) 00.1 N HCl.

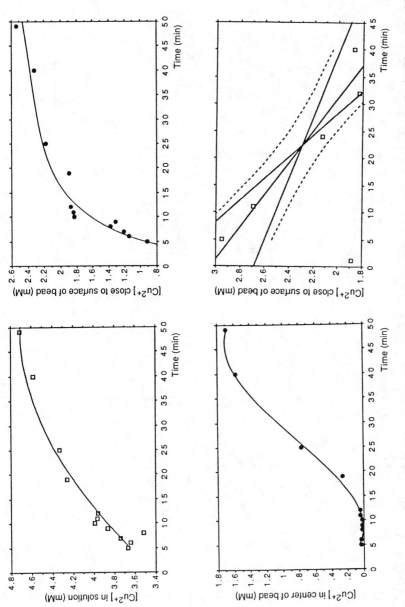

Figure 6 MRI of CU^{+2} surrounding and within the bead of P. ochro-chloron shown schematically in Figure 3. (a) Cu^{+2} concentration in regions 1 and 2 as a solution of CuSO$_4$ is eluted through the column. (b) Cu^{+2} concentration in region 3. (c) Cu^{+2} concentration in the core of the bead - region 4. (d) Cu^{+2} in region 3 of the bead after initiation of 0.1 N HCl flow. The t-value for the slope of the line is 6.578, p<0.0071 and y = 0.034 + 3.046, R^2 = 0.935.

REFERENCES

1. Blanchard, G., Maunaye, M., and Martin, G., "Removal of Heavy Metals from Waters by Means of Natural Zeolites,," *Water Res.* Vol. 18, pp. 1501-1507, 1984.

2. Coughlin, R.W., Deshaies, M.R., and Davis, E.M., "Chitosan in Crab Shell Wastes Purifies Electroplating Wastewaters," *Environ. Prog.* Vol. 9, pp. 35-39, 1990.

3. Crusberg, T.C., Weathers, P., and Baker, E., "Fungal Biotrap for Retrieval of Heavy Metals from Industrial Wastewaters," pp. 196-203, *Proc. Third Intern. Conf. on New Frontiers for Haz. Waste Management*, U. S. Government Printing Office, Washington D.C., EPA600/9-89/072, 1989a.

4. Crusberg, T.C., Weathers, P., and Cheetham, R.D., "Biotraps for Heavy Metal Recovery from Electroplating Wastewaters," pp. 283-306, In Y.C. Wu (ed.), *Proc. Third Intern. Conf. on Physiochemical and Biological Detoxification of Hazardous Wastes, Vol. I*, Technonmic Publ. Co., Lancaster, Pennsylvania, 1989b.

5. Crusberg, T.C., Weathers, P.J., and Baker, E.F., "Biotrap for Heavy Metal Removal and Recovery from Industrial Wastewater," pp. 91-96, In H.M. Freeman and P.R. Sferra, eds., *Innovative Hazardous Waste Treatment Technology Series, Vol. 3, Biological Processes*, Technomic Publ. Co. Lancaster, Pennsylvania, 1991a.

6. Crusberg, T.C., Mayer, J.A., and Weathers, P.J., "Comparison of Several Biotraps for Heavy Metal Removal and Recovery from Wastewaters," pp. 266-270, In *Proc. HMC-Northeast '91*, Hazardous Materials Control Research Institute, Greenbelt, Maryland, 1991b.

7. Darnall, D.W., and Gabel, A., "A New Biotechnology for Recovering Heavy Metal Ions from Wastewater," pp. 217-224, *Proc. Third Intern. Conf. on New Frontiers for Haz. Waste Management*, U. S. Government Printing Office, Washington, D.C., EPA600/9-89/072, 1989.

8. deRome, L., and Gadd, G.A., "Use of Pelleted and Immobilized Yeast and Fungal Biomass for Heavy Metal and Radionuclide Recovery," *J. Indust. Microbiol.*, Vol. 7, pp. 97-104, 1991.

9. Dixon, R.L., and Ekstrand, K.E., "The Physics of Proton NMR," *Med. Phys.*, Vol. 9, pp. 107-108, 1982.

10. Dudley, L.M., McLean, J.E., Furst, T.H., and Jurinak, J.J., "Sorption of Cadmium and Copper from an Acid Mine Waste Extract by Two Calcareous Soils: Column Studies," *Soil Sci.*, Vol. 151, pp. 121-135, 1991.

11. Freeman, H.M., ed., *Innovative Hazardous Waste Treatment Technology Series, Vol. I, Thermal Processes*, Technomic Publ. Co., Lancaster, Pennsylvania, 1990.

12. Freeman, H.M., ed., *Innovative Hazardous Waste Treatment Technology Series, Vol. II, Physical/Chemical Processes*, Technomic Publ. Co., Lancaster, Pennsylvania, 1990.

13. Fukami, M., Yanayaki, S., and Toda, S., "Distribution of Copper in the Cells of the Heavy Metal Fungus *Penicillium ochro-chloron* Cultured in Concentrated Copper Medium," *Agricultural Biochem.*, Vol. 47, pp. 1367-1369, 1983.

14. Giesy, J.P., and Paine, D., "Uptake of Americum-241 by Algae and Bacteria," *Prog. Water Technol.*, Vol. 9, pp. 845-857, 1977.

15. Kuhn, S.P., and Pfister, R.M., "Accumulation of Cadmium by Immobilized *Zoogloea ramigea 115*," *J. Indust. Microbiol.*, Vol. 6, pp. 123-128, 1990.

16. Lindsay, A.W., "Demonstrating Technologies for the Treatment of Hazardous Wastes at EPA," pp. 17-34, In Y.C. Wu, ed., *International Conf. on Physiochemical and Biological Detoxification of Hazardous Wastes, Vol. I*, Technomic Publ. Co., Lancaster, Pennsylvania, 1989.

17. Mahan, C.A., . Majidi, and Holcombe, J.A., "Evaluation of the Metal Uptake of Several Algae Strains in a Multicomponent Matrix Utilizing Inductively Coupled Plasma Emission Spectroscopy," *Anal. Chem.*, Vol. 61, pp. 624-627, 1989.

18. Nishi, N., Maekita, Y, S.I. Nishimura, Hasegawa, O., and Tokura, S., "Highly Phosphorylated Derivatives of Chitin, Partially Deacylated Chitin and Chitosan as New Functional Polymers: Metal Binding Property of the Insolubilized Materials," *Internat. J. Macromol.*, Vol 9, pp. 109-114, 1987.

19. Norris, P.R., and Kelly, D.P., "Accumulation of Cadmium and Cobalt by Saccharomyces cerevisiae," *J. Gen. Microbiol.*, Vol. 99, pp. 317-324, 1977.

20. Stokes, P.M., and Lindsay, S.E., "Copper Tolerance and Accumulation in Penicillium ochro-chloron Isolated from Copper Plating Solution," *Mycologia*, Vol. 71, pp. 788-806, 1979.

18

DETOXIFICATION TREATMENT OF CHROME SLUDGE

Chin-Tson Liaw, Wen-Shung Ma Lin, Tzong-Tzeng Lin
UCL, ITRI
Hsinchu, Taiwan, R.O.C.

INTRODUCTION

The paper describes a method to detoxification of chrome sludge by calcination at high temperature with Na_2CO_3, the resulting chromate may be then recovered.

The chrome sludge containing 30% to 80 chromium hydroxide was dried, crushed, passed through a 100 mesh sieve and then mixed with one to two times stoichiometric amount of sodium carbonate (Na_2CO_3). The mixture was calcined for one hour at various temperatures between 400°C-1000°C. It was then extracted extensively with water.

The conversion temperature of chrome within the sludge depends on the concentration of sludge impurities. The reaction temperature of the sludges with high content of $Al(OH)_3$ and $Fe(OH)_3$ is higher than the one with lower content. The alkali amount comes close to the stoichiometric as long as the sludge contains no $Al(OH)_3$ as impurities.

Since the chrome sludges are in hydroxide form, which is in fine particle with large surface area, the reaction can be accomplished under the temperature of 700°C -950°C which is 300°C lower than the required temperature for chrome ore reaction.

The conversion rate of chrome sludge can reach 95% at proper conditions regardless of the amount of impurities.

Chrome is widely used in industries, electroplating, tannery, dyeing as well as finishing factories. However, chromium ion, particularly the Cr(6), is one of the most toxic chemical substances, chronic inhalation particles of chromic salt dust may result in lung cancer.

The traditional method for waste water treatment that containing Cr(6) components was to convert the Cr(6) into Cr(3) by reaction with $FeSO_4$ or $NaHSO_3$ to form $Cr(OH)_3$ precipitate in alkali condition (i.e NaOH). However, the resultant chrome sludge, in addition to $Cr(OH)_3$ also contains impurities such as $Al(OH)_3$, $Fe(OH)_3$, $Ni(OH)_2$, $Cu(OH)_2$, $Zn(OH)_2$ and Na_2SO_4. For final disposal, the $Cr(OH)_3$ sludge was further solidified with cement and then landfill. The disadvantage of this method is that the Cr(3) salt when reacted with strong alkali may be converted into water soluble salt $Na3Cr(OH)_6$. Thus, it may result in the pollution of underground water. Thus reutilization of chrome sludge is much favorable than disposal. It may be reused as following:

(1) chromemagnesia refractory,
(2) pigment for ceramic, and
(3) chrome alloy, etc.

This study suggests a new approach of chrome sludge treatment, that is the metallurgy method. Sludges with $Cr(OH)_3$ are considered as metallurgic materials, from which Na_2CrO_4, $Na_2Cr_2O_7$, $K_2Cr_2O_7$ can be recovered. This is because Cr(3) can be converted to Cr(6) when reacted with strong alkali under high temperatures. This method of treatment produces useful material thus fulfilling the response of resource recycle, waste minimization and pollution control.

This article also discusses the effects of reaction temperature, amount of sodium carbonate used, and the sludge impurities effect on conversion rate.

EXPERIMENTAL SECTION

Three different kinds of chrome sludges were selected for the study of recovery rate. The compositions are identified in Table 1. Samples No.1 and No.2 are the

Table 1. Composition of Different Sludge Samples (unit: %)

Composition	No. 1	No. 2	No.3
$Cr(OH)_3$	34.59	36.11	52.03
$Al(OH)_3$	28.36	54.51	trace
$Fe(OH)_3$	15.12	0.73	0.18
$Cu(OH)_2$	5.30	0.64	0.39
$Zn(OH)_2$	3.21	0.05	0.01
$Ni(OH)_2$	1.51	0.10	0.37
Na_2SO_4	11.90	7.85	43.67
Other	2.4	4.20	3.35

sludges from the dilute plating waste which were treated by reduction and neutralization methods, sample No.3 is the sludge from concentrate aging plating liquid.

Three kinds of sludges were dried, crushed, and passed through 100 mesh sieve.

The resulting materials were examined for the effects of reaction temperature, amount of sodium carbonate used, and impurities content on the conversion rate of chrome. Detoxification process of the chrome sludges are shown in Table 2.

REACTION TEMPERATURE

100g of the chrome sludge and sodium carbonate mixture was calcined at various temperatures between 400°C ~ 1000°C in an alumina crucible for one hour, then it was thoroughly extracted with water. The chrome content of this material was determined by atomic absorption.

AMOUNT OF ALKALI USED

Sludge which has been dried, crushed, passed through 100 mesh sieve was first mixed with powder Na_2CO_3 and then calcined in an alumina crucible. The weight percentage is as following:

$$\frac{Na_2CO_3}{Chrome\ sludge\ +\ Na_2CO_3} = 26 \sim 50\%\ (wt\%)$$

RESULTS AND DISCUSSION

The conversion rates of the three different chrome sludges at various temperatures are shown in Figure 1. The results indicate the conversion rate increases as temperature increases, each at different temperatures, all reached the maximum conversion rate of 95%, beyond this point further increase in temperature resulted in a drop of the conversion rate. No. 1 sample reaches maximum conversion at 900°C, No. 2 at 950°C while No. 3 at 750°C, which is the lowest among three. This indicates the higher concentration of chrome content from concentrated plating solution has a lower conversion temperature.

The composition of the three chrome sludges are shown in Table 1, sample 1 contains more $Al(OH)_3$ and $Fe(OH)_3$, while sample 2 is dominated by $Al(OH)_3$. Thus it seems to indicate the higher the chrome content, the lower the conversion temperature, further, higher in $Al(OH)_3$, increases the converting temperature. However, the final conversion rate does not seem to be effected by impurity of chrome content.

Figure 2 shows the amount of Na_2CO_3 used in the conversion of all three samples. It indicates that the Na_2CO_3 needed for maximum conversion are different. The optimum condition of conversion rate (i.e, 95%), for temperatures,

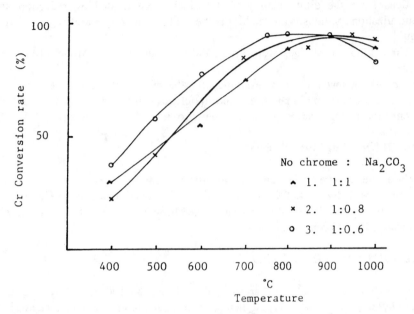

Figure 1 The conversion rate of the different chrome sludges at various temperature.

% of Na_2CO_3 used for the three tested materials are: Sample No.1: 900°C, 50%; Sample No. 2: 950°C, 44%; and Sample No. 3: 750°C, 37.5%. The amount needed for Na_2CO_3 depends on the amount of such impurities as $Al(OH)_3$, $Fe(OH)_3$, $Cu(OH)_2$, and $Ni(OH)_2$, the sludge contained.

Figure 3 indicates the different conversion rates at different temperatures for various Na_2CO_3 % added into sample No. 1. It is clear that 50% of Na_2CO_3, and 900°C results in the highest conversion rate. It is interesting to note that the stoichiometric amounts of Na_2CO_3 for the 3 samples are 26%, 27% and 34% respectively, which are much lower than the maximum conversion (Na_2CO_3) concentration, as 50% for sample 1, 44% for sample 2 and 37.5% for sample 3.

In chrome ore, it takes 1000°C ~ 1200°C to obtain chrome salt, that is 250°C ~ 450°C higher than from chrome sludge, this may be due to the fact that in chrome ore, the chrome is in Cr_2O_3, and the Cr-O chemical bond is stronger than Cr-OH in the sludge, thus it takes higher temperatures to break the Cr-O bond.

The main composition for chrome sludge $Cr(OH)_3$ which was formed by coprecipitation method had the advantages of fine particle and very large surface area which results in being able to react easily with Na_2CO_3. These two elements, chemical bond and surface area, provide sludge for better reactivity than chrome ore.

Table 2. Detoxification Process of the Chromium Sludge

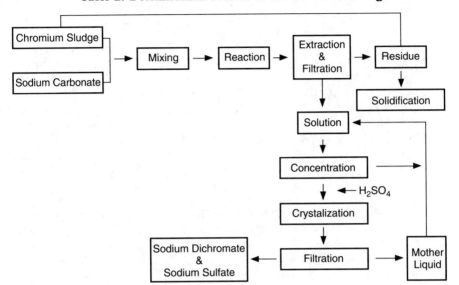

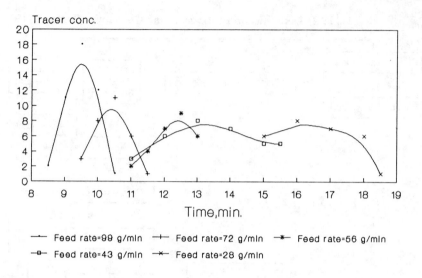

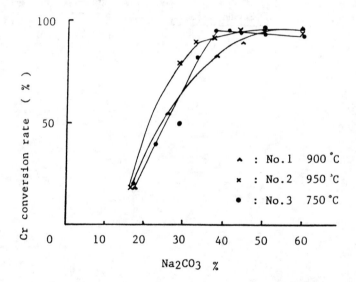

Figure 2 The relationships between chrome rate and amount of Na_2CO_3.

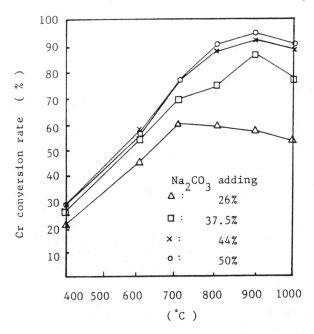

Figure 3 The relationships among conversion rate, temperature, and the amount of Na_2CO_3 for no. 1 material.

CONCLUSIONS

The conventional method of chrome sludge treatment although already change the highly toxic Cr(6) to Cr(3), however landfill of Cr(3) product may result secondly ground water pollution. The new detoxification treatment of chrome sludge adopts the concept of metallurgy. Thus not only has the advantage of waste minimization but resource recovery and pollution control of chrome sludge.

REFERENCES

1. M. Grayson, *Encyclopedia of Chemical Technology*, 3rd ed., Vol.6 John Wiley & Sons, Inc., New York, N.Y. 1979, pp. 82-116.

2. *Treatment and Disposal Methods for Waste Chemicals*, International Register of Potentially Toxic Chemicals United Nations Environment Programme, Geneva, Switzerland, TRPTC, No.5, 1985, pp. 134-137.

3. Daiichi Nenryo Kogy, KK JP 52, 128, 879 (Oct. 28, 1977).

4. Nakashima Mitsuo, (Toyota Motor Co., Ltd) JP 8, 014, 868 (Feb. 01, 1980).

19

EFFECTS OF CHROMIUM CONTAMINATION ON CONCRETE

Dorairaja Raghu and Hsin-Neng Hsieh
Department of Civil and Environmental Engineering
New Jersey Institute of Technology
Newark, NJ

INTRODUCTION

During the period from 1890 to 1964, huge quantities (over 1.5 million metric tons) of chromates and dichromates were produced by some chemical industries in Hudson County, New Jersey. This activity generated approximately 2.75 million tons of an alkaline waste residue containing from 2% to 5% chromium (Raghu and Hsieh). The waste material was subsequently used as fill and diking material in what are now commercial, industrial, and residential areas. Over a period of time, this fill material undergoes complex physical and chemical changes resulting in ground and surface water pollution, deterioration of structures and parking lots, and contamination of soils.

In their description of the phenomenon of structural distress, Raghu and Hsieh noted the following effects:

1) Mortar between the brick becomes contaminated with chromium. Over a period of time (usually 5 to 10 years), the mortar joints flake out and bricks fall out of the wall.
2) Structural cracks developed in the walls along concrete seams and/or mortar joints.
3) Walls tended to shift off plumb.
4) Floor slabs exhibited heaving and buckling.

New Jersey Institute of Technology initiated a study into the mechanisms causing structural distress as a part of an overall investigation involving chromium contamination. A report was prepared on the above subject (Deng). Some salient aspects of the above study are reported in this paper.

DETERIORATION OF CONCRETE

Generally, concrete deterioration can be attributed to physical and chemical effects. Physical causes of concrete deterioration have been grouped into two categories, one due to surface wear and the other due to cracking (Mehta). The former could be caused by the loss of mass due abrasion, erosion, and cavitation. Volume changes due to normal temperature and humidity gradients and crystallization pressures of salts in pores, structural loading, and exposure to temperature extremes such as freezing or fire could result in cracking.

The chemical causes of deterioration may be due to hydrolysis of the cement paste components by soft water, cation-exchange reactions between aggressive fluids and the cement paste and reactions leading to formation of expansive products, such as in sulfate expansion, alkali-aggregate expansion, and corrosion of steel in concrete. The authors believe that the deterioration of concrete by chromium could be caused by the crystallization pressures of salts in the pores and the reactions involving formation of expansive products.

INTERNAL STRUCTURE OF CONCRETE

In order to further understand the damage mechanisms in concrete due to chromium contamination, it is necessary to study its internal structure. The gross structure of a material that is visible to unaided human eye is called macrostructure. In the macrostructure of concrete, two phases are readily distinguished: aggregates of varying shapes and size, and binding medium, which consists of an incoherent mass of the hydrated cement paste.

At the microscopic level, the complexities of the concrete structure begin to show up. It becomes obvious that the two phases of the structure are neither homogeneously distributed with respect to each other, nor are they themselves homogeneous. Besides the aggregate and cement paste, there is a third phase, called transition zone, which represents the interfacial region between the particle of coarse aggregate and hydrated cement paste (HCP). Existing as a thin shell, typically 10 to 50 μm thick and around large aggregate, the transition zone is generally weaker than either of the main components of concrete, and therefore, it exercises a far greater influence on the mechanical behavior of concrete than is reflected by its size. Since the HCP and transition zone are subject to change with time, environmental humidity, and temperature, the structure of concrete does not remain stable.

Observing the HCP and transition zone under scanning electron microscope, hexagonal crystals of monosulfate hydrate and needle like crystals can be seen, (most of them exist in the transition zone). Those needle-like crystals are called ettringite. The main components of ettringite are calcium aluminate and calcium sulfate.

PORE STRUCTURE OF CEMENT

The pore structure of cement consists of clinker grains that separated from each other by a hydrated mass of calcium hydrosilicates, in the central portion of which there run thin veins of capillaries which form an interconnected network. According to those, the pore structure in cement is interconnected and permeability is very low. The nature of the pore size distribution also will affect the properties of cement (Meyer). So far, an overview of the internal structure of concrete has been presented. As a matter of fact, the structure-property relationships in concrete are not yet well developed; however, some researchers in this field have proven that the properties of the concrete can be affected and modified by changes in the structure (Wakeley).

ETTRINGITE

Ettringite is a kind of C-S-H, calcium-silicatehydrate gel, which exists in concrete at microscopic levels, and can be represented as $C_6S_3H_{32}$. The formation of ettringite and hydration of hard-burnt calcium oxide are the two phenomena known to cement chemists that can cause disruptive expansive cements. Also, there is general agreement that sulfate-related expansions in concrete are associated with ettringite; however, the mechanisms by which ettringite formation causes expansion is still a subject of controversy. Exertion of pressure by growing ettringite crystal, and swelling due to adsorption of water in alkaline environment by poorly crystalline ettringite are two hypotheses that are supported by most researchers.

Ettringite will form when alumina, silica and water are present. To a certain extent, the properties of concrete will be unaffected if both the size and quantity of ettringite stay stable. Some cases have been recorded that related to the growth of ettringite. Cohen explained the kinetics of hydration and formation of ettringite crystals. He created a model of expansive cement, which was divided into three sub-systems: the solution, which includes water, dissolved sulfates and lime, the matrix containing the unhydrated calcium silicate particles surrounded by the solution and the expansive particles, consisting of spherical cement particles. Each particle is surrounded uniformly by a film of solution which is of the same thickness for each particle regardless of its diameter.

The spherical C_4A_3S particles can be assumed to have equal diameters if the range of the particle sizes is small (i.e. 2-5 μm, 3-8 μm, 5-10 μm and 10-15 μm). The reaction between the C4A3S particles and the solution to produce ettringite can follow the stoichiometric reaction shown in the next page, where

$$C = CaO, \quad A = Al_2O_3, \; S = SO_3 \wedge H = H_2O$$
$$\text{(Solid spheres)} \qquad \text{(Solution)} \qquad \text{(Ettringite)}$$

$$C_4A_3S + 8CS + 6C + 96H \; \text{---} \; 3C_6AS_3H_{32}$$
$$\text{(or 8CSH}_2\text{)} \qquad \text{(or 80H)}$$

As soon as the reaction starts, the spherical C_4A_3S particles become covered uniformly with ettringite crystals. During the topochemical reaction of the reactants and the formation of the product, ettringite crystals, the lengths of these crystals will increase but their number and thickness will remain constant. This process will continue as long as necessary reactants are available in stoichiometric quantities. Moreover, the ettringite crystals are considered to have equal average cross-sectional areas during the period of expansion.

The expansion starts when the length of these crystals become larger than the solution film thickness, so that the crystals begin to exert pressure against the surrounding matrix. During the expansion, the ettringite crystals will have spaces or pores between them and the volume of these pores will increase as the length of crystals increase. Figure 1, in the next page illustrates the expansion process.

PRIOR RESEARCH ON THE EFFECT OF CHROMIUM ON CONCRETE

Craig found that the compressive strength decreased from 5.9 ksi with no chromium added to less than 4.5 ksi with 6% potassium chromate. Craig reported a decrease in splitting tensile strength in concrete with 6% potassium chromate. Tashiro reported that chromium oxide and some other metal oxides could promote the growth of ettringite. By performing a pore size distribution analysis of a specimen containing 5% chromium oxide additive, he also found that chromium oxide generally increases the total pore volume of hardened pastes and shifts the pore size distribution to the larger pore sizes between three days to twenty eight days. Test results indicated that the compressive strength decreased about 50%.

Wakely studied the expansion of a salt-saturated concrete and pointed out that considerable expansion could result if ettringite continues to form after the rigid structure is established. The growth of ettringite may be accompanied by

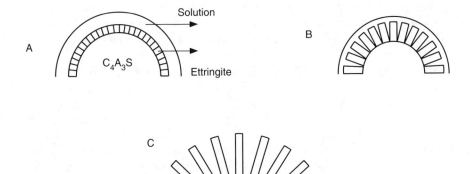

Figure 1 Expansion process.

expansion (over 200%) and gradual disintegration of concrete (Nicholls). There is ample evidence to indicate that the strength of concrete decreases with increasing quantities of potassium chromate.

EXPERIMENTAL INVESTIGATIONS PERFORMED FOR THIS STUDY

The purpose of this research is to study the effects of chromium on concrete strength, and the mechanisms responsible for causing structural distress in the contaminated sites. For this purpose, two additives, namely chromium nitrate and potassium chromate were selected to represent the trivalent and the hexavalent chromium species present in the contaminated sites.

Two types of experiments have been performed. One was to confirm that chromium had an influence in the growth of ettringite by observing the specimen under a scanning electron microscope. The other one was to determine the compressive strength with the chromium additives.

OBSERVATION UNDER SCANNING ELECTRON MICROSCOPE

For this study, three sets of samples were prepared by mixing C_3A synthesized from reagents and $CaSO_4.2H_2O$ of reagent grade at a mole ratio of 1:3 to an ettringite composition. One set of samples were prepared with no additives and the other two sets were made by mixing chromium additives in the proportion of 5% by weight. The samples were observed after three days, fourteen days and twenty-eight days of preparation and curing.

COMPRESSIVE STRENGTH TESTS

The mixing proportions of water: cement: sand utilized for this study were 1:1:2 by weight. The chromium compounds were added to the mix after dissolving them in water first. Two-inch cubical specimens were cast and cured at 70°F in water. All compressive strength tests were performed at the same loading rate.

DISCUSSION OF RESULTS: OBSERVATION UNDER MICROSCOPE

The scanning electron micrographs of hardened cement paste after three days, fourteen days of curing with no additives are shown in Figure 2. The corresponding micrographs for samples prepared with chromium additives are presented in Figures 3 and 4 respectively. Figure 5 and Figure 6 are micrographs of samples which were picked randomly from the broken chips of the compressive strength tests. It can be observed from these pictures that the needle-like crystals of ettringite grow with time and that this growth is very pronounced for specimens

Figure 2(a) Paste with no addition. Curing, 3 days.

Figure 2(b) Paste with no addition. Curing, 14 days.

Figure 2(c) Paste with no addition. Curing, 28 days.

Figure 3(a) Paste with Cr(NO$_3$)$_3$. Curing, 3 days.

Figure 3(b) Paste with Cr $(NO_3)_3$. Curing 14 days.

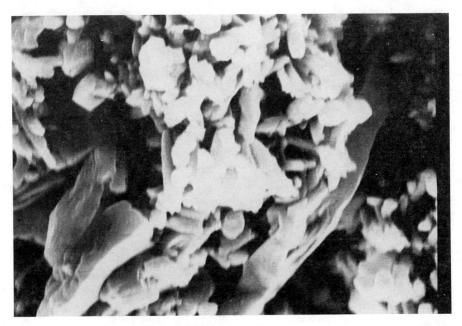

Figure 3(c) Paste with Cr $(NO_3)_3$. Curing, 28 days.

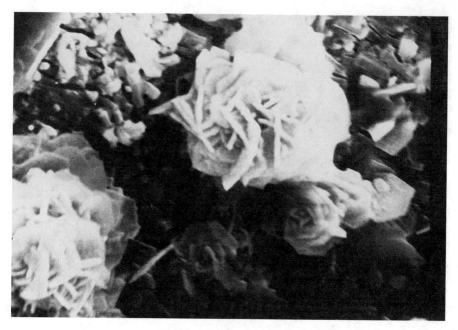

Figure 4(a) Paste with K_2CrO_4. Curing, 3 days.

Figure 4(b) Paste with K_2CrO_4. Curing, 14 days.

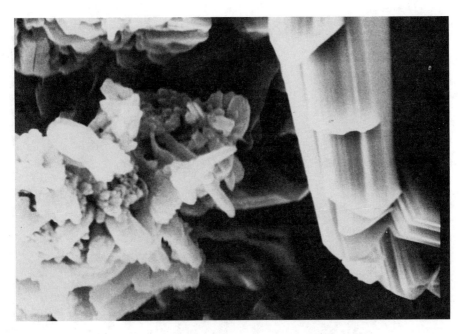

Figure 4(c) Paste with KCrO$_4$. Curing, 28 days.

Figure 5(a) Curing time 60 days. No addition.

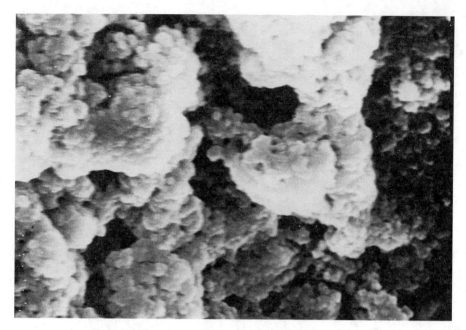

Figure 5(b) Curing time 60 days. Cr(NO$_3$)$_3$ added.

Figure 5(c) Curing time 60 days. K$_2$CrO$_4$ added.

Figure 6(a) Curing time 3 days. No addition.

Figure 6(b) Curing time 3 days. $Cr(NO_3)_3$ added.

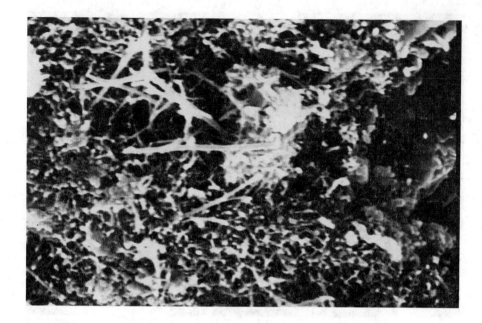

Figure 6(c) Curing time 3 days. K_2CrO_4 added.

with hexavalent chromium additives. Similar observations were made by Tashiro in his research regarding formation of ettringite due to the addition of heavy metal oxides.

It can also be seen from Figures 2, 3 and 4 that the microstructure changed due to the addition of chromium compounds. The structure was tight and void spaces were small with no chromium addition. As the ettringite crystals grow, the structure becomes loose and the void spaces enlarged. The reduction in strength could be related to this phenomenon.

DISCUSSION OF RESULTS: COMPRESSIVE STRENGTH TESTS

The results are shown in Figure 7. These indicate that all groups exhibited an initial increase in compressive strength with increase in curing time, up to a period of seven days. Compressive strength of groups with additives showed a decrease till the twenty-eighth day followed by an increase thereafter. It was also observed that the compressive strength on the twenty-eighth day was the lowest.

CONCLUSIONS

The results of this study showed that two chromium compounds, chromium nitrate and potassium chromate would affect the formation of ettringite from hydration of $C_3A + 3CaSO_4.2H_2O$ cement and microstructure of hardened cement by

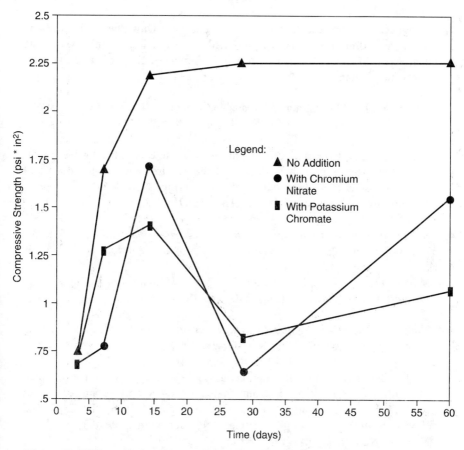

Figure 7 Effects of chromium admixtures on compressive strength of concrete.

prompting the growth of the ettringite crystals. And the compressive strength of mixes decreased with the addition of chromium compounds.

There is a conclusion that the growth of the crystal of ettringite is one of the causes for the expansion of concrete, and since this phenomenon occurs at microscopic levels, it will take a long time to determine expansion rates by general measurement involving mechanical instrumentation until the microcrack develops. On the other hand, the enlargement of the pore sizes of cement may also play an important role in the expansion of cement, which can seen by the naked eye only after the crack occurs. Additional studies consisting of a long term experiment should be performed to investigate as to whether there is a stable tendency of the compressive strength and to determine between the growth of ettringite and large pore size the main cause of the expansion of concrete.

REFERENCES

1. Craig, R. J.," Physical Properties of Cement Paste with Chemical Corrosion Reducing Admixtures," Thesis presented as partial fulfillment towards the degree of Master of Science in Civil Engineering, Purdue University, Lafayette, Indiana, June 1969.

2. Cohen, M. D., "Modelling of Expansive Cements," *Journal of Concrete and Cement Research*, Volume 13, No. 4 , pp. 519- 528, 1983.

3. Deng, Li, "The Effects of Chromium on Concrete Strength," Project report submitted towards partial fulfillment of the requirements for the degree of Master of Science in Environmental Engineering, New Jersey Institute of Technology, Newark, New Jersey, April 1989.

4. Mehta, P. Kumar, *Concrete: Structure, Properties and Materials*, Prentice Hall Publishing Company, Englewood Cliffs, New Jersey, 1986.

5. Meyer, H. I., "Pore Distribution in Porous Media," *Journal of Applied Physics*, Volume 24, No. 3, pp. 510-520, 1953.

6. Nicholls, Robert, *Inorganic Cement Chemistry*, Prentice Hall Publishing Company, Englewood Cliffs, New Jersey, 1959.

7. Raghu, Dorairaja and Hsieh, Hsin-neng, "A Feasibility Study of the Removal of Chromium from Selected Contaminated Sites," Report Submitted as part of funded research Contract to the City of Jersey City by the Civil And Environmental Engineering Department, New Jersey Institute of Technology, Newark, New Jersey, August 1967.

8. Raghu, Dorairaja and Hsieh, Hsin-neng, "Performance of Some Structures Constructed on Chromium Ore Fills," American Society of Civil Engineers, *Journal of the Performance of Constructed Facilities Division*, Volume 3, No. 2, 1989, pp. 113-120.

9. Tashiro, C., Oba, J. and Akama. K., "The Effects of Several Metal Oxides on the Formation of Ettringite and the Microstructure of Hardened Ettringite," *Journal of Cement and Concrete Research*. Volume 9, No. 3, pp. 723- 733, 1987.

10. Wakeley, L. D., " Optimizing Workability and Expansion of a Salt-saturated Concrete," *Journal of Cement and Concrete Research*, Volume 17, No.5, pp. 723-733, 1987.

20

REMOVAL OF CHROMIUM FROM A HIGHLY CONTAMINATED SOIL/SLAG MATRIX BY WASHING AT LOW pH

Erez Gotlieb, Victor Ososkov, and Joseph W. Bozzelli
NJ Institute of Technology
Newark, NJ

Itzhak Gotlieb
CHEA Associates
Roseland, NJ

Ed Stevenson
NJ Department of Environmental Protection
Trenton, NJ

INTRODUCTION

Chromium and its compounds are widely used by modern industries, resulting in large quantities of this element being discharged into the environment[1]. Chromium is one of the major soil contaminants of industrial sites in New Jersey as well as in many of the superfund sites throughout the United States.

Most adverse health effects due to exposure to chromium are associated with hexavalent chromium or Cr(VI). It has been identified as being mutagenic and carcinogenic. Cr(III), or trivalent chromium, can be metabolized in the human body and, therefore, poses less of a risk than does Cr(VI)[1,2]. Routes of entry are by ingestion, inhalation, or through the skin[2]. The maximum concentration limit set for total chromium in drinking water is 50 ppb[3].

Chromium contamination of soil poses risk to the general welfare by its potential for leaching into ground water supplies. The various remediation options available include stabilization and land application, and biological and chemical treatments. Although stabilization and land application methods have shown effectiveness in controlling the leaching of chromium [4,5], they are only temporary measures. Unless the contaminant is removed, its migration is always a concern.

The weakness of biological treatment in removing hexavalent chromium limits its applicability[6]. Detoxification of chromium contamination by biological methods requires chemical reduction of Cr(VI) to Cr(III) as a preliminary step. Biological treatments are also limited by the sensitive requirements of the microorganisms.

The promising technologies for chromium decontamination are chemical treatments. Barlett and Kimble[7] experimented with $Na_4P_2O_7$, NH_4OAc, 0.1M NaF and 1M HCl by adding them to soils with various organic content previously impregnated with Cr(III). Only $Na_4P_2O_7$ and HCl removed significant (up to 65%) amounts of chromium. HCl was shown to be capable of removing both inorganic and organic complexes of chromium, while $Na_4P_2O_7$ is only effective for removal of organic complexes. Hsieh et al[8] studied the effectiency of chromium extraction from soil by washing with sodium hypochlorite and EDTA. Ten successive cycles of washing with hypochlorite yielded 46% of chromium removal, and with EDTA 58% removal. Peters and Elliott[9] reported removal efficiencies 40-60% for chromium and lead using EDTA solutions as extragent. Tan[10] treated a range of clay and sand media impregnated with chromium with acid solution and achieved mixed results. The strongest acid concentration used had pH 1.5.

So an effective method for chromium extraction from soil has not been found. Promising results have been achieved with acid extraction, thus warranting further study. The objectives to the study were to evaluate several different extraction parameters associated with the removal of chromium from soil by washing at low pH. The extraction parameters are acid concentration, contact time, temperature and the effect of two stage extraction.

All the tests were proceeded by a water wash step (time 60 min, temperature 95°C, water/soil ratio 75/1 ml/g) to remove water soluble chromium. Water washed samples were oven dried overnight at 85°C. A series of tests were run to examine the various parameters. After the extraction run was completed, both the extract and residue were analyzed for total chromium by atomic absorption spectroscopy (AAS). Extraction efficiency were calculated as the fraction of chromium extracted from total chromium.

SOIL/SLAG CHARACTERISTICS

A sample of chromium-laden industrial slag mixed with soil was received via the NJ DEPE from a contaminated site in Kearny, NJ. This material will be referred to as "soil/slag" because it has the unique properties of industrial slag and cannot be properly termed soil.

The soil/slag was dry screened through 1/8" pore stainless steel screen. Typical total chromium concentration in the slag was 21,000 ppm, or 2.1% by weight, as determined by AAS following acid digestion and dilution.

The content of some metals in the soil/slag after performing digestion procedure was estimated by inductive couple plasma mass spectrometry (ICPMS) method (Table 1).

Table 1. Concentration of Some Metals in Soil/Slag, g/kg

Ca	Fe	Mg	Al	Cr	Ti	Mn	V	Ni	Pb	Zn	Cu
228	66	61	60	21	2.9	0.9	0.5	0.3	0.2	0.2	0.2

The water fraction was determined by placing a sample of material in the oven to dry overnight at 85°C. Bulk density was determined by measuring the volume of a given dried sample in a graduated cylinder and weighing the content of the cylinder. Real density was arrived at by filling the graduated cylinder contacting the sample with water and weighing the sample in water. pH was measured by a glass electrode in a soil- lM KCl slurry (the ratio soil/solution 1:20).

Particle size analysis was done on a dried 200g sample in a analytical sieve shaker with standard Tyler wire mesh screen. Total organic extractables were determined by soxhlet extraction analysis using dichlormethane as a solvent. The characteristics of the soil/slag sample are shown in Table 2.

Table 2. Soil/Slag Characteristics

Sieve #	>12	12	20	60	100	200	325
Wt.Percent	0.02	1.4	18.7	12.3	20.0	13.0	34.5

Bulk density	1.12 g/ml (dry)
Real density	2.24 g/ml (dry)
Water fraction	30.6%
Total organic extractables	0.1%
pH	9.8

EXPERIMENTAL PROCEDURE

Prior to acid extraction, soil/slag samples of 20g each were water washed at a 75:1 v/w water/soil ratio at 95°C for one hour as a preliminary step to remove water soluble chromium and then dried overnight at 85°C. Acid concentrations were measured as weight/volume rations. Solutions containing soil/slag samples of 1-3g (dry basis) each were heated to 95°C for one hour and filtered under vacuum. The residue cake was rinsed with an equal volume of extracting solution to remove residual acid and chromium. Unless otherwise noted, solvent/soil ratios were 75:1 v:w (ml/g). When acid concentration was not used as variable, it was maintained at 2% weight to volume extraction solution. The sample residue was digested after drying and both the filter extract and digestate were analyzed for total chromium.

Digestion was done by EPA method 3050 for heavy metals digestion[11]. Residue and digestate were separated by gravity filtration. The residue cake rinsed with 100 ml 30% nitric acid to remove residual chromium.

Samples were analyzed for total chromium using a Thermal Jarrel Ash model 1200 atomic absorption flame spectrometer at a wavelength of 357.9 nm with an acetylene/air flame and Smith Hieftje background correction. The content of some metals in the soil/slag was estimated by inductive couple plasma mass spectrophotometr VGPQ-2 following its digestion by EPA method 3050. Extraction efficiencies were calculated by mass balance; chromium content was determined as the sum of chromium removed by extraction and chromium removed by digestion. Extraction efficiency is the mass of chromium removed by digestion divided by total chromium removed.

No differentiation was made between Cr(VI) and Cr(III) in the study. Hexavalent chromium has a higher water solubility and is, therefore, less likely than trivalent chromium to be found in soil after hot water extraction. It was assumed that the majority of Cr(VI) was removed in the water wash and the acid extraction step was devoted to removal of Cr(III).

RESULTS AND DISCUSSION.

Acid Strength

The chromium extraction efficiency of three acids (sulfuric, hydrochloric and nitric) were studied. The results indicated that extraction efficiency at the same pH is practically not dependent upon which of the three acids is used. Selection of sulfuric acid as the extractant is based upon its lower cost. It is also less corrosive and doesn't produce fumes.

Chromium extraction as a function of sulfuric acid concentration is shown in Figure 1. The extraction parameters of time, temperature, and solvent/soil ratio are held constant at one hour, 95°C, and 75:1 v:w, respectively. There is a steep increase in extraction efficiency between 0.5% and 2% acid concentration and efficiency levels out 95% removal at concentrations higher than 2%.

Matrix Solubility

There is considerable dissolution of sample matrix as a result of acid extraction. Figure 2 illustrates that matrix weight loss is nearly linear as a function of acid concentration. This phenomena is a result of a unique character of soil/slag matrix. The main reason for the significant weight loss is the transition of the soil/slag macro elements (Ca, Fe, Mg, Al) to the extract during washing with acid

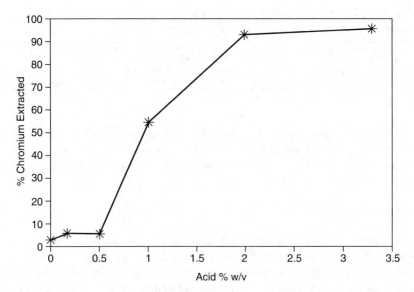

Figure 1 Effect of acid concentration on extraction efficiency.

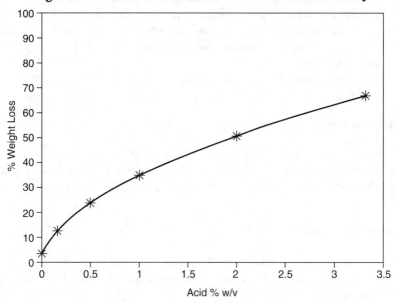

Figure 2 Matrix weight loss.

solution. It is clear from data received by ICPMS method and represented in Table 3 that more then 300 mg of metals is extracted by 2% sulfuric acid from 1 g of soil/slag. The separation of chromium from this extract is a special problem and is not investigated in this article.

Table 3. Mass of Some Metals Extracted from 1 g of Soil/Slag by 2% Sulfuric Acid at 95 C, Solvent/Soil Ratio 75:1, Time 30 min.

Ca	Fe	Mg	Al	Cr	Ti	Mn	V	Ni	Pb	Cu	Zn
185	50.5	45.2	27.8	18.2	1.9	0.7	0.4	0.2	0.15	0.15	0.15

Kinetics

A study of extraction kinetics indicates that extraction efficiency with 2% sulfuric acid at 95°C is about 85% within 5 minutes and achieves 92% for 1 hour treatment, as seen in Figure 3. So the basic mass of chromium is extracted from soil/slag for the first 5 minutes of washing.

Temperature

The effect of temperature on extraction efficiency was assessed. The results are shown in Figure 3. The curve shown in Figure 3 is meant to be illustrative only and doesn't represent a speculation of the true form of the curve. A comparison of efficiencies at three different sulfuric acid concentrations (extracted for one hour) at boiling temperature (20°C) shows that extraction increases significantly when it is carried out at boiling temperature over room temperature. At higher temperatures the solubility of chromium compounds in acid solutions as well as the rate of desorption and chemical reactions increases and efficiency of the processes is much more.

Previous investigations of chromium extraction from soil[7-10] were carried out at room temperature.

Solvent/Soil Ratio

As seen in Figure 4 at 75:1 solvent/soil ratio peak extraction is achieved at a lower concentration than at 25:1. Tests indicate that at 75:1 solvent/soil ratio 95% extraction is achievable, but at 25:1, the maximum chromium removal levels off at 80%.

Second Stage Extraction

Analysis of two stage extraction (Figure 5) shows that a second extraction following acid washes of 2% sulfuric acid concentration yields little additional chromium removal. At lower concentrations of the acid (0.5-1%) two stage extraction significantly increases the efficiency of chromium removal.

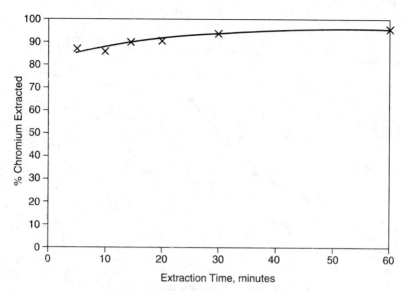

Figure 3. Chromium extraction kinetics

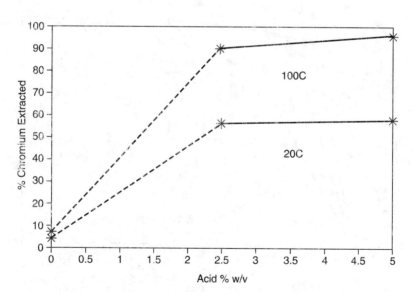

Figure 4. Effect of temperature on extraction efficiency

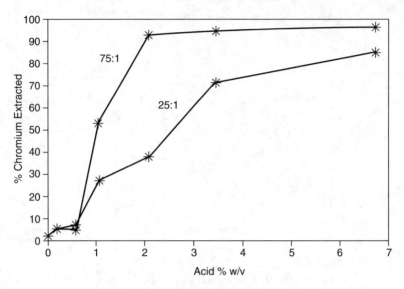

Figure 5. Effect of solvent/soil ratio on extraction efficiency

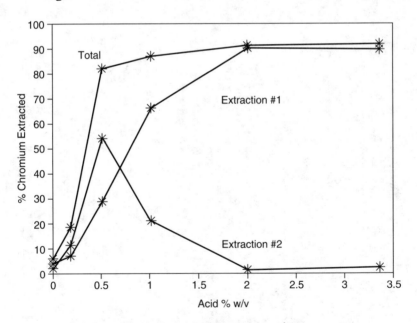

Figure 6. Second stage extraction

CONCLUSIONS

Acid extraction of the soil/slag samples yielded 95% chromium extraction efficiency, but about 50% matrix weight loss. All leachable chromium is believed to be removed. The matrix weight loss is accounted mostly by the transition of the slag macro elements (Ca, Fe, Mg, Al) to the extract during washing with acid solution.

We have found that there is a steep increase in chromium extraction efficiency between 0.5% and 2% sulfuric acid concentration and efficiency levels out at 95% removal at concentrations higher than 2%. Extraction is almost completed (about 85%) within 5 minutes and is much more efficient when carried out at boiling temperature rather than at room temperature. At 75:1 v:w solvent/soil ratio, peak extraction is achieved at a lower concentration than 25:1 and the efficiency of the process is more. Two stage extraction with 2% or higher sulfuric acid concentration yields little additional chromium removal.

REFERENCES

1. *Chromium in the Natural and Human Environments*, ed. by J.O. Nrigau, and E. Nieboer. John Wiley & Sons, New York, 1988, p. 512.

2. Sitting, M., *Handbook of Toxic and Hazardous Chemicals and Carcinogens*, Noyes Publ., Park Ridge, NJ, 2nd Ed., 1985, pp. 243-247.

3. "Environmental Reporter," The Bureau of National Affairs, 7/19/91, p. 132:0105.

4. Rinaldo-Lee, M., Hagarman, J., Diefendorf, A., "Siting of a Metals Industry Landfill on Abandoned Soda Ash Waste Beds," Hazardous and Industrial Waste Management and Testing: Third Symposium, ASTM STP 851, 1984, pp. 171-192.

5. Dreiss, S., "Chromium Migration Through Sludge-Treated Soils," *Ground Water*, V. 24, No.3, pp. 312-321.

6. Lester, J., "Biological Treatment," *Heavy Metals in Wastewater and Sludge Treatment Processes*, Volume II, CRC Press, Boca Raton, 1987, pp. 21-23.

7. Barlett, R., Kimble, J., "Behavior of Chromium in Soils: I. Trivalent Forms," *J. Environ. Qual.*, V. 5, 1976, pp. 379-386.

8. Hsieh, H., Rahgu, D., Liskowitz, J., Grow, J., "Soil Washing Techniques for Removal of Chromium Contaminants from Soil," Proceedings of the

Twenty-first Midlantic Industrial Waste Conference, ed. C. Cole, D.Long, pp. 651-660.

9. Peters, R., Elliott, H., American Institute of Chemical Engineers, 1991 Annual Meeting, Nov.17-22, 1991, Los-Angeles, CA.

10. Tan, K., "Adsorption and Desorption of Chromium on Clayey Soils," NJ Inst. of Tech., MSc. Thesis, 1989.

11. "Test Methods for Evaluating Soil Waste," United States Environmental Protection Agency, SW 846, Nov. 1986, 3rd Edition.

21

BIODEGRADATION OF POLYCYCLIC AROMATIC HYDROCARBON DEPENDING UPON OXYGEN TENSION IN UNSATURATED SOIL

Hun Seak Park
Washington State Pollution Liability Insurance Agency
Olympia, WA

Ronald C. Sims
Utah Water Research Laboratory
Utah State University
Logan, Utah

INTRODUCTION

Oxygen is known to be a critical factor influencing the functioning of oxygenases in the initial cooxidative attack on the reduced polycyclic aromatic hydrocarbon (PAH) molecule which is a group of hazardous organic substances of critical public health and environmental concern due to the chronic health effects (e.g., carcinogenicity), microbial recalcitrance, high bioaccumulation potential, and low removal efficiency in traditional waste treatment processes (Herbes et al., 1976; McGinnis et al., 1988). Low concentrations of oxygen may limit the cooxidative process as discussed by Sims and Overcash (1983), when applied to in situ biodegradation of relatively insoluble high-molecular weight mutagenic PAHs present in wood preservative waste impacted soil. Soil tends to be more anoxic with depth because oxygen is consumed rapidly relative to the rate of diffusion of oxygen from the soil surface (Lee, 1986; Stetzenbach, 1986). Thus, the study of oxygen-limited biodegradation of PAHs in soil under cooxidation condition would have significant implications for the engineering design of in situ biological treatment of PAHs. Specific objectives of the research that address the above hypothesis are described as follows: develop a mathematical model incorporating the effect of O_2 tension; evaluate the sensitivity of the model; identify the critical O_2 tension level; investigate the relationship among degradation rates, microbial density, and O_2 consumption rates as a function of O_2 tension.

MATERIALS

7,12-Dimethylbenz(a)anthracene (7,12-DMBA) was chosen based on reported genetic toxicity and rate and extent of degradation measured in laboratory studies (Park, 1987; Bigger et al., 1983). 7,12-Dimethylbenz(a)anthracene (97%; m.p. 121-123°C; F.W. 256.35; Lot# JM02013HM; $C_{20}H_{16}$) was purchased from Aldrich Chemical Co. (Milwaukee, Wis). N-hexadecane was chosen as a growth substrate for the cooxidation of 7,12-DMBA in soil. N-hexadecane has been previously demonstrated to affect cooxidation of hydrocarbons under laboratory conditions (Davis and Raymond, 1961). N-hexadecane ($CH_3(CH_2)_{14}CH_3$; F.W. 226.45; 99+ %; ASTM grade) was purchased from EM Science Co. (Cherry Hill, NJ). Uncontaminated Durant clay loam soil was collected from a U.S. EPA Research Facility, the Robert S. Kerr Environmental Research Laboratory, Ada, Oklahoma. The soil was characterized with regard to its properties (Table 1) and was demonstrated to be microbially active (Sims et al., 1986). Soil was sieved to pass a 2-mm screen. Soil moisture was maintained at 80% of soil moisture content at -1/3 bar tension throughout the study.

METHODS

Experimental Design

The effect of O_2 tension on the degradation of 7,12-DMBA was assessed on the basis of experimental conditions shown in Table 2. For a 7,12-DMBA degradation kinetic study, 40g (wet weight, or 26.7g dry weight) of soil at the incubation water potential of 80% of soil moisture at -1/3 bar tension were placed in 500-ml glass Erlenmeyer flask. Soils were then thoroughly stirred to mix with water. Following four weeks of preincubation at the desired O_2 tension, 7,12-DMBA was added to each soil in methylene chloride solution.

After the methylene chloride solution evaporated from the soil (approximately 24 hours), water was added to adjust the soil moisture. Soil flasks were sealed by rubber stoppers to reduce the loss of soil water and seal the system while maintaining the desired O_2 tension. Soil microcosms were stirred at one week intervals. Head space O_2 was regulated by gassing flasks with nitrogen gas and readjusting to the desired O_2 tension with a plastic and gas-tight syringe. Rubber stoppers were removed periodically (every 4 or 5 days) to replenish moisture and to allow gas exchange. Evaporative water losses during incubation were compensated by adding water to the flasks approximately every 2 weeks to maintain soil-water potential. All gases (air, oxygen, and nitrogen) were humidified through the fritted glass gas dispersion tubes. All incubations, sample

Table 1. Characterization of Durant Clay Loam Soil

Characteristics	Value
Bulk Density	1.59 g/cm^3
Texture	Clay Loam
pH	6.6
Electrical Condu.	20.5 me/100g
Organic Carbon	2.88%
Cation Exchange Cap.	0.5 mhos/cm
Total Phosphorus	0.03%
Total Nitrogen	18 ppm
Iron	28 ppm
Zinc	3.8 ppm
Potassium	177 ppm
Bacteria	5.1 × 10^7 cfu/g
Funi	2.6 × 10^5 cfu/g

cfu: colony-forming units/g of soil by plate count agar

Table 2. Experimental Design for 7,12-Dimethylbenz(a)anthracene Degradation Kinetic Study

Items	Explanation
Type of experiment:	Semi - Continuous System for oxygen
Type of Extraction:	Sacrifice of Entire Sample in Flask
Oxygen Tensions:	21.0%, 7.00%, 2.33%, 0.78%, 0.26%, 0.00%
Oxygen Readjusting Period:	Every 4 to 5 days
7,12-DMBA Concentration:	Initially One Time Spike 40 µg/g of wet soil
7,12-DMBA Carrier:	CH_2Cl_2
Growth Substrates:	N-Hexadecane (1%) and Indigenous Organic Carbon
Sampling Events:	Seven
Monitored Parameters:	O_2, CO_2, 7,12-DMBA, and Biomass Density
Duration of Experiment:	54 days
Control Sample:	Sterile (2% $HgCl_2$) @ 21% O_2 Tension
Weight of Soil:	40 g wet weight (or 26.7g dry weight) per Flask
Replicates:	Three
Temperature:	25 ± 3°C

handling, and chemical analysis were strictly conducted under the Gold Bulb due to the photosensitivity of 7,12-DMBA. Poisoned controls (2% $HgCl_2$) with and without 7,12-DMBA were also prepared to account for possible abiotic reactions. In order to obtain biomass density information under reduced O_2 tensions, sets of duplicate soil samples were incubated as described previously except that 500 g of soil were placed in a 3-L glass flask. To ensure that head space O_2 was maintained at the desired levels throughout the experiment, head space gas subsamples were analyzed for O_2 on a gas chromatograph equipped with a thermal conductivity detector (TCD). Every 4 or 5 days head space gas samples were removed from the flasks using a gas-tight syringe, and the composition of the gas phase was determined with gas chromatography. A Perkin-Elmer Sigma 4 isothermal gas chromatograph, equipped with TCD and an Alltech CTR1 6 ft × 1/4 inch activated molecular sieve column (Deerfield, IL), was used to measure gas composition and component concentrations (O_2, CO_2, N_2). The detector and oven temperatures were maintained at 50°C and ambient temperature, respectively. Helium was used as a carrier gas at a flow rate of 65 mVmin. Sample injection was 0.5 ml. The gas chromatograph was calibrated using laboratory air and compressed gas standards obtained from Scott Specialty Gases (Plumsteadville, PA).

Under identical conditions of incubation, the initial, substrate-induced, maximal respiratory responses were correlated to the actual size of the living, non-resting soil microflara as introduced by Anderson and Domsch (1978). Thus a constant had to be derived which would allow conversion into microbial biomass weight. Glucose was chosen as a substrate suitable for Substrate Induced Response (SIR) method based on the following criteria: water-soluble to allow its rapid dispersion to as many microhabitats as possible; relatively complex in structure, to prevent its mineralization by bound or free soil enzymes; and non-toxic, to avoid adverse side effects when added to soil in quantities in excess of those which can be immediately metabolized. The amount of microbial carbon in soil was calculated on the basis of the carbon fraction (53%) in a characteristic heterotrophic microbial cell ($C_5sH_7O_2N$). Based on experimental data, CO_2 measurements between the first 5 and 6 hrs were optimum for representing the initial substrate-induced maximal respiratory rate of a specific microcosm. An infrared (IR) detector (Oceanography International, College Station, Texas) was used as a CO_2 monitoring system. Procedures were as follows: (1) Equilibrated and moistened 10 g soil samples were placed in 25 ml empty flasks with rubber stoppers; (2) Flasks were amended with glucose to 4 mg/g of soil; (3) Flasks were incubated at constant temperature (35°C) and dark environment; (4) One ml subsample of each flask atmosphere was injected into the IR detector injection port of IR detector for the measurement of CO_2 after 5 and 6 hrs of incubation; (5) Hourly production rates of CO_2 were converted to biomass C size. Substrate induced response (SIR) measurement was calibrated against the known biomass size added to soil in terms of Volatile Suspended Solid (VSS), and used as a measure of microbial biomass in soil. The simultaneous measurement of VSS in the added inoculum and SIR showed a highly significant correlation ($r^2 = 0.979$) between VSS and SIR. From this correlation it could be concluded that a substrate induced maximal respiratory rate of 1 μmol CO_2/h corresponds to approximately 3.47 mg microbial biomass carbon, or 1 mg biomass carbon produces approximately 0.288 μmol CO_2/hr.

Chemical Analysis

Soil beakers were withdrawn from the incubation units at 0, 9, 18, 27, 36, 45, and 54 days after 7,12-DMBA addition. The schedule of sampling times ensured that samples would be taken beyond the published 7,12-DMBA half-life (reportedly, 20 to 28 days) in soil (Park, 1987). Each beaker was extracted with 170 ml of methylene chloride using a homogenization technique, then the extracts were dried over anhydrous sodium sulfate and evaporated to 1 ml in the K-D apparatus. A Tekmar Tissuemizer was used for extraction of 7,12-DMBA from each soil sample (Coover and Sims, 1987). 7,12-DMBA was determined by reverse-phase HPLC using a Shimadzu Model SCL-6A liquid chromatograph. An aliquot of the extract was injected into a HPLC system fitted with 4.6 mm I.D. × 250 mm 5 μ octadecylsilance column and eluted with a water/acetonitrile gradient mobile phase program (from 35 to 100% of acetonitrile) at a flow rate of 1.5 ml/min (injection volume of 15 μl). The extracts were immediately placed in dark bottles with Teflon caps in a freezer at -20 °C. The extracts were analyzed for 7,12-DMBA in the peak area mode at a UV detector of 254 nm.

Statistics and Model Development

Analysis of Variance (ANOVA) was used to compare half-life values among different treatments based on O_2 tension. It was necessary to determine statistically whether or not the observed differences between slopes of regression lines (or degradation rate, k), were significant (Kleinbaum and Kupper, 1978). The appropriate null hypothesis for a comparison of slopes, which can be converted to half-life, (i.e., for a test of parallelism) was given by a weighted average of the two separate slope estimates, and tested by the Student's "t" table with the two-sided test. In addition, first- and zero- order regression plots were prepared to calculate an apparent half-life for various O_2 tension treatments.

In soil, the introduction of a nongrowth substrate (7,12-DMBA) and growth substrate (n-hexadecane + soil organic carbon) would be expected to induce qualitative and quantitative variations in different segments of the microbial population. To model the biodegradation of compounds that do not contribute appreciably to the growth of the microbes responsible for their metabolism and cooxidation in the soil system, the logistic expression for growth of the active microbes could be used (Simkins and Alexander, 1984; Schmidt and Alexander, 1985; Lambrecht et al., 1988) as follows:

$$\mu = \mu_m \left(1 - \frac{M}{M_{max}}\right) \quad (1)$$

where,

M_{max} = Maximum Microbial Density achievable (MIM)
μ_m = Maximum Growth Rate (1/T)

The specific growth rate (μ) of microbes in the concentration of limiting substrates is likely to follow some form of saturating kinetics. To express this concept, the Michaelis-Menten relationship is used. It is thought that the enzyme responsible for 7,12-DMBA oxidation would be active when growth substrate is available on a long term basis. However, the rate and extent of 7,12-DMBA disappearance would directly depend on the cooxidizing microbial population density and O_2 concentrations. If one does not allow for maintenance consumption that is assumed to be provided entirely by growth substrate, the terms can be written as:

$$-\frac{dS}{dt} = k_{max} M \left(1 - \frac{M}{M_{max}}\right) \left(\frac{O}{K_O + O}\right) \quad (2)$$

where,

S = 7,12-DMBA Concentration (M/M)
K_O = Half Saturation Constant for Oxygen (M/M)
k_{max} = Maximum 7,12-DMBA Degradation Rate (1T)
O = Oxygen Tension (M/M)

If microbial growth during the incubation is neglected, Eq. 2 can be reduced to:

$$-\frac{dS}{dt} = k_{max} \left(\frac{O}{K_O + O}\right) \quad (3)$$

Eq. 3 was designated as Model #1, whereas Eq. 2 as Model #2 to be simulated by the experimental design and data collection.

Parameter Estimation and Sensitivity Analysis

Once data were available relating degradation rate to O_2 tension, several different techniques were used to estimate k_{max} and K_o by using three linear transformations (Lineweaver-Burk, Hanes, and Hofstee plots). All nonlinear regression analysis of data were performed with a microcomputer and the SANREG statistical package developed by Utah State University. SANREG provides weighed or unweighted least squares parameter estimation on a nonlinear mathematical model. SANREG uses the modified Gauss-Newton method for minimizing the sum of squares function. This version of the program was written by Dr. David K. Stevens (Utah State University) for use on IBM compatible microcomputers. Three major parameters (K_o, M_{max}, and k_{max}) were estimated from laboratory studies. The differential equation of Model #2 previously described is intrinsically nonlinear. It is therefore not possible to transform Model #2 into a linear form, so a nonlinear least square regression method was required, which allowed estimation of these parameters by a computer program. To determine which parameters influenced the system most, a parameter sensitivity analysis was conducted for the Models #1 and #2. These sensitivity equations can be used to predict whether the parameters may be uniquely estimated and the range of the dependent variables over which the model is most sensitive to changes in the parameters. Model #1 (Eq. 3) can be integrated with respect to time to form the 7,12-DMBA depletion equation as follows:

$$S = S_o - tk_{max}\left(\frac{O}{K_O + O}\right) \quad (4)$$

By differentiation of S in Eq. 4 with respect to kma, and Ko, the sensitivity coefficient equations were obtained (Eqs. 5 and 6).

$$\frac{dS}{dk_{max}} = -\frac{tO}{K_O + O} \quad (5)$$

$$\frac{dS}{dK_O} = \frac{t\,O\,k_{max}}{(K_o + O)^2} \quad (6)$$

For Model #2, Eq. 2 was integrated with respect to time as follows:

$$S = S_O - k_{max}\, t\, M \left(\frac{O}{K_O + O}\right)\left(1 - \frac{M}{M_{max}}\right) \quad (7)$$

The sensitivity equations of Model #2 were obtained from the differentiation of Eq.7 as follows:

$$\frac{dS}{dK_{max}} = -tM\left(\frac{O}{K_O + O}\right)\left(1 - \frac{M}{M_{max}}\right)$$

$$\frac{dS}{dK_O} = \frac{t\,k_{max}\,OM}{(K_O + O)^2}\left(1 - \frac{M}{M_{max}}\right) \tag{9}$$

$$\frac{dS}{dM_{max}} = -\frac{t\,k_{max}\,OM^2}{(K_O + O)M_{max}^2} \tag{10}$$

Eqs. 5, 6, 8, 9 and 10 were drawn against a range of O_2 tensions for the purpose of the parameter sensitivity test.

RESULTS AND DISCUSSION

As shown in Table 3, 7,12-DMBA was observed to have wide range of half-lives (from 645 to 109 days) depending on O_2 tension with the lowest half-life associated with the highest O_2 tension of 21%. Degradation rates ranged from 0.00 to 0.183 µg/g of soil/day for zero-order kinetic and 0.001 to 0.006 /day for first-order kinetic, respectively. The greatest increase in loss rate occurred between 2.33 and 7.0% O_2 tensions.

Anoxic samples (0% O_2) showed no difference in k from the poisoned control sample. Loss of 7,12-DMBA from soil samples poisoned by 2% $HgCl_2$ was statistically not significant. Generally, apparent degradation did not show clearly that first- or zero-order was a more appropriate model. Results indicated that 7,12-DMBA losses over the course of the experiment with varying O_2 tension were sufficient to provide a clear conclusion that soil treated with higher O_2 tensions (7.0 and 21% O_2) had a higher 7,12-DMBA apparent degradation rate than in the soil treated with lower O_2 tensions (0.0, 0.26, 0.78, and 2.33% O_2). These results agree well with those of DeLaune et al. (1981) and Bauer and Capone (1985) who found (^{14}C)-naphthalene mineralization to be directly related to sediment redox potential (or E_h), which in turn is largely controlled by oxygen availability. Hambrick et al. (1980) found 0 and 35% of (^{14}C)-naphthalene mineralized at -250 and +250 mV, respectively, after 35 days. This has several implications for both the fate of aromatic compounds in systems of variable redox and their long-term effects on resident microbiota. Figure 1 shows the O_2 cumulative consumption data in 7,12-DMBA applied soil as a function of various O_2 tension maintained. The rates of O_2 consumption and CO_2 production in closed soil flasks containing 7,12-DMBA amended soil were measured to determine the microbial activity. The maximum O_2 consumption rate was found to be 27.23 moles O_2/g of soil/day at

Table 3. Degradation Kinetic Information for 7,12-Dimethylbenz(a)-anthracene in Soil (Zero-Order Kinetics)

Oxygen Tension (%)	Total DF[#]	F-test (p)	k*	Half-Life[#]	r^2	95% Confidence Interval			
						Lower Limit		Upper Limit	
						k*	Half-Life[#]	k*	Half-Life[#]
0	20	0.171	-0.031	645	0.096	-0.077	260	0.015	—
0.26	20	0.827	0.007	—	0.003	-0.058	345	0.072	—
0.78	20	0.064	-0.049	408	0.169	-0.1	200	0.003	—
2.33	20	0.028	-0.069	290	0.229	-0.129	155	-0.01	2500
7	17	0.003	-0.149	134	0.426	-0.24	83	-0.06	351
21	17	1E-04	-0.183	109	0.794	-0.233	86	-0.13	149

*ug/g of soil/day
[#]Days

the 21% O_2 tension. A peak in the rates of O_2 consumption and CO_2 production were observed after 20 days of incubation. Lower rates were measured following this peak in activity after 30 days of incubation. Approximately 70% of O_2 consumed in the soil microcosm was converted to CO_2. The molar ratio of CO_2 produced/O_2 consumed was constant throughout the incubation period. Independent estimates of biomass size were obtained using the live and non-resting biomass measurement technique. To determine if the microbial measures employed were indicative of apparent degradation rates of 7,12-DMBA in soil, apparent degradation rates were regressed (least-squares linear regression) against biomass density (Figure 2) and O_2 consumption rate - the underlying assumption being that the relationship between rates and biomass density could be approximated by a linear function. Apparent degradation rates of 7,12-DMBA (zero-order reaction rate constants), k were directly correlated (r^2 = 0.803) to microbial density, i.e., the living and non-resting biomass as measured by the SIR method, and O_2 consumption rate (r^2 = 0.859). The regression analysis showed that a direct relationship generally existed between apparent degradation
rates and biomass densities measured by SIR method. The results were analyzed with models to obtain values for the kinetic parameters. Using Eq. 3 (Model #1), linear transformations of the data were performed to determine the maximum degradation rates (k_{max}) as well as the halfsaturation constants for O_2 tension (K_o) and turnover times (Tt) under O_2 limitation (Table 4). The Hanes transformation (Figure 3) was chosen to compare the measured and predicted data because it gives the best distribution of data along the X axis and therefore, provides a good estimate of slope values (Cornish-Bowden, 1976). The reasonable agreement observed over the entire range of oxygen tension suggests that the average values

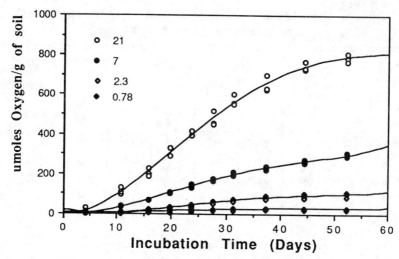

Figure 1 Cumulative oxygen consumption in 7,12-Dimethylbenz(a)anthracene applied soil

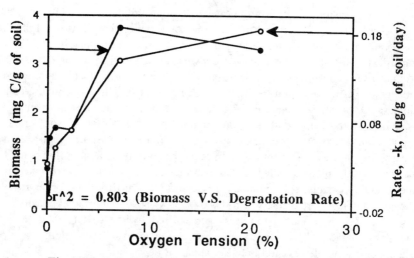

Figure 2 Relationship between biomass density and 7,12 Dimethylbenz(a)anthracene degradation rate (k).

adequately describe the kinetics of 7,12-DMBA degradation. During computer analysis, the initial parameters were varied individually to obtain the best fit of the experimental data. For Model #1 the validity of the average values was tested by simultaneously applying the coefficients to describe the kinetics of 7,12-DMBA degradation over the entire range of O_2 tension. Figure 4 is a plot of the predicted 7,12-DMBA degradation as a function of the O_2 tension tested in this soil study.

For Model #1, when determining by nonlinear regression technique, Ko and kma, were higher and T_t was lower than by linear transformation method due to the intrinsic different data analysis methods. However, K_o (approximately 15% O_2 tension) estimated by nonlinear regression was found to be five times higher than Ko (approximately 3% O_2 tension) estimated by linear transformation. Confidence interval (95%) for K_o estimated by nonlinear regression were relatively large beyond the experimental conditions conducted for O_2 tension. The Model #2's estimates of k_{max}, M_{max}, and K_o values by the nonlinear regression technique are presented in Table 4. Kinetic parameter values (k_{max} and K_o) estimated by nonlinear regression technique were almost identical regardless of kinetic Models (Model #1 and 2)- M_{max} estimated by nonlinear regression was found to be 8.54 mg C/g of soil which was a reasonable value when comparing with the results of biomass density measurement.

A model input can be varied in order to examine the degree of its influence on the system response. By varying one input parameter while holding other constants, insight can be gained on the relationship between that parameter and 7,12-DMBA concentration. The results of sensitivity analysis are shown in Figure 5. k_{max} was the parameter for which the model is most sensitive. Sensitivity test was conducted when considering O_2 tensions of 21, 5.25, 1.312%. As the value of K_o increases, the O_2-limitation of degradation increases, as would be expected. It is recognized that, under field conditions, where O_2 is replenished by dispersion/diffusion from the atmosphere this rate limitation will be less severe over the long term. However, for short term dynamics, such as immediately after a waste application in land treatment of PAHs, O_2 limits may be very important and warrant inclusion. Eqs. 8, 9, and 10 are plotted in the Figure 6 to show sensitivity of S (7,12-DMBA concentration) with respect to the parameters estimated by nonlinear regression technique. The relative sensitivity analysis based on the best estimated parameters with Model #2 are of the order $k_{max} \times 10 > M_{max} \times 10^{-1} \geq K_o \times 10^{-1}$ implying that K_o was the least sensitive parameter.

The validity of the kinetic parameters was confirmed by simultaneously applying the coefficients to predict the rates of 7,12-DMBA degradation over a wide range of O_2 tensions. The rate of 7,12-DMBA degradation was shown to be influenced by the availability of molecular O_2, which is consistent with the role of molecular O_2 as a cosubstrate for the initial cleavage of the ether side chain and both the hydroxylation and subsequent fission of the aromatic nucleus. Analysis of the experimental data indicated that the rate of 7,12-DMBA apparent degradation was a hyperbolic function of the molecular O_2 tension. According to the statistical significance test results, O_2 tensions in the range of 3 to 7% may be rate limiting for the microbial degradation of high molecular weight aromatic compounds which have a requirement for molecular O_2 as a cosubstrate.

The availability of O_2 has been shown to influence the rates of hydrocarbon degradation which is consistent with the role of molecular O_2 as a cosubstrate for microbial hydrocarbon degradation (Hambrick et al., 1980). The results obtained from this investigation provide some insight into the degradation of PAHs within

Table 4. Kinetic Parameters of 7,12-Dimethylbenz(a)anthracene Degradation in Soil as a Function of Head Space Oxygen Tension

Parameters Estimated	Model #				
	1				2
	Data Analysis Method				
	Lineweaver-Burk	Hanes	Hofstee	Nonlinear	Nonlinear
k_{max}	0.163	0.214	0.187	0.343	0.221
T_t	245	187	213	116	181
M_{max}	NA	NA	NA	NA	8.540
K_o	1.933	3.545	2.421	14.99	15.00
95% C.I. of K_o					
Lower	0.0033	-0.082	-1.61	-16.27	-14.54
Higher	3.864	7.185	6.453	46.27	44.55

k_{max}: Maximum 7,12-DMBA Degradation Rate (μg/g of soil/day).
T_t: 7,12-DMBA Turn Over Time (Days).
K_o: Half-Saturation Constant for Oxygen Tension (% Headspace).
M_{max}: Maximum Microbial Density Achievable (mg C/g of soil).
NA: Not Applicable

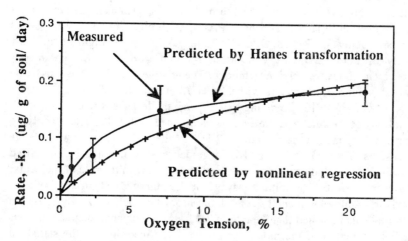

Figure 3 Comparison of Measured (with One Standard Error Bar) and Predicted (by HANES Transformation and Nonlinear Regression) 7,12-Dimethylbenz(a)anthracene Degradation Rate (k) (with Model #1).

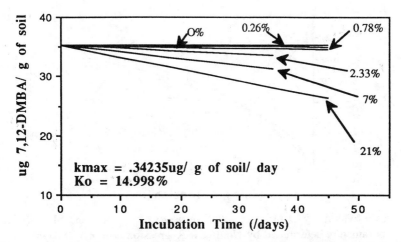

Figure 4 Predicted 7,12-Dimethylbenz(a)anthracene Degradation in Soil with Model #1 using Nonlinear Regression Technique.

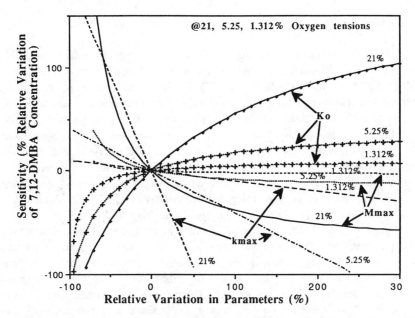

Figure 5 Sensitivity of the Model #2 for a change in the parameters (with constants determined by nonlinear regression). The relative variation of the 7,12-DMBA concentration at t=100 days was defined as the sensitivity.

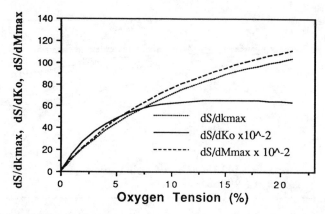

Figure 6 Sensitivity of Model #2 with respect to k_{max}, K_o, and M_{max} (with parameters determined by Nonlinear Regression at t = 100 days).

a vadose zone soil microcosm and could have implications for the potential reclamation of contaminated sites.

CONCLUSIONS

The effect of O_2 tension on PAH degradation in unsaturated soil under cooxidation conditions was evaluated for one PAH compound: 7,12Dimethylbenz(a)anthracene (7,12-DMBA). Results indicated that 7,12-DMBA losses over the course of the experiment with varying O_2 tension were sufficient to provide a clear conclusion that soil treated with higher O_2 tensions (7.0 and 21% O_2) had a higher 7,12-DMBA apparent degradation rate than in the soil treated with lower O_2 tensions (0.0, 0.26, 0.78, and 2.33% O_2)- Mean degradation halflives decreased from 645 days for 0% head space O_2 tension to 109 days for 21% O_2 tension in laboratory soil microcosm. Anoxic samples showed no difference in apparent degradation rates, k, from controls. Apparent degradation rates (k) of 7,12-DMBA under O_2 limitation ranged from 0.00 to 0.180 µg/g of soil /day for a zero-order kinetic model and 0.001 to 0.006 /day for a first-order kinetic model, respectively (@ initial concentration, 40 µg/g of wet soil; 25 °C). Data obtained from experiments with 7,12-DMBA were evaluated using a hyperbolic saturation type - Michaelis Menten - enzymatic kinetic model under O_2 limitation. The kinetic parameters kma, (maximum degradation rate) and K_o (half-saturation constant for O_2) were best estimated with a Hanes linear transformation and a nonlinear regression technique. Values of k_{max} and K_o for the Hanes linear transformation were found to be 0.214 mg/g of soil/day and 3.6%, and for the nonlinear regression k_{max} and K_o values were 0.221 mg/g of soil/day and 15%, respectively. M_{max} (maximum microbial density achievable) was found to be 8.54 mg C/g of soil using the nonlinear regression technique. Apparent degradation rates (zero-order reaction rate constants) were directly correlated to microbial density, i.e., the living and non-resting biomass as measured by the SIR method, and O_2 consumption rate. The relative sensitivity analysis showed that Ko was the least sensitive parameter.

ACKNOWLEDGEMENTS

This work was sponsored by the United States Environmental Protection Agency Grant No. R14475-01, Office of Research and Development, Exploratory Research Program, Washington, D.C., (Mr. Donald Carey, Project Officer). This work has not been subjected to the Agency's peer review and therefore does not necessarily reflect the views of the Agency, and no official endorsement should be inferred.

REFERENCES

1. Anderson, J. P. E. and Domsch, K. .H., "A Physiological Methods for the Quantitative Measurement of Microbial Biomass in Soil," *Soil Biol. Biochem.*, 10, pp. 215-221, 1978.

2. Bauer, J. E. and Capone, D. G., "Degradation and Mineralization of the Polycyclic Aromatic Hydrocarbons Anthracene and Naphthalene in Intertidal Marine Sediments," *Appl. Environ. Microbiol.*, 50, pp. 81-90, 1985.

3. Bigger, C. A., Sawicki, J. T., Blake, D. M., Raymond, L. G., and Dipple, A., Products of Binding of 7,12-Dimethylbenz(a)anthracene to DNA in Mouse Skin, *Cancer Res.*, 43, pp. 5647-5651, 1983.

4. Coover, M. P. and Sims, R. C., "The Effect of Temperature on Polycyclic Aromatic Hydrocarbon Persistence in an Unacclimated Agricultural Soil," *Hazardous Waste & Hazardous Materials*, 4(1), pp. 69-82, 1987.

5. Cornish-Bowden, "Principles of Enzyme Kinetics," Butterworths, London, 1976.

6. Davis, J. B. and Raymond, R. L., "Oxidation of Alkyl-Substituted Cyclic Hydrocarbons by a Nocardia During Growth on A-alkanes," *Appl. Microbiol.*, 9, pp. 383-388, 1961.

7. Delaune, R.D., Hambrick, G. A., and Patick, W. H., "Degradation of Hydrocarbons in Oxidized and Reduced Sediments," *Mar. Poll. Bull.*, 11, pp. 103-106, 1981.

8. Hambrick, G. A., Delaune, R. D., and Patrick, W. H., "Effect of Esturine Sediment pH and Oxidation-Reduction Potential on Microbial Hydrocarbon Degradation," *Appl. Environ. Microbiol.*, 40, pp. 365-369, 1980.

9. Herbes, S. E., Southworth, G. R., and Gehrs, C. W., "Organic Contaminants in a Aqueous Coal Conversion Effluents: Environmental Consequences and Research Priorities," *Trace Substances in Environmental Heal-X*, A Symposium D. D. Hemphill (ed.), University of Missouri, Columbia, 1976.

10. Kleinbaum, D. G. and Kupper, L. L., "Applied Regression Analysis and Other Multivariable Methods," Duxbury Press, 1978.

11. Lambrecht, R. S., Carriere, J. F., and Collins, M. T., "A Model for Analyzing Growth Kinetics of a Slowly Growing *Mycobacterium sp.*," *Appl. Environ. Microbiol.*, 54(4), pp. 910-916, 1988.

12. Lee, M. D., "Biodegradation of Organic Contaminants in the Subsurface of Hazardous Waste sites," Rice University, Ph.D. Dissertation, 1986.

13. McGinnis, G. D. et al., "Characterization and Laboratory Soil Treatability Studies for Creosote and Pentachlorophenol Sludge and Contaminated soil," US EPA, Washington D.C., EPA 600/S2-88/055, 1988.

14. Park, K., "Degradation and Transformation of Polycyclic Aromatic Hydrocarbons in Soil Systems," Utah State University, Ph.D. Dissertation, 1987.

15. Schmidt, S. K. and Alexander, M., "Effects of Dissolved Organic Carbon and Second Substrates on the Biodegradation of Organic Compounds at Low Concentrations," *Appl. Environ. Microbiol.*, 49(4), pp. 822-827, 1985.

16. Simkins, S. and Alexander, M., "Models for Mineralization Kinetics with the Variables of Substrate Concentration and Population Density," *Appl. & Environ. Microbiol.*, 47(6), pp. 1299-1306, 1984.

17. Sims. R. C., et al., "Contaminated Surface Soils In-place Treatment Techniques," Noyes Publications, New Jersey, 1986.

18. Sims, R. C. and Overcash, M. R., "Fate of Polynuclear Aromatic Compounds (PNAs) in Coal Systems," *Residue Reviews*, 88, pp. 1-68, 1983.

19. Stetzenbach, L. D. A., "The Degradation and Utilization of Polycyclic Aromatic Hydrocarbons by Indigenous Soil Bacteria," The University of Arizona, Ph.D. Dissertation, 1986.

22

APPLICABILITY OF BIODEGRADATION PRINCIPALS FOR TREATMENT OF SOILS AND GROUNDWATER

Robert D. Norris, Ph.D. and W. Wesley Eckenfelder, Jr., P.E. DSC.
ECKENFELDER INC.
Nashville, TN

INTRODUCTION

Microbiological processes for treating contaminated soils and groundwater are generally referred to as bioremediation. The development of bioremediation technology has benefited from the incorporation of the engineering, hydrogeology, and chemistry developed for other environmental technologies such as groundwater production, pump and treat technology, and in situ vapor stripping. The relatively rapid development of bioremediation technology has been possible in part because of the extensive understanding of biooxidation that has evolved over the long history of wastewater treatment and interest in biooxidation processes by the academic community. As a result, the development of commercial bioremediation technology has been able to focus on the critical engineering and hydrogeological aspects of bioremediation while relying on established microbiological principles and data concerning both aerobic and anaerobic biodegradability of hundreds of compounds (Gibson, 1984; Munnecke, 1982; Pitter, 1990; Reineke, 1988, and USEPA, 1986). The large body of biodegradation data provides details of degradation pathways, including the intermediate products that are formed from many common contaminant constituents.

Years of biological wastewater treatment experience have identified critical parameters that affect the rate and extent of biodegradation of organic compounds. While there exist many subtleties concerning how these parameters affect degradation under various applications, the general effects of these parameters hold across all technologies as discussed in the following sections.

Nutrient Requirements

Several mineral elements are essential for the metabolism of organic matter by microorganisms. For soil and groundwater biooxidation treatment processes, all elements except nitrogen and phosphorus are nearly always present in sufficient quantities. This is because these elements are found in at least trace levels in nearly all soils and groundwaters and are required in extremely low amounts relative to the mass of biodegradable organic constituents. Table 1 shows the ratio of trace nutrients to biological oxygen demand (BOD) as determined for wastewater applications (Eckenfelder, 1967). Nitrogen and phosphorus requirements were the subject of much research. Early work (Helmers, 1951) suggested a nitrogen requirement of 4.3 pounds per 100 pounds of BOD and a phosphorous requirement of 0.6 pounds per 100 pounds of BOD.

Nutrient requirements for bioremediation of soils and groundwater also depend on the extent to which nutrients are recycled from dead microorganisms, the fraction of the contaminant that is converted to carbon dioxide, and the mass of nutrients already present in the soils and/or groundwater. Various practitioners in the industry have suggested rule of thumb nutrient requirements from 3 to 10 pounds of nitrogen and from 0.3 to 1 pounds of phosphorous per 100 pounds of biodegradable organic carbon. Nutrient requirements can be estimated from the contamination concentrations using such ratios or determined from the results of laboratory treatability tests. In some instances most or all of the nutrient requirements can be met without addition of nutrients.

Nutrients consist of a phosphorus source that is most typically a salt of phosphoric acid and a nitrogen source that may be an ammonium salt, a nitrate

Table 1. Trace Nutrient Requirements for Biological Oxidation

	mg/mg BOD
Mn	10×10^{-5}
Cu	15×10^{-5}
Zn	16×10^{-5}
Mo	43×10^{-5}
Se	14×10^{-10}
Mg	30×10^{-4}
Co	13×10^{-5}
Ca	62×10^{-4}
Na	5×10^{-5}
K	45×10^{-4}
Fe	12×10^{-3}
CO_3	27×10^{-4}

salt, urea, or a combination of sources. Nutrient selection is more important for in situ systems than for the ex situ methods. Selection of nutrients must take into account the degree of retention of the nutrient components by the soils. While some retention is beneficial, the degree of retention directly impacts the volume of water and time required to distribute the nutrients. Other considerations are also important. If sodium salts are used, clayey soils may swell and reduce permeability to both water and air. Use of nitrate as the nitrogen source may result in concern for increased nitrate levels in the groundwater as nitrate will be readily leached from the soils. Residual ammonium ion levels and potential conversion of ammonium ions to nitrate may also be of regulatory concern.

Oxygen Requirements

Although other electron acceptors such as nitrate can be used for microbial based destruction of organic compounds, most commercial bioremediation projects have used oxygen as the electron acceptor. The stoichiometric amount of oxygen required for conversion of hydrocarbons to carbon dioxide and water can be estimated from the equation

$$2CH_2 + 3O_2 \rightarrow 2CO_2 + 2H_2O$$

For most hydrocarbons the oxygen requirement can be conveniently approximated as three pounds of oxygen for every pound of degradable hydrocarbon. Since not all of the transformed organic substances will be converted to carbon dioxide and water, the microorganisms will require less than the stoichiometric amount of oxygen. Wastewater treatment processes typically require 0.8 to 1.7 pounds of oxygen per pound of carbon. In most bioremediation processes, either oxygen is present in large excess as in landfarming or the efficiency of the oxygen distribution process is likely to require that excess oxygen needs to be introduced. In the latter case, preliminary estimates should be made using the 3 to 1 ratio.

Effect of Temperature

Variations in temperature effect all biological processes. Most aerobic biological-treatment processes operate within a temperature range of 4 to 39°C. Within this range the rate of biological reactions will increase with temperature to a maximum value at approximately 31°C. At temperatures above 35°C, it is highly likely that biological process rates will begin to decrease. In cold climate areas the indigenous microorganisms are likely to be more active at the lower end of the acceptable temperature range than are bacteria indigenous to more temperate climates.

Fortunately, in situ bioremediation processes can frequently be conducted within the acceptable temperature range without any active steps being taken to control the temperature. Above ground treatment systems may not be practical under very hot or very cold weather conditions unless steps are taken to control the temperature within the soils. For instance, soil cells can be operated during cooler seasons by operating the blower only during the warmest part of the day or, when appropriate, by blowing in heated air.

Effect of pH

Biooxidation processes are typically effective over a limited pH range. Most processes are practically limited to a pH range of 5 to 9 with optimum rates occurring over the pH range of 6.5 to 8.5. In soils and groundwater where the pH has historically been outside the these values due to natural effects, biooxidation may proceed quite rapidly. Groundwater in the New Jersey Pinelands can be in the pH range of 4.5 to 5.0 and still support active biodegradation (Brown, 1991). The carbon dioxide generated from biooxidation forms carbonic acid when dissolved in water. It is thus necessary to be aware of pH changes during treatment. The use of phosphate salts as nutrients can help buffer the soils and groundwater during treatment.

Toxicity

Toxicity to biological oxidation may result from one of several causes. Organic substances, such as phenol, which are biodegradable at low concentrations, may be present at higher concentrations that are inhibitory to the microorganisms. Substances such as heavy metals may be present above their toxic threshold. Inorganic salts which may be beneficial at low concentrations may retard biooxidation at high concentrations. Fortunately, microorganisms can, under some conditions, become tolerant of otherwise toxic levels of some substances. Toxicity problems can generally be detected from standard plate count methods or through respirometer tests.

THE BIOREMEDIATION PROCESSES

The goal of commercial bioremediation processes is to modify environmental conditions such that specific constituents are converted to species of lesser environmental concern at acceptable rates. The parameters that are important for bioremediation are basically those that were discussed above; temperature, pH, nutrient availability, oxygen levels, and absence of toxic or inhibitory species as well as the presence of a consortium of bacteria that is capable of effecting the desired conversions and, for some processes, soil moisture level. In most cases pH, soil moisture, and temperature are within acceptable levels and an effective microbial consortium is present.

As a result, bioremediation can be viewed as an engineered process for efficiently providing adequate amounts of phosphorus, nitrogen, and oxygen. Nutrients are added either continuously or in batches. Oxygen is provided continuously in either a passive form or through active transport of a liquid phase (water) or a gas phase (air). How this is accomplished depends on a number of factors including types of contaminants, soil properties, contaminant distribution, and hydrogeology.

Effect Of Physical Properties

Various bioremediation processes incorporate pump and treat techniques to introduce oxygen and nutrients through movement of groundwater or in situ vapor stripping techniques to move air to introduce oxygen. The movement of fluids to transport an oxygen source and/or nutrients causes other changes as well as biodegradation to occur. The movement of air or water can result in the transport of volatile or water soluble species, thus removing some portion of these species from the contaminated zone.

As a result, the physical properties of the contaminant(s) will affect the rate at which remedial goals are reached and the amount of recovered groundwater or air requiring treatment. When air is the fluid, the ease of vaporization of the contaminant(s) and the air flow as well as many other factors will determine the extent to which the contaminant is transferred to the vapor phase. When the fluid is water, the ease of transport of the individual constituents and the rate of groundwater movement as well as many other factors will determine the extent to which the contamination is transported and partially removed in the dissolved phase.

In designing a bioremediation system, the impact of transport of contaminants needs to be understood. The proportions of the constituents of interest that will be addressed through biodegradation versus physical removal from the soils will affect the demand for nutrients and oxygen and for treatment of recovered air and/or water. Since treatment of water and air can represent a significant portion of total remediation costs, a properly designed system will balance the benefits and potential costs resulting from contaminant transport.

Relevant physical constants such as shown in Table 2 are available for many compounds of environmental interest (Montgomery, 1990; Howard, 1989; Howard, 1990; and Verschueren, 1983) or can be estimated from properties of compounds with similar structures and molecular weight. For conceptual planning purposes, it is adequate to utilize available physical properties to estimate the relative importance of transport of dissolved or vapor phase constituents to the overall process, the extent to which nutrient and oxygen demands will be reduced, and the extent of treatment of recovered water or air that will be required. The same physical properties also provide good insight into which methods of air or water treatment are likely to be feasible. For instance, highly water soluble compounds are not effectively removed from the aqueous phase by either air stripping or carbon adsorption.

Table 2. Physical Properties Important To Bioremediation Processes

Compound	Water Solubility (mg/L)	Log K_{ow}	Vapor Pressure (mm Hg)	Henry's Law Constant
Benzene	1,791	2.13	95	5.4×10^{-3}
TCE	1,100	2.40	69	1.0×10^{-2}
Acetone	Miscible	-0.24	231	3.7×10^{-5}
Butanol	77,000	0.88	7	5.6×10^{-6}
Phenol	87,000	1.46	0.5	4×10^{-7}
Naphthalene	31	3.28	0.2	4.6×10^{-4}
Pyrene	0.1	5.20	6.9×10^{-7}	1.1×10^{-5}
DEHP	0.3	5.11	5.0×10^{-6}	1.1×10^{-5}
Phenanthrene	0.9	4.46	6.8×10^{-4}	3.9×10^{-5}

Data obtained from: Howard, Phillip H. et al., Fate and Exposure Data, Vol. I, 1989 and Vol. II, Lewis Publishers, New York, 1990, and Montgomery, John H. and Linda M. Welkins, Groundwater Chemicals Desk Reference, Vol. 1, Lewis Publishers, New York, 1990.

Each of the bioremediation processes described below incorporate the potential for physical removal of contaminants from the soils as well as the primary mechanism of biodegradation. Each process contains procedures to create conditions favorable to microbial growth and generally use the contaminant(s) as food and energy sources for the bacteria in order to promote conversion of the contaminant(s) to substances of lesser or no environmental concern. In some processes degradation of the contaminant occurs through co-oxidation as a result of exocellular enzyme activity in conjunction with another substance that serves as the metabolite. The specific procedures in each process are utilized to take advantage of a variety of potential site conditions and metabolic behavior.

Treatment Of Excavated Soils

Land Treatment. Land treatment, sometimes referred to as land farming, incorporates batch addition of nutrients using physical mixing of the nutrients into the soils and continuous passive introduction of oxygen (air) (Loehr, 1986). Excavated soils are spread over a relatively large area (approximately 0.5 acres per 1,000 cubic yards of soil) to a depth of 6 to 18 inches depending on soil type, land availability, and time constraints. Nutrients are added in a dry form or as a dilute solution which also provides moisture. The soils are periodically tilled and/or plowed and nutrients added as required. Ten to fifty percent of the total anticipated nutrient requirement is added during the initial batch addition. Subsequent nutrient additions are made over the course of the project with the amounts estimated based on consumption during the process as determined from periodic analysis of soil samples.

Monitoring includes sampling of the treatment soils prior to, during, and after treatment for contaminant parameters, pH, nutrients, soil moisture, and bacteria counts. Additionally, the soils beneath the treatment system are sampled and analyzed prior to and after remediation to document the lack of impact on the surficial soils.

The construction and operation of a land treatment system exposes the soils to large volumes of air (and thus oxygen) which allows volatile species to evaporate. Unless the system is contained within a tent-like structure, volatiles will not be contained and thus will be discharged to the atmosphere (Yare, 1991). Uncontrolled discharge of large amounts of volatile species is typically inappropriate and nonconforming with most state regulations. For soils contaminated with less volatile contaminants such as diesel fuel and especially the heavier heating oils, the volatile losses may be acceptable depending on the location relative to human activities and on state regulations.

High degrees of conversion and/or low residuals relative to remediation standards are possible with low molecular weight contaminants and mixtures such as gasoline. However, the levels achievable with other contaminants such as polyaromatic hydrocarbons (PAHs) may be limited by their rate of solubilization (Brubaker, 1991).

Soil Cell Treatment. Excavated soils are mixed with nutrients in a single batch addition and placed on a liner (Jacobson, 1987). Air is provided continuously through a series of slotted PVC pipes that are manifolded to a blower (see Figure 1). A treatment area of approximately 0.1 acres per 1,000 cubic yards is required. The area is prepared by removing debris and grading to provide a shallow slope to collect any water that may drain from the soils. The area is covered with a suitable synthetic liner which is then covered with approximately six inches of gravel or coarse sand. A geotechnical fabric is placed over the gravel or sand layer to prevent intrusion of the soils. Nutrients of a similar composition to those used for land treatment are mixed with the contaminated soils which are then placed over the geotechnical fiber in two to three foot lifts. Slotted PVC pipes are placed at each lift. The PVC pipes are manifolded to a blower or vacuum pump. The soils are covered with an impervious material which extends over shallow berms located around the pile. The cover controls moisture in the soils, prevents runoff, and minimizes both dust and vapor emissions. The PVC pipe/blower system moves air through the soil pile to provide one to three pore volumes of air per day. The off-gas from the blower is treated if necessary.

In addition to providing oxygen, the movement of air through the soils will capture volatile species. The extent to which the process physically removes specific contaminants depends on the lumped partitioning coefficients and the airflow through the soils. Operating at a low flow rate favors degradation over

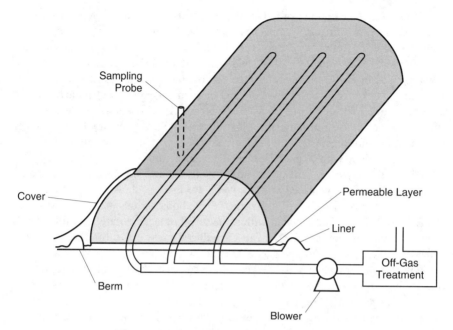

Figure 1 Bioremediation using a soil cell.

volatilization. Flow rates that exchange two to three pore volumes of air per day will probably support the maximum degradation rates. The flow rate of air through the soils should be designed based on the type of vapor treatment system, the mass of volatile species, and the importance of the treatment time.

Operation is relatively simple. In addition to routine maintenance of the blower and off-gas treatment system, vapor samples from probes placed within the cell, the PVC manifold, and the blower off-gas are analyzed for temperature, carbon dioxide, oxygen and volatile contaminants. Once the carbon dioxide and oxygen data indicate that remediation is nearing completion (as indicated by either material balance estimates based on initial contamination levels and the total carbon dioxide produced or by a reduced rate of oxygen consumption and carbon dioxide production), additional soil samples are collected and analyzed to determine the residual levels of contamination.

In Situ Treatment

In situ treatment methods offer the advantage of minimal disruption to the site. In some instances in situ treatment is the only approach that can result in site remediation without demolishing buildings or other structures. In situ treatment methods that destroy or convert the contaminants to innocuous materials mitigate long term liability and maintain property values.

In Situ Bioremediation of Unsaturated Soils. In situ bioremediation of unsaturated soils utilizes systems for batch addition of aqueous solutions of nutrients and continuous addition of oxygen through induced air movement (Norris, 1989 and Downey, 1991). Nutrients, if required, are added by one of several types of surface or near surface percolation systems. If the surface soils are exposed or covered with gravel or sand or even vegetation, nutrients are added to the surface in a dry form or as a dilute solution. Water is sprayed on the surface to carry the nutrients through the soils. Soils that are located under pavement or buildings are more difficult to address. If the pavement or building floor is underlain with a gravel or sand layer, a nutrient solution is carefully forced into this layer and will spread horizontally before percolating into the soils. In other cases, closely spaced shallow wells are used as shown in Figure 2.

Oxygen is provided as air utilizing the methods of in situ vapor stripping (USEPA, 1991). Wells screened within the unsaturated zone are manifolded to a blower or vacuum pump which creates a reduced pressure within the well bore. The partial vacuum causes air to enter the soils from the surface and sweep through the contaminated soils. Exchanging the air in the unsaturated zone from one to three times per day provides adequate oxygen for the bacteria. Air flow through the soils may be reduced as a result of the water added to introduce nutrients, particularly in finer grained soils. The design of the system must balance the need for both oxygen and nutrients.

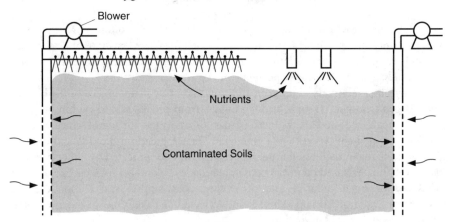

Figure 2 In situ bioremediation of unsaturated soils.

The movement of air and water through the contaminated soils may mobilize some of the contaminants. Water soluble compounds may be carried downward with the percolating water. The extent of vertical migration will depend on the properties of the contaminant most appropriately represented by the octanol/water coefficient, the organic content of the soils, and the volume and percolation rate of the water as well as the rate of degradation under the established conditions. The degree to which volatile components are removed are, as for a soil cell, dependent upon the lumped partitioning coefficient, the rate of air flow through the contaminated soils, the rate of degradation, and to some extent, the solubility of the contaminant and the amount of water present. Volatilization and vertical migration can also be competitive.

While removal of volatiles eliminates the contaminants from the treatment zone, downward migration may result in the contaminants reaching the aquifer. If the aquifer is already contaminated and has a treatment system in place, the leachate compounds will be addressed by the aquifer treatment system. However, if the groundwater has not been previously impacted, careful consideration should be given before selecting this method. Groundwater monitoring should be implemented and possibly, groundwater recovery wells installed. Minimization of downward migration can be achieved by operating the air system for a period before nutrients are added even if few volatile constituents are present. Most soils contain sufficient nutrients for biodegradation to occur.

Although the removal of volatile contaminants contributes to the cleaning of the soils, it may require the inclusion of a treatment system to prevent discharge of these constituents to the atmosphere. Treatment of air can be accomplished with activated carbon, a catalytic converter, an incinerator, or a vapor phase bioreactor.

Monitoring consists of measuring pressure readings at monitoring probes and sampling of air from the probes and the recovered air for analysis for oxygen, carbon dioxide, and specific volatile constituents. As for soil cells, once the carbon dioxide and oxygen levels approach ambient air levels, soils samples are obtained for analysis of the constituents of interest. Carbon dioxide levels in the off gas can provide an indication of the rate of biodegradation, but may be misleading because of other carbon dioxide sources and the complex equilibrium between carbon dioxide vapors and the dissolved and precipitated forms of carbon dioxide. Carbon isotope ratios can provide a more accurate estimation of biodegradation of petroleum derived substances. Monitoring may also include groundwater sampling for contaminants and nutrient constituents.

In Situ Bioremediation of Saturated Soils. Groundwater is recovered from the aquifer, treated, and reinjected into the formation after amendment with nutrients and an oxygen source (Norris, 1987). Groundwater recovery is designed to capture the contaminant plume and the injected nutrients and oxygen amended water. The recovered water is, in most cases, treated at the surface using activated carbon, an air stripper, or a fixed film bioreactor. A portion of the treated groundwater is amended with nutrients and an oxygen source and introduced into the formation through a series of injection wells or trenches (see Figure 3).

Nutrient selection is more critical for this type of bioremediation than for the other processes. Sodium salts can cause swelling of clays and should be avoided for soils with appreciable clay contents. Orthophosphates can cause precipitation of iron, calcium, and magnesium and thus should be used with great care. Tripolyphosphate salts solubilize iron, calcium, and magnesium and are much less likely to result in blockage of the formation. Nitrogen is usually provided as an ammonium salt. Nitrate has some advantages but is usually restricted by state regulations.

Oxygen can be provided by sparging air into the injection well, but this can provide only 8 ppm of dissolved oxygen (plus some entrained gas) unless injected at considerable depth below the water table. Because the remediation process cannot proceed any faster than oxygen is provided, other sources of oxygen are typically used. Sparging of pure oxygen gas instead of air into the injection water will result in a five fold increase in the rate of oxygen introduction into the aquifer. However, unless oxygen gas is already available, this method is usually not practical. Hydrogen peroxide, which undergoes conversion to oxygen and water, can be introduced at up to levels of 500 to 1,000 ppm corresponding to 250 to 500 ppm of oxygen. In practice some of this oxygen is likely to be lost to the unsaturated zone due to too rapid abiotic or microbial induced decomposition (Lowes, 1991). Hydrogen peroxide is more costly than air and should be used only where the increased rate of remediation more than offsets the added chemical costs. Recently, there has been a trend to use the direct injection of air into the aquifer. Air is forced into the formation several feet below the water table using sparging wells which are screened over a one or two foot interval. The air moves radially outward before reaching the unsaturated zone, thus providing dissolved oxygen for aerobic biodegradation.

In designing an in situ bioremediation system for saturated soils it is necessary to understand the distribution and total mass of the contamination so that the amount of nutrients and oxygen can be estimated. The practicality and time for remediation then depends on how long it will take to introduce sufficient nutrients and oxygen to degrade the constituents of concern.

Generally, it is not practical to implement in situ bioremediation if more than intermittent pockets or very thin films of free phase - lighter than water material is present. Free phase material should be removed using skimmers or dual phase pumps. Residual or thin layers of free phase volatile contaminants such as gasoline can be removed using in situ vapor stripping which will also serve to provide oxygen oxygen in the capillary zone and to soils exposed to air during periods of low groundwater levels.

The nutrient and oxygen requirements as well as the remediation time are further dependent on the extent to which the degradable species are removed from the aquifer. Soluble species are removed with the captured groundwater at rates which are a function of the octanol/water coefficient of the contaminants; the percentage of carbon in the soils; the rate of groundwater recovery; and the location of the recovery wells relative to the contaminant mass. If air sparging is used to provide oxygen, volatile constituents will be transferred to the unsaturated zone where they are captured by an in situ vapor stripping system. The extent to which they are removed will depend on the Henry's Law Constant of the volatile constituents and the air flow volumes and patterns.

For treatment of lighter petroleum hydrocarbons and many easily degraded compounds, the critical issues of feasibility are not rate or extent of biodegradation

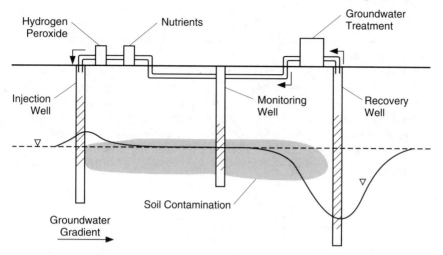

Figure 3 Bioremediation in the saturated zone.

that might be indicated from laboratory experiments, but rate of injection of amended water and groundwater velocities under pumping and injection conditions. For many sites it is better to first determine the optimum pumping and injection scenario using simple modeling techniques and aquifer test data to determine if sufficient groundwater recirculation flow rates and pathways can be established (Falatico, 1990). From these an estimate of the time of remediation can be made. If the aquifer data are already available, the modeling effort can be done much more quickly and at lower cost than can laboratory tests.

During operation of an in situ bioremediation system, groundwater is sampled from recovery wells and monitoring wells located within and outside of the treatment area. Groundwater is analyzed for nutrient constituents, bacteria populations, dissolved oxygen, pH, nitrate, calcium, iron, magnesium and contaminant levels. Groundwater flow rates and piezometric surfaces are determined and used to compare system operation to predictions of the model to design the system. When necessary, the model can be used to determine changes in flow rates needed to provide a more advantageous distribution of nutrients and/or oxygen.

EFFLUENT TREATMENT

The same biological water treatment methods that have been developed for wastewater treatment can be utilized to treat recovered groundwater during in situ bioremediation or other aquifer remediation processes. The various biological water treatment processes in practice today are shown in Figure 4. Selection of the most appropriate process requires an understanding of the significant differences in the treatment of groundwaters compared to industrial wastewaters. In most cases, groundwater contamination levels are low compared to industrial wastewaters, e.g., 10 mg/L, and frequently the majority of the constituents are volatile. The concentration and volatility of the contaminants frequently limits the choices of processes.

Biologically Activated Carbon Beds. These systems can be successfully employed for very low levels of organic constituents, i.e., less than 5 mg/L. In these systems, oxygen is added to the water either by aeration or as hydrogen peroxide in solution with nutrients to provide one milligram of oxygen for each milligram of carbon. Biological growth occurs on the carbon and organic removal results from both adsorption and biodegradation resulting in typical retention time requirements of 5 to 15 minutes.

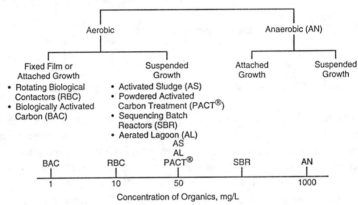

Figure 4 Biological water treatment processes.

Granular Activated Carbon Fluid Bed Reactors. This type of reactor has recently been developed for groundwater treatment. In this process high purity oxygen is dissolved in the water prior to passing through the reactor. The use of pure oxygen as the oxygen source eliminates any off-gas from the system. Organics resistant or refractory to biological treatment can be removed by adsorption on the carbon. Hydraulic retention times are on the order of five minutes with organic loadings (chemical oxygen demand) of up to three kg COD/cubic meter-day.

Fixed Film Processes. Waters with organic contents of 5 to 50 mg/L can be treated with fixed film reactors. Organic slime layers are grown on plastic surfaces. The organic constituents in the water diffuse into the slime layer where they undergo oxidation. Traditionally, rotating biological contactors (RBC) consisting of plastic disks that rotate through the water exposing the microbial slime to both the organic containing water and air (as the oxygen source) have been widely used. Fixed film reactors consisting of a series of aerated flow-through chambers containing sheets of slime coated static plastic media have become common in the United States (Skladany, 1989).

SLURRY REACTORS

A number of systems that provide nutrients and oxygen to soil/water mixtures are referred to as slurry reactors. These can be engineered reactors capable of treating a few hundred cubic yards at a time or simply large lagoons or lined ponds. The smaller systems provide constant agitation while lagoons may depend on aeration to provide some mixing. Because the systems incorporate a liquid phase and some mixing, addition of nutrients, surfactants, cometabolites, and additional bacteria sources, can be easily accomplished. Nutrient selection is less important than with in situ bioremediation because there are no transport issues. A compressor can be used to supply air or pure oxygen can be used to maintain high dissolved oxygen levels while minimizing loss of volatile constituents.

These systems permit a large degree of control over reaction conditions and thus can attain faster rates of biodegradation and lower residual contamination levels than is typically attained with some of the other bioremediation methods. The faster rates of degradation may be offset by the much smaller reactor size and thus treatment times for large soil volumes can be relatively slow. Further, the cost of treatment is typically much higher using slurry reactors than for the other methods. The time of treatment, however, can be reduced if the soils are pretreated using soil washing techniques to reduce the volume to be treated. Soil washing separates the fine soil particles from the coarse soil particles. Since the fine particles are likely to contain a much higher fraction of the contamination mass than are the coarse particles, only the fine particles may need treatment.

Since slurry reactors use some form of aeration, volatiles can be transferred to the air phase. If volatile emissions surpass the applicable air regulation limits, some form of air treatment will be required. For small systems this is an added cost, for large lagoons, it may not be possible to effectively capture emissions.

Remediation Times

Estimating remediation times can be difficult, especially for an in situ system where the heterogeneity of the soils and contaminant distribution will not be well defined. For in situ treatments of aquifers, treatment times are most appropriately estimated by calculating the time it will take to distribute sufficient nutrients and oxygen across the site. Typically, oxygen distribution is the limiting factor. In these cases, remediation times can be estimated from the total oxygen demand (three pounds of oxygen for each pound of biodegradable organic), the groundwater reinjection rate, and the concentration of oxygen in the reinjected groundwater. Making reasonable estimates of remediation times requires adequate modeling of the aquifer and a reasonable estimate of the contaminant mass. Appropriate and cost effective numerical models are available for this purpose. These models can also be used for developing conceptual and detailed designs and making adjustments to operating systems (Falatico, 1990). More detailed models which incorporate transport and degradation rates are appropriate for more in-depth analysis (Borden, 1991).

For ex situ treatment of soils, remediation times are generally not dependent on transport of nutrients or oxygen and rough estimates can be made from degradation rates determined from laboratory tests conducted with site soil samples. Estimations from literature degradation rates are useful but less accurate because they do not take into account soil specific effects, temperature, moisture content, and, frequently, acclimation of the bacteria to the specific contaminant(s). However, such estimates may be adequate for preliminary evaluations.

Biodegradation kinetics can be approximated by first-order kinetics. This allows the concentration at any time to be predicted from the biodegradation reaction half-life. The half-life is the time that it takes for the concentration to decrease by one half. Thus if the half-life was ten days and the initial concentration was 1,000 ppm, the concentration would be 500 ppm after 10 days, 250 ppm after 20 days, and 125 ppm after 30 days. Provided that optimum conditions are maintained, the remediation time will depend on the half-life of degradation, the initial concentration of the most highly contaminated soils, and the final concentration that must be obtained. This approach is least reliable at very high and very low contaminant concentrations.

Extent Of Remediation

The above discussion on estimation of remediation times assumes that the remediation goals can be met. There has been considerable discussion about the extent of remediation that can be obtained. Unfortunately, there exists meager good quality case history data. In some instances, low ppm levels or levels below the detection limit have been attained. Typically laboratory treatability tests will yield lower levels than will full scale systems where control of critical parameters is more difficult. For highly degradable constituents such as gasoline and many non-chlorinated solvents, low ppm or lower levels are reasonable to expect provided adequate control can be maintained over the system. For slower-to-degrade constituents such as heavier petroleum hydrocarbon mixtures, it is best to consider the percent conversion required to attain the remediation goals. There are no hard and fast rules for estimating the extent of conversion that might occur. The less readily degradable the target constituents, the greater care that needs to be exerted when high percent conversions are required.

Process Selection

Which bioremediation process, if any, is the most appropriate solution to environmental problems at a specific site depends on several factors. Selection of bioremediation as the sole method of remediation requires that all of the contamination of interest be biodegradable or capable of being physically removed by the specific process under consideration. Otherwise bioremediation must be used in conjunction with other remediation technologies (Norris, 1990) or not used at all.

At many sites, excavation is not practical due to site activities or structures and thus in situ methods offer the most attractive approach to remediation. In situ methods may also provide the most cost effective solution. Other advantages include avoidance of soils being classified as hazardous waste upon excavation, minimum disruption of the site, minimal public exposure during treatment, and, in some cases, provision of a ready remedy for potential future releases. On the other hand, in situ methods are applied under frequently highly heterogeneous conditions and provide the least control over the treatment conditions.

Ex situ methods require excavation of soils which is a significant cost factor and either require the excavation to be left open for a period of time or clean soils to be obtained to fill the excavation. However, ex situ treatment is generally fast compared to in situ methods, allowing the contaminated area to be returned to unimpaired or less restricted uses or to facilitate sale of the property. Ex situ methods also allow some level of blending of soils to reduce heterogeneity and simplify treatment and monitoring. Land treatment in particular provides a great deal of control over the soils. Nutrients can be readily added and the soils can be thoroughly mixed and worked. It is also relatively easy to monitor progress. Land treatment does have the disadvantages that a large area is required unless treatment is done sequentially, and volatile emissions can not be easily controlled. Soil cell treatment has the advantage that a smaller area is required, that emissions can be controlled, and that operations are relatively simple. Costs per unit soil volume, however, tend to be higher than for land treatment.

In selecting a biological treatment process for treating groundwater, it is necessary to take into account the concentration of the organic constituents, the presence of volatile compounds, and the changes in dissolved constituents that will occur over the lifetime of the project. The change in constituent level and composition will occur much more rapidly for systems where active in situ treatment such as bioremediation is employed than when only pump and treat methods are used. The selection of a biological groundwater treatment system also needs to take into consideration the availability of trained personnel at the site. Many in situ remediation systems require only limited presence of trained personnel. In such cases, the surface treatment system should also require a low level of maintenance.

REFERENCES

1. Borden, R.C., "Simulation of Enhanced In Situ Biorestoration of Petroleum Hydrocarbons," *In Situ Bioreclamation: Application and Investigation for Hydrocarbons and Contaminated Site Remediation*, Eds., Hinchee, R.E. and Olfenbuttel, R.F. Butterworth-Heinemann, 1991, pp. 529-534.

2. Brown, R.A., Dey, J.C. and McFarland, W.E., "Integrated Site Remediation Combining Groundwater Treatment, Soil Vapor Extraction, and Bioremediation," *In Situ Bioreclamation: Application and Investigation for Hydrocarbons and Contaminated Site Remediation*, Eds., Hinchee, R.E. and Olfenbuttel, R.F. Butterworth-Heinemann, 1991, pp. 444-459.

3. Brubaker, G.R., "In Situ Bioremediation of PAH-Contaminated Aquifers," *Proceedings of the Petroleum Hydrocarbons and Organic Chemicals in Ground Water: Prevention, Detection, and Restoration.* Houston, 1991, pp. 377-390.

4. Downey, D.C., and Guest, P.R., "Physical and Biological Treatment of Deep Diesel-Contaminated Soils," *Proceedings of the Petroleum Hydrocarbons and Organic Chemicals in Groundwater: Prevention, Detection and Restoration.* Houston, 1991, pp. 361-376.

5. Eckenfelder, W. Wesley, Jr., *Industrial Water Pollution Control*, McGraw-Hill Book Company, New York, 1967.

6. Falatico, R.J. and Norris, R.D., "The Necessity of Hydrogeological Analysis for Successful In Situ Bioremediation," *Proceedings of the Haztech International Pittsburgh Waste Conference.* Pittsburgh, PA, October 2-4, 1990.

7. Gibson, D.T., *Microbial Degradation of Organic Compounds*, Microbiology Series, Marcel Dekker, Inc., New York, 1984.

8. Helmers, E.N., Frame, J.D., Greenberg, A.F. and Sawyer, C.N., *Sewage Industrial Waste*, Vol. 23, Pt 7, 1951, p. 834.

9. Howard, P.H., *Handbook of Environmental Fate and Exposure Data for Organic Chemicals: Volume I Large Production and Priority Pollutants*, Lewis Publishers, Chelsa, Michigan, 1989.

10. Howard, P.H., *Handbook of Environmental Fate And Exposure Data For Organic Chemicals: Volume II Solvents*, Lewis Publishers, Chelsa, Michigan, 1990.

11. Jacobson, J.E. and Hoehn, G.D., "Working with Developing Air Quality Regulations: A Case Study Incorporating On-Site Aeration Standards for Vadose Zone Remediation Alternative," *Proceedings from Petroleum Hydrocarbons and Organic Chemicals in Groundwater - Prevention, Detection and Restoration*, NWWA API Conference, Houston, TX, 1987.

12. Loehr, R.C., *Land Treatment as a Waste Management Technology: An Overview. Land Treatment: A Hazardous Waste Management Alternative.* R.C. Loehr, et al., eds. Center for Research in Water Resources, The University of Texas at Austin, 1986, pp. 7-17.

13. Lowes, B.C., "Soil-Induced Decomposition of Hydrogen Peroxide," *In Situ Bioreclamation: Application and Investigation for Hydrocarbons and Contaminated Site Remediation*, Eds., Hinchee, R.E. and Olfenbuttel, R.F., Butterworth-Heinemann, 1991, pp. 143-156.

14. Montgomery, J.H. and Wilkins, L.M., *Groundwater Chemical Desk Reference*, Lewis Publishers: New York, Vol. I, 1990.

15. Munnecke, D.M., Johnson, L.M., Talbot, H.W., and Barik, S., "Microbial Metabolism and Enzymology of Selected Pesticides," *Biodegradation and Detoxification of Environmental Pollutants*. A.M. Chakrabarty, ed. CRC Press, Boca Raton, Florida, 1982.

16. Norris, R.D., and Brown, R.A., "In Situ Bioreclamation - A Complete On Site Solution," *Proceedings of the Hazardous Waste Management Conference - Hazmat West*, Long Beach, CA, December 1987.

17. Norris, R.D., Muniz, F.P., and Crosbie, J.R., "A Survey of Current Groundwater Remedial Methods", *Proceedings of the Construction and the Contaminated Site, Connecticut Society of Civil Engineers and Connecticut Groundwater Association*, November 2-3, 1989, Berlin, CT.

18. Norris, R.D., Sutherson, S.S., and Callmeyer, T.J., "Integrating Different Technologies to Accelerate Remediation of Multiphase Contamination," *Proceedings of NWWA Focus Eastern Regional Groundwater Conference*, Springfield, MA, October 17-19, 1990.

19. Pitter, P. and Chudoba, J., *Biodegradability of Organic Substances in the Aquatic Environment*. CRC Press, Boca Raton, Florida, 1990.

20. Reineke, W. and Knackmuss, H.J., "Microbial Degradation of Haloaromatics." *Ann. Rev. Microbiol.* 42, 1988, pp. 263-287.

21. Skladany, G.J., "Onsite Biological Treatment of an Industrial Landfill Leachate: Microbiological and Engineering Considerations," *Hazardous Waste and Hazardous Materials*. Vol. 6, 1989, pp. 213-222.

22. U.S. Environmental Protection Agency, *Microbiological Decomposition of Chlorinated Aromatic Compounds*. EPA 600/2-86/090, 1986.

23. U.S. Environmental Protection Agency, *Soil Vapor Extraction Technology; Reference Handbook*, EPA/540/2-91/003, 1991.

24. Verschueren, K., *Handbook of Environmental Data on Organic Chemicals*, 2nd Edition, VanNostrand Reinhold, New York, 1983.

25. Yare, B., "A Comparison of Solid-Phase and Slurry-Phase Bioremediation of PNA-Containing Soils." Presented at *In Situ and On-Site Bioremediation - an International Symposium*, San Diego, CA, March 19-21, 1991.

23

KINETIC EVALUATION OF ANAEROBIC BIODEGRADATION OF TRICHLOROETHYLENE USING ACTIVATED CARBON FLUIDIZED BED

SUXUAN HUANG
Keystone Environmental Resources, Inc.
Monroeville, PA

YEUN C. WU
New Jersey Institute of Technology
Newark, NJ

INTRODUCTION

Biodegradation of trichloroethylene (TCE) has been a major focus of investigations for the past decade. Under aerobic conditions, TCE was found to be biodegraded in the presence of methane gas, toluene or phenol. Recently, TCE was reported to be biodegraded by *Nitrosomonas* at concentrations of approximately 1 ppm.[1] TCE was shown to be readily biodegraded under anaerobic conditions, but with the production of vinyl chloride (VC) under incomplete biodegradation process. This paper describes a TCE biodegradation pathway under anaerobic conditions that has led to the production of both VC and dichloroethane (DCA), and further breakdown of these compounds to chloroethane (CA), methane (CH_4) and chloride ion (Cl^-). Kinetic data were evaluated and a general correlation was derived.

Several years ago, we reported our preliminary findings on the operating characteristics of the fluidized bed bioreactor systems. This study is a continuation of that work. The objectives of this study were: 1) to further investigate the possibility of attaining complete mineralization of TCE under anaerobic conditions; 2) to evaluate the relative distribution of TCE, and the intermediate and end products in the various media of the system; 3) to determine the effects of certain operating variables on process performance; and 4) to derive kinetic data and a correlation pertinent to design and scale-up purposes.

MATERIAL AND METHOD

The experiments were conducted in the Department of Civil & Environmental Engineering, New Jersey Institute of Technology using two-stage granular activated carbon (GAC) fluidized bed systems. Description of the experimental set-up has been presented in a previous paper.[2] Three systems were operated: Systems 1 and 2 were test systems which were seeded with a mixed microbial culture collected from an anaerobic digester, and fed with TCE solution and a defined medium.[2] Glucose was used as a co-substrate. The third system (System 0) was a control which was fed with the same solution as the test systems but not seeded. Sodium azide was periodically added to System 0 to inhibit microbial growth.

The objective of setting up System 0 was to serve as a baseline to which results obtained from the test systems were to be compared. Systems 1 and 2 were identical in all respects except that System 2 was fed at half the rate of System 1. By operating Systems 1 and 2 simultaneously, more data could be collected in a limited time frame.

Influent and effluent samples of both the first and the second stages were collected and analyzed for TCE, and the potential intermediate chlorinated products using EPA 601. Intermediate compounds that could potentially be produced were *trans*-1,2 dichloroethylene (TDCE), *cis*-1,2 dichloroethylene (CDCE), 1,1-dichloroethylene (1,1-DCE), vinyl chloride (VC), 1,2-dichloroethane (DCA) and chloroethane (CA). Possible end products were methane and Cl^-. Gas samples from the headspace were also collected periodically and analyzed for VOCs, as well as methane. The GAC in both stages were concurrently sampled and analyzed for the chlorinated compounds. A mass balance was then performed based on the organic loading applied to the system.

All samples were analyzed using a gas chromatography (GC - Hewlett Packard 5890) equipped with a flame ionization detector, which was kept at 250°C and a 8' × 1/8" o.d. stainless steel column packed with 60/80 Carbopack B/1% SP-1000 (Supelco, Inc. Bellefont, P.A.). The nitrogen carrier gas flowrate was 40 ml/min. The flow rates of hydrogen and air were 30 and 70 ml/min, respectively. The column temperature was programmed to start at 45°C, followed by a linear temperature gradient of 12°C/min to 200°C. Peak integrations were obtained with a Hewlett-Packard Integrator. The GC was coupled with a purge and trap system, Tekmar Model 4000 (Tekmar Co., Cincinnati, Ohio) and the analyses were performed in accordance with EPA Method 601. The conditions employed in this procedure were as follows: Purge gas nitrogen flow rate = 40 ml/min; purge time = 11 minutes; desorb time = 4 minutes; and bake time = 10 minutes.

The other parameters measured during the study include: glucose, NH_4^+-N, pH, PO_4^{-3}-P and alkalinity as $CaCO_3$. Standard test methods were used for these measurements. Glucose concentration was determined by Anthrone test.[2] Table 1 presents the ranges of values for each of the variables tested during the study.

Table 1. Scope of Study for Variables Tested

Variable	Range
Initial TCE Concentration	0.3 - 480 mg/L
Feed Rate	1.8 - 30.6 ml/min
Recycle Flow Rate	666 - 921 ml/min
Glucose/TCE Ratio	0.63 - 1,000

RESULTS AND DISCUSSIONS

Effect of Loading

In a fluidized bed bioreactor, loading is one of the most important parameters used in describing the process performance since it accounts for the effects of both detention time and influent concentration. Figure 1 depicts the effect of loading on the effluent composition of System 1. As expected, an increase in the loading has resulted in an increase in both the effluent TCE and TDCE concentrations. It is also noted that while the effluent TCE concentration continued to increase with loading, the effluent TDCE concentration seemed to level off beyond a TCE loading of 10 mg/L-hr.

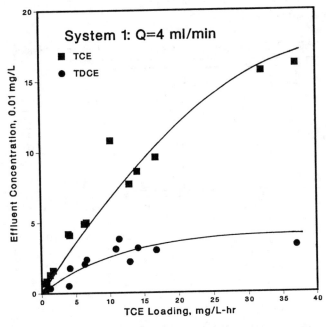

Figure 1 Effect of TCE loading one effluent composition.

The same trend was observed with the VC and DCA in the headspace (Figure 2). In Figure 2 methane was seen to increase initially with TCE loading and then decline beyond a loading of approximately 12 mg/L-hr. This may be attributed to the inhibitory effect exerted on the microbial activity at high TCE loading.

Figure 3 depicts the effect of TCE loading on the removal rate of TCE. It is interesting to note that, in the range of loading investigated, the removal rate increases linearly with the loading and does not level off. This lack of slowed down reaction with increased loading is probably because maximum loading has not been applied, or continual renewal and thus exposure of adsorption sites on the GAC as a result of microbial activity has been taking place[3] or both.

Evidence of Biologically-Mediated Transformation

In order to prove that the TCE removal attained in this study was biologically mediated, tests were conducted on System 0 (control) containing only GAC and no microorganisms. Headspace gas samples were collected and analyzed for VOCs and methane. Concentrations of potential intermediate products in effluents were also measured.

Table 2 presents the methane concentrations obtained from the headspace of Systems 0 and 1. It must be noted that under similar conditions, no methane was produced in System 0, while methane was detected in all gas samples collected from System 1. This indicates that without the presence of microbial activity, there could not be any biotransformation reaction and thus production of any end product such as methane.

Table 2. CH_4 Production in Seeded and Unseeded Systems

TCE Loading mg/L-hr	Unseeded µg/L	Seeded µg/L
10.81	ND	6.83
18.67	ND	5.91
27.90	ND	4.52
43.72	ND	3.19

Note: ND - not detected; individual concentrations are averages of two to three data points.

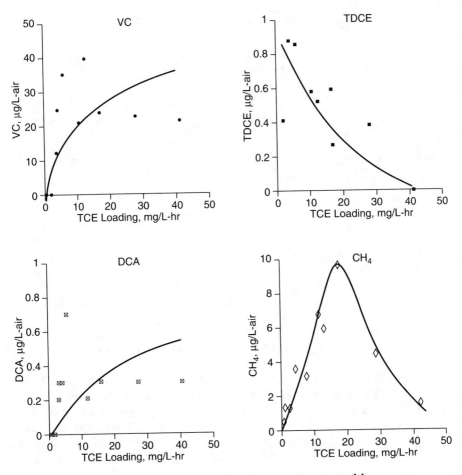

Figure 2 Effect of TCE loading on gas composition.

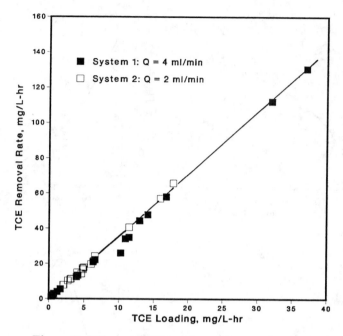

Figure 3 Effect of TCE loading on removal rate.

Figure 4 depicts the effluent TDCE concentrations for Systems 0 and 1. As shown, TDCE was not detected in System 0 until the visual observation indicated apparent growth of microorganisms on the GAC due to cross-contamination. The TDCE concentrations in System 0 during the experiments, however, were orders of magnitude lower than those in System 1. The effluent TDCE decreased to undetectable levels when high dosage of sodium azide was introduced to System 0 in an attempt to kill the microorganisms.

Chloride ions (Cl^-) were occasionally detected in the effluent from System 1. Stoichiometric conversion from influent TCE was observed in several samples. Interferences were encountered during analysis of certain headspace gas samples by GC. Hydrochloric acid was implicated for these interferences. This substantiates, in part, the production of Cl^- as a result of biological mineralization of TCE.

The foregoing discussion demonstrated that the removal of TCE observed in the test systems was biologically mediated, and not just a chemical reaction or volatilization.

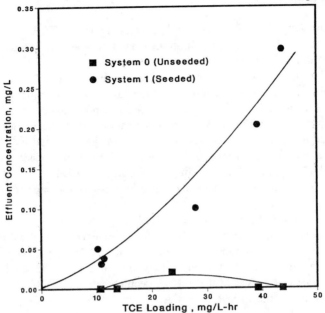

Figure 4 Comparison of effluent TDCE in seeded and unseeded systems.

Relative Importance of Removal Pathways and Mass Balance

As stated before, there are three pathways whereby TCE could be removed from the aqueous phase in this process. The influent TCE first encountered the GAC bed where most of it was presumably adsorbed. This initial adsorption step served to concentrate the TCE for subsequent microbial utilization. However, due to the nature of the reaction, complete transformation of TCE could not be achieved in the limited time of the experiment. Therefore some TCE would remain unchanged and be detected as residual TCE on the GAC particles. The fraction of TCE that did not adsorb would escape to the headspace of the column and thus could hardly be available for microbial action. Still another fraction of TCE might escape adsorption and elute with the effluent stream from the system. In order to obtain an assessment of the relative importance of the three removal pathways, the relative amounts of TCE in the different phases were determined.

Using TCE loading as the parameter, a mass balance accounting for the relative distribution of TCE in the various phases was carried out as follows:

(1) The difference between the influent and effluent TCE concentrations was calculated and designated as the overall TCE depleted in the reactor.

(2) The TCE concentration in the headspace of the column was converted to the corresponding hypothetical liquid phase concentration using Henry's law. These concentrations were found to be approximately 1% of the total TCE fed to the system.

(3) The difference between (1) and (2) is the amount of TCE adsorbed on the GAC and subsequently biotransformed to the intermediate and end products, or remained unaltered on the GAC.

(4) The amounts of TCE that remained unaltered were accounted for by analyzing the GAC. The results (not shown) indicated that about 2% of the influent TCE was not transformed and remained on the GAC.

(5) As shown in Figure 5, biotransformation to TDCE accounts for the majority of TCE removal in the first stage. At relatively low TCE loading, further dechlorination appeared to have taken place, as discussed later. Beyond a TCE loading of about 35 mg/L-hr, only TDCE was produced as the intermediate product because the loading was excessively high.

(6) The difference between the ordinates of the two curves in Figure 5 represents the fraction of TDCE that underwent further transformation to the subsequent intermediate products, such as DCA, VC and CA. Figure 6 presents the concentrations of these intermediates as detected during the analyses. It is evident that further dechlorination of TDCE has primarily resulted in the formation of CA.

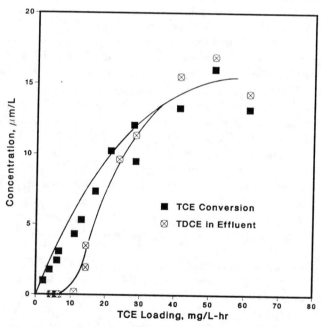

Figure 5 Comparison between TCE conversion and effluent TDCE.

The results obtained from the mass balance revealed that, in the range of TCE loading of approximately 1 to 60 mg/L-hr, about 1 percent of the total mass of TCE introduced into the system volatilized to the reactor headspace, a maximum of 2% remained unaltered on the GAC particles, and a maximum of 35% escaped biotransformation and eluted with the first-stage effluent. Over 62% was biotransformed to the DCEs, 75% of which was further degraded to CA.

This part of the experiment clearly illustrated that the process performance of the present system can well be approximated, without undue generality, by considering only the difference between the concentrations of TCE in the influent and effluent streams.

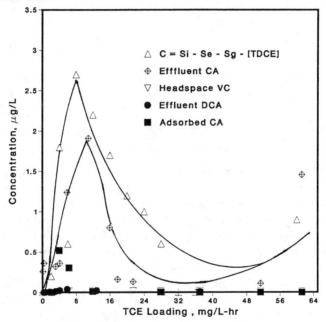

Figure 6 Comparison of conversions.

Biokinetic Parameters

In order to derive kinetic parameters, the effluent TCE concentrations were plotted against the rates of TCE biodegradation in Figure 7. As shown in the figure, the overall rate of depletion of TCE from the aqueous phase follows a Michaelis-Menten type of kinetics. This is consistent with the foregoing mass balance computation. As discussed in the previous analysis, biotransformation was the most important among the three pathways operative in the removal of TCE for the present work. Therefore, in the overall kinetics, the biological behavior would be controlling. The Michaelis-Menten equation describes enzymatic kinetics and applies as well to surface-catalyzed reactions. The latter, in turn, characterizes most biological reactions.

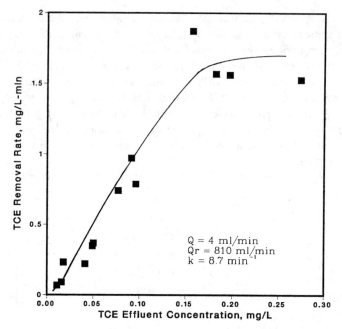

Figure 7 Evaluation of first-order rate constant for TCE.

The linear portion of the curve was then used in evaluating the first-order rate constant, k. The value obtained for k was 8.7 min^{-1}, which was orders of magnitude greater than values reported in the literature.[4] This might be attributable to the difference in the removal mechanism as glucose was used as co-substrate in our system but not in other's work.

Figure 8 is a Lineweaver-Burke plot of the TCE removal data. The maximum removal rate, V_m, and the half-velocity constant, K_m, calculated by linear regression, are 1.63 mg/L-min and 0.11 mg/L, respectively.

Kinetic evaluation of TDCE transformation was performed in a similar manner. The reaction rate constant (k) was found to be 0.3 min^{-1}, which was orders of magnitude smaller than that of TCE. It is well-documented in the literature that the rates of transformation decrease as chlorine is removed from the chlorinated compounds during the reductive dechlorination process. Using a Lineweaver-Burke plot, V_m and K_m for TDCE were found to be 0.059 mg/L-min and 0.096 mg/L, respectively.

Proposed Biodegradation Pathway

Based on our experimental observation and data, the anaerobic biodegradation pathway of TCE in the two-stage fluidized bed process used in this study is proposed as shown in Figure 9. Accordingly, the intermediate products from the initial reductive dechlorination of TCE involved all three geometric isomers 1,1-

DCE, TDCE and CDCE. Our data indicated that TDCE was the predominant species based on peak identification using standard solution of TDCE. No attempt was made in this study to differentiate between TDCE and CDCE. Parsons et al.[5].

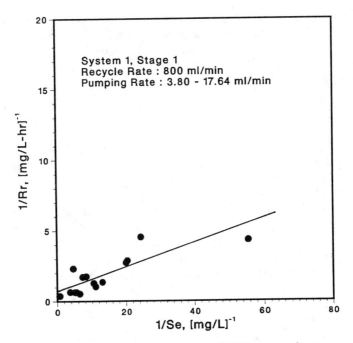

Figure 8 Lineweaver-Burke plot of TCE removal rate.

were able to show isomer specificity, i.e., CDCE was favored over TDCE. It is therefore suspected that CDCE was also produced in our system. 1,1-DCE was occasionally detected in our system, especially in the latter part of the study, i.e., at relatively high TCE loading.

In Vogel's work[6] where partial mineralization of TECE under anaerobic environment to CO_2 was observed, it was speculated that the intermediate product, VC, formed during the process was degraded in a way similar to that of DCA. Accordingly, DCA was initially oxidized yielding 1,2-dichloroethanol, which decomposed spontaneously to HCl and 2-chloroacetaldehyde. The latter compound was then further oxidized to chloroacetate, which was dehalogenated to glycolate, a metabolite readily utilizable as a carbon source by many bacteria. The suggested mechanism was based on the detection of a NAD-dependent 2-chloroacetaldehyde dehydrogenase activity in the extracts of cells grown on DCA. Baek and Jaffe [7] had shown the anaerobic biodegradation of TCE to CH_4 via the formation of VC and CA, and had also suggested a similar pathway. In our study, no attempt was made in identifying the intermediate products in the transformation from DCA to

CA. However, production of methane and chloride was noted in several occasions, as discussed earlier, which indicated that mineralization of TCE had occurred.

Correlation

In an attempt to derive a general correlation that would be useful for future scale-up and design purposes, the experimental data were arranged in terms of dimensionless parameters $F/V_r G_m$ and C_i/K_m.[8] A computer program written in Fortran was developed to derive the following polynomial equation with a correlation of 0.94:

$$\frac{C_e}{C_i} = 6.9 - 0.12\left(\frac{C_i}{K_m}\right) - 182\left(\frac{F}{V_r G_m}\right) + 2.9\left(\frac{C_i}{K_m}\right)\left(\frac{F}{V_r G_m}\right) + 1173\left(\frac{F}{V_r G_m}\right)^2$$

where C_e and C_i are the effluent and influent concentrations, respectively; V_r is the reactor volume; F is the feed rate; G_m is the maximum growth rate. A good agreement between the experimental and the calculated data is indicated by Figure 10. With a given set of F, G_m, V_r and K_m, the TCE level in the effluent could be predicted if the influent TCE concentration is known. On the other hand, for designing a new reactor, the polynomial could be solved for V_r using known C_e, C_i, F, K_m and G_m. However, this correlation was based on a narrow range of $F/V_m G_m$, as shown in Figure 10, due to the limited time frame for the study. Further work is therefore required to extend the data base for this correlation.

CONCLUSIONS

Based on the data generated from this study, the following conclusions can be drawn:

(1) The percent reduction of TCE ranged from 65% to 98% for the first stage and more than 98% for the entire system at an initial TCE level of about 0.3 to 480 mg/L.

(2) The TCE removal observed in the present work is biologically mediated. The distribution of TCE in the various media is as follows: 1% volatilized to the head space, a maximum of 2% remained unaltered on the GAC particles, and a maximum of 35% eluted from the first stage. A minimum of 62% was converted to intermediate products, the majority of which was CA.

(3) Production of CH_4, Cl^- and appreciable amounts of CA is an indication that attainment of complete mineralization of TCE is possible with this process.

(4) Both the rates of removal of TCE and appearance of TDCE in the effluent follow Michaelis-Menten type of kinetics. For TCE, $V_m = 1.63$ mg/L-min and $K_m = 0.11$ mg/L. For TDCE, $V_m = 0.06$ mg/L-min and $K_m = 0.1$ mg/L.

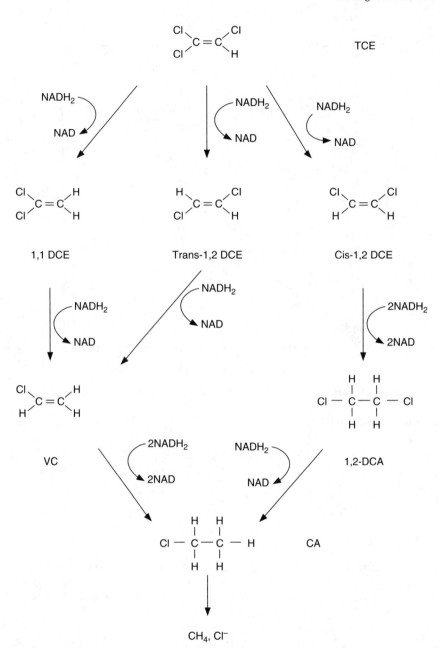

Figure 9 Proposed anaerobic biodegradation pathway of TCE.

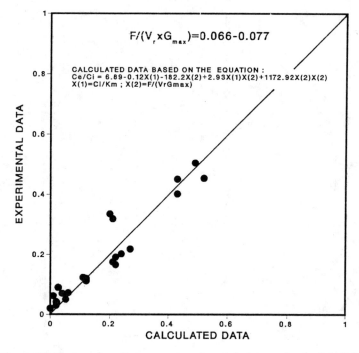

Figure 10 Comparison between experimental data and calculated data.

ACKNOWLEDGEMENT

The authors acknowledge the support granted by the Department of Civil & Environmental Engineering of the New Jersey Institute of Technology and Keystone Environmental Resources, Inc. during the preparation of this paper.

REFERENCES

1. Vannelli, T., Logan, M., Arciero, D.M. and A.B. Hooper, "Degradation of Halogenated Aliphatic Compounds by the Ammonia-Oxidizing Bacterium *Nitrosomonas europaea*," *Appl. Environ. Microbiol.*, Vol. 56, No. 4, pp. 1169-1171, 1990.

2. Huang, S. and Wu, Y.C., "Co-Metabolism of Trichloroethylene with Sugar Using Two-Stage Upflow Activated Carbon Fluidized Beds," pp. 783-802, In Y.C. Wu (ed.), *International Conference on Physiochemical and Biological Detoxification of Hazardous Wastes*, Volume Two, Technomic Publishing Co., Inc., Lancaster, PA, 1988.

3. Kinner, N.E. and Eighmy, T.T., "Biological Fixed-Film System," *J. Wat. Pollut. Contr. Fed.*, Vol. 58, No. 6, pp. 498-501, 1986.

4. Barrio-Lage, G., Parson, F.Z., Nassar, R.S. and Lorenzo, P.A., "Sequential Dehalogenation Of Chlorinated Ethenes," *Environ. Sci. Technol.*, Vol. 20, pp. 96-99, 1986.

5. Parsons, F.Z. and Lage, G.B., *J - Am. Water Works Assoc.*, Vol. 77, pp. 52-59, 1985.

6. Vogel, T.M. and McCarty, P.L., "Biotransformation of Tetrachloroethylene to Trichloroethylene, Dichloroethylene, Vinyl Chloride, and Carbon Dioxide Under Methanogenic Conditions," *Appl. Environ. Microbiol.*, Vol. 49, No. 5, pp. 1080-1083, 1985.

7. Baek, N.H. and Jaffe, P.R., "Anaerobic Mineralization of TCE," Paper presented at the International Conference on Physiochemical and Biological Detoxification of Hazardous Wastes, Atlantic City, N.J., May 3-5, 1988.

8. In P.P. Cooper and B. Atkinson (ed.), *Biological Fluidized Bed Treatment of Water and Waste Water*, Ellis Horwood Ltd., Chichester, England, 1981.

24

TREATMENT PROCESS EVALUATION FOR THE REMEDIATION OF HERBICIDE-CONTAMINATED GROUNDWATER: PERCEPTION AND REALITY

Rong-Jin Leu and Chen-Yu Yen
Gannett Fleming, Inc.
Baltimore, Md

INTRODUCTION

Drake Chemical, Inc. (Drake) located in Lock Haven, Pennsylvania, was a specialty chemicals manufacturer for the production of dyes, pharmaceuticals, cosmetics, herbicides and pesticides. After Drake was cited several times for violations of environmental and health and safety regulations, and failed to respond for voluntary cleanup, the U.S. Environmental Protection Agency (EPA) initiated emergency cleanup activities. Three phases of Superfund remedial investigation/feasibility studies (RI/FS) have been conducted by EPA since 1983. A Record of Decision (ROD) was signed to include pumping and treating the contaminated groundwater to an acceptable level for discharge, and considering biological activated carbon (BAC) as an alternative treatment technology for the groundwater at the Drake Chemical Superfund Site.

The major contaminants in the groundwater are Fenac® (2,3,6-trichlorophenyl acetic acid, an herbicide), chlorobenzene, 1,2-dichloroethane (DCA), and several heavy metals (lead, chromium, cadmium, and arsenic). Risk analysis performed for this Superfund site identified Fenac® as one of the most significant risk contributors. A nearby Resource Conservation and Recovery Act (RCRA) treatment facility is using granular activated carbon (GAC) to treat chlorobenzene, DCA, and other organic compounds, but not the herbicide Fenac®.

The objective of this paper is to evaluate the effectiveness of pretreatment processes (coagulation and aeration) and treatment processes (GAC and BAC) for the removal of the groundwater contaminants based on a series of treatability studies.

PRETREATMENT

Perception

Previous studies indicated that adsorption/coprecipitation should be an effective means of reducing trace metal concentrations to regulatory levels (Leckie, et al, and Merrill, et al.). This is usually accomplished by using jar tests to select an appropriate coagulant and operating conditions for the coprecipitation process.

The purpose of groundwater pretreatment is to reduce metal contamination to levels that would not inhibit the growth of bacteria in the BAC study. Two stages of pretreatment were performed—bench-scale testing and pilot-scale operation. The bench-scale testing was to determine the required operating parameters to be applied in the pilot-scale operation; the pilot-scale operation was to prepare the amount of groundwater required for the GAC and BAC studies.

Jar Test

In this study, four coagulants—alum, lime, ferric chloride and a polymer—were screened; then, the selected coagulant was tested with various dosages ranging from 1.0×10^{-4} M to 2.0×10^{-3} M and pH values ranging from 3.5 to 9. Each jar test was conducted using one liter of groundwater and consisted of the following steps:

- Rapidly mixing the groundwater at 200 rpm for one minute after adding the coagulant and adjusting the pH.
- Slowly mixing the groundwater at 30 rpm for 20 minutes to initiate the coagulation process.
- Allowing the coagulated suspension to settle under quiescent conditions for 60 minutes. The supernatant was then sampled for analysis.

The most effective coagulant and its optimum operating conditions were selected based on turbidity instead of the heavy metal concentrations because the concentrations were detected below their respective detection limits after the jar tests. Jar testing indicated that ferric chloride was the best coagulant because of its effectiveness in turbidity removal. A concentration of 5×10^{-4} M of ferric chloride proved to be the most effective in removing turbidity over a wide pH range of 4.5 to 6.5. With this pH operating range, pH adjustment was not necessary because the pH of the groundwater was about 5.5 after the addition of 5×10^{-4} M ferric chloride.

The treatability test was performed for groundwater pumped out from a new interceptor well. Chemical data from other wells were reviewed before the jar test.

Because iron was not considered a priority pollutant in the earlier studies, data were scanty and variable for groundwater throughout the site. During the selection of ferric chloride through jar testing, full chemical analyses were performed on the groundwater. Results indicated iron levels of 200 to 300 ppm, approximately 10 times the ferric chloride concentration of 5×10^{-4} M. This new information necessitated a change in treatment strategy to take advantage of the high levels of iron already present in the groundwater.

Onsite Pretreatment

After the operating parameters for the coagulation/precipitation process were determined in the bench-scale study, a pilot-scale operation including aeration, coagulation/precipitation, and pH neutralization was conducted to pretreat the groundwater at the Drake Chemical Site. The purpose of aeration was to simulate the iron oxidation in the bench-scale testing, and the pH neutralization was to provide a suitable living environment for bacteria for the upcoming BAC study.

The groundwater was pumped from an interceptor well of the Drake Chemical Site and was collected in two 4-foot high, 150-gallon tanks and subjected to the following treatment:

- Aerating the groundwater using fine-bubble diffusers. The pH was adjusted to about 8 before the experiment and was maintained at approximately 7 throughout the test.
- Adding an anionic polymer (Magnifloc 835A at 1 mg/L) and slowly mixing the groundwater for 20 minutes.
- Allowing the coagulated suspension to settle overnight under quiescent conditions.

Figure 1 presents the iron oxidation kinetics which could be used to design the aeration tank. The results indicate two or three hours of aeration of the Drake groundwater would oxidize more than 90 percent of ferrous iron to ferric iron.

Reality

It was not anticipated that the total iron concentration of the groundwater would be one order of magnitude more than that of the added coagulant, ferric chloride. Because most iron in the groundwater is in ferrous form instead of in ferric form, aeration was then considered as a pretreatment option to convert ferrous iron into ferric iron which can be used as an *in situ* coagulant for the coprecipitation of the groundwater contaminants. Aeration, with the addition of a polymer as a coagulant aid, was considered as the groundwater pretreatment process instead of the addition of coagulant. A valuable lesson learned is to test the groundwater first to see if the needed coagulant is already present.

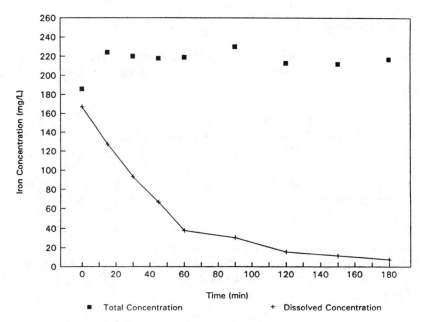

Figure 1 Aeration experiment results.

TREATMENT PROCESSES

Perception

BAC was considered to be the most effective treatment technology for the Drake groundwater because it has the advantage of biodegradation of organics and bioregeneration of exhausted carbon beyond adsorption mechanism. Both biodegradation and bioregeneration would have greatly extended the service life of the column, resulting in a savings in operating and regeneration costs. Theoretically, regeneration is not needed for a BAC column operation. In order to verity this perception, the BAC column experiment was conducted using the same activated carbon and same operating conditions as those of the GAC column experiment.

Because of our earlier perception of the pretreatment based on previous groundwater data, groundwater treated with 5×10^{-4} M ferric chloride was used for this portion of the study. The GAC experiment was conducted to compare column performance to BAC column performance in removing volatile and refractory organics. Adsorption isotherms were established for both total organic carbon (TOC) and Fenac® for different activated carbons in order to select an appropriate carbon and to estimate carbon adsorption capacity. Column experiments were then conducted to investigate column breakthrough curve characteristics.

Carbon Preparation

Carbons have to be prepared for both the isotherm analysis and the column experiments. The procedures for preparing activated carbons for the isotherm analysis were as follows:

- Grind GAC with a mortar and pestle.
- Sieve the ground GAC through a 200-mesh screen.
- Rinse the sieved GAC with high performance liquid chromatography (HPLC) water.
- Dry GAC in a 105°C oven to constant weight (less than 0.1% relative percent difference).
- Store in a desiccator.

The procedures for preparing carbon for the column studies, including the GAC and BAC, were similar to those for the isotherm analysis except that carbon was directly sieved between 20- and 40-mesh screens without going through the grinding step.

Isotherm Study

Three commercially available activated carbons (React-A and Filtrasorb-400 from Calgon, and HD-3000 from American Norit) were screened in the isotherm study. The following procedures for isotherm analysis were used:

- Add varying dosages of the carbons to four-liter plastic bottles.
- Add pretreated groundwater to each bottle.
- Place bottles in a rotary agitation apparatus and agitate at 30 rpm for at least seven hours.
- Vacuum filter the samples and analyze the filtrates.

Experiment results of the isotherm analysis were used to establish adsorption equilibria for TOC and Fenac®. The equilibrium relationship can be expressed by either the Freundlich or Langmuir isotherm which can then be used to estimate carbon adsorption capacity. From the isotherm analysis, Filtrasorb-400 proved to be more effective than the other two carbons in TOC adsorption, however; no significant difference was observed between Filtrasorb-400 and React-A in Fenac® adsorption. React-A was selected as the carbon to be used in the GAC and BAC column studies because it is used by the nearby RCRA facility for carbon adsorbers and it is a regenerated carbon, satisfying both environmental and cost concerns.

GAC Column Study

In the column study, two sets of design parameters including bed size, flow rate, and running time, were evaluated to investigate the breakthrough characteristics for TOC and Fenac®. The small column was designed to study the complete breakthrough characteristics; the big column was designed to mimic the operating conditions of the nearby RCRA treatment facility due to the possible use of that facility to treat the contaminated groundwater. The configuration for the GAC experiment is presented in Figure 2 which shows a five-gallon cubitainer was used as the influent container and a peristaltic pump with a speed controller was used to maintain the flow rate. The pretreated groundwater was directed down through a sand filter and then up through the GAC column without disturbing the bed. Samples were collected at the effluent of the GAC bed through a preset auto sampler.

Experiment results of the small column study showed that the 50 percent breakthrough of TOC and Fenac® occurred much earlier than what was estimated from the carbon capacity. In addition, a symmetrical breakthrough curve was not

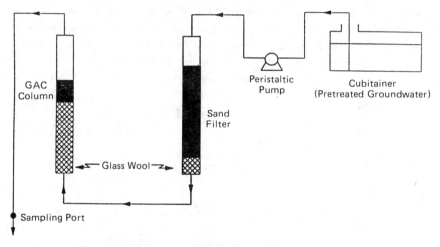

Figure 2 Column configuration for GAC experiment.

observed because of either short empty bed contact time (EBCT) or high linear velocity for the operation. From the experiment results of the big column study, the breakthrough curves for both TOC and Fenac® were not fully developed in the 10 day column operation. However, the combination of the breakthrough curves from both the small and the big column operations show the detailed, complete breakthrough characteristics for both TOC and Fenac® in GAC column operations. A computer model was established to describe and predict the GAC column performance under various operating conditions (Yen and Leu). Figures 3 and 4 present the model prediction and experiment results of TOC and Fenac® breakthrough curves for the GAC big column study.

Bacteria Culturing

The purpose of the bacteria culturing is to grow a sufficient quantity of bacteria and select an appropriate seed for the BAC column study. Activated sludge samples were obtained from three established cultures: the Kalamazoo Powdered Activated Carbon Treatment (PACl-) Plant, the Baltimore Back River Waste Water Treatment Plant (WWTP), and the duPont Chamber Works PACT Plant. The activated sludge seeds were cultured by maintaining a temperature of 20 to 22°C, a pH between 6.0 and 8.0, and dissolved oxygen (DO) greater than 7.0 mg/L with the stepwise increase of the groundwater intensity and stepwise decrease of the nutrient amounts over an eight-week period. Well water from Pennsylvania was used to initiate the culturing so that no residual chlorine would be present, and extract beef solution was used as nutrient. During the culturing period, mixed-liquor suspended solids (MLSS) and mixed-liquor volatile suspended solids (MLVSS) were monitored and the sludge was microscopically observed to ensure bacteria survival. At the end of the period, appropriate activated sludge seeds were selected based on their performances on BOD_5 tests. As shown in Figure 5, the test results not only prove that the best bacteria sources can be as diverse as the duPont Chamber Works PACT plant as well as the Baltimore Back River WWTP, but indicate that activated sludge may be a promising process to remove organic contaminants in the groundwater. Before the BAC column study was initiated, the selected carbon from the GAC study—Calgon React-A—was inoculated with the selected activated sludge seeds over a 10-day period. During

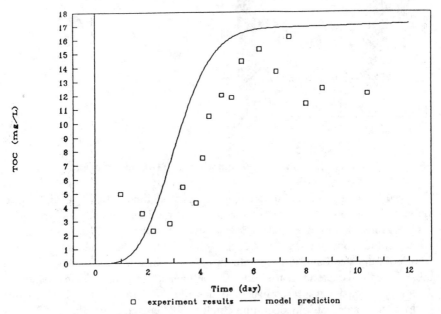

Figure 3 TOC breakthrough curve for GAC column.

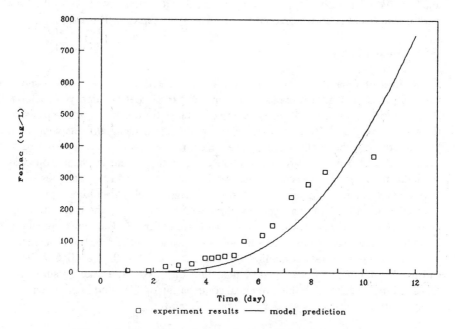

Figure 4 Fenac® breakthrough curve for GAC column.

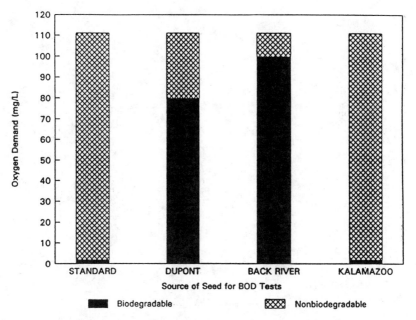

Figure 5 Biodegradation potential test results.

the inoculating period, gentle shaking by hand a few times each day mixed the carbon and bacteria and promoted the growth of bacteria on the carbon.

BAC Column Study

Experiment apparatus of the BAC study is presented in Figure 6 which shows the post-BAC sand filter was placed to trap fine particulates and decayed bacteria. The operating parameters of the GAC big column study was used for comparison. Temperature, DO, pH, and turbidity of both the influent and effluent were monitored from the BAC column. The first three parameters were measured to ensure an environment that would promote the growth of bacteria; the last one was to indicate when to backwash the sand filter. The effluent samples were periodically collected for analyses during the eight weeks of column operation. Results of the BAC column experiment showed no decline in the concentrations of both TOC and Fenac®, which implies the failure of the BAC column operation. In addition, the BAC column did not show significantly better removal efficiencies for organic column test contaminants when compared with the GAC column during the running period. Finally, modeling the BAC experiment results as a GAC column indicate that the biological contribution of TOC and Fenac® removal exists, but is minimal, as shown in Figures 7 and 8.

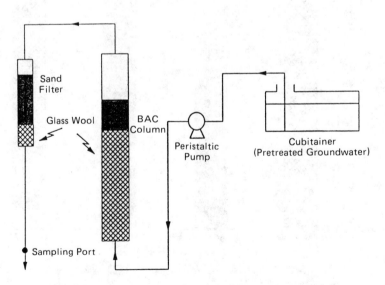

Figure 6 Experiment configuration for BAC column test.

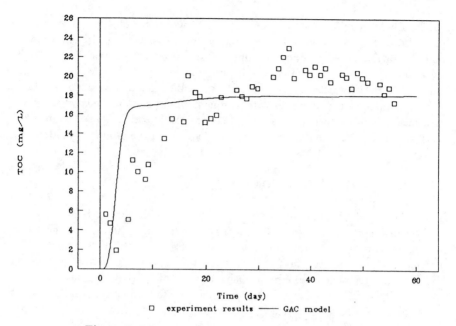

Figure 7 TOC breakthrough curve for BAC column.

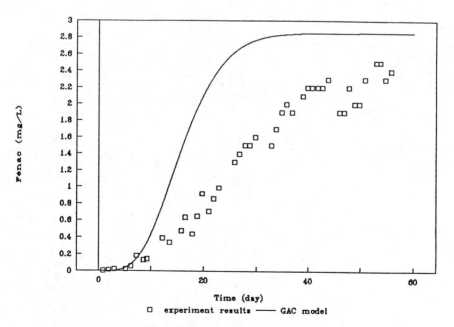

Figure 8 Fenac® breakthrough curve for BAC column.

Reality

The EBCT required for the BAC column to work is probably too long to be cost-effective. In addition, other technical problems such as insufficient biomass and the difficulty of forming biofilm on activated carbon would have to be overcome for BAC to be effective. In order to increase biological activities, a suspended-growth treatment process such as traditional activated sludge might be more promising than attached-growth treatment processes. The Drake experiments indicated that although an innovative technology may theoretically appear to be the most promising treatment process, the tried-and-true technologies should not be ignored.

SUMMARY AND CONCLUSIONS

In the GAC study, Calgon React-A was selected out of the three activated carbons as the most appropriate carbon for both GAC and BAC column experiments based on the established adsorption isotherm analyses for TOC and Fenac®. The combination of two breakthrough curves from the two sets of design parameters exhibited the complete breakthrough characteristics of the GAC column operation. GAC is effective in removing organic contaminants following metal precipitation, however; the operation cycle is short and related regeneration costs are consequently high.

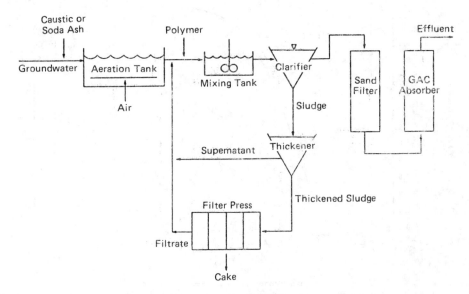

Figure 9 Schematic of GAC treatment train.

RECOMMENDATIONS

The proposed GAC treatment train for the Drake groundwater is shown in Figure 9 and the cost estimate for the treatment train is presented in Table 1.

After eight weeks of operation, the BAC column operation did not show significantly better removal efficiencies for organic contaminants than the GAC column. The TOC and Fenac® concentrations did not decrease toward the end of the experiment as was expected (Bouwer, 1982). Limited study of BOD5 tests using the three acclimated sludges as seeds indicates that activated sludges or similar biological treatment processes could be very effective in treating organic contaminants. However, the effectiveness of Fenac® removal needs to be studied further to assess its treatability. Currently, a treatability test of activated sludge processes is being conducted at the Gannett Fleming laboratory. If the activated sludge process does not provide adequate removal of Fenac® a post-activated sludge GAC is recommended to complete the removal.

Table 1. Summarized Cost Estimate for Proposed Treatment Train

Major Equipment Cost	$402,000
Site Preparation	$220,000
Total Capital/Construction Cost	$625,000
Total Project Cost	$780,000
Amortized Capital Cost	$130,000
Annual Operating and Maintenance Cost	$1,030,000
Total Annual Cost	**$1,155,000**

ACKNOWLEDGEMENT

We thank Mr. Gerry Slattery, Plant Manager of the Back River Waste Water Treatment Plant at Baltimore, Maryland, Mr. Rohel Amundson, General Superintendent of the Kalamazoo Water Reclamation Plant, and Mr. Harry Heath, Chief Consultant at the duPont Chamber Works Wastewater Treatment Plant for their generous gifts of sludge seeds. We also appreciate our text editor, Ms. Carol Royal, who word processed and provided comments on this manuscript. This study was conducted under Contract Number DACW45-90-C-0117 from the U.S. Army Corps of Engineers, Omaha District, with the funding indirectly from the U.S. Environmental Protection Agency. The authors want to thank these agencies for their support. This paper, however, does not necessarily reflect the views and policies of these agencies.

REFERENCES

1. Bouwer, E.J. and McCarty, P.L. (1982). "Removal of Trace Chlorinated Organic Compounds by Activated Carbon and Fixed-Film Bacteria." *Environmental Science and Technology*, 16(12), 836-843.

2. Leckie, J.O., et al. (1980). "Adsorption/Coprecipitation of Trace Elements From Water With Iron Oyhydroxide." EPRI CS-1513, Project 910-1, Final Report.

3. Merrill, D.T., Maroney, P.M., and Parker, D.S. (1985). "Trace Element Removal by Coprecipitation with Amorphous Iron Oxyhydroxide: Engineering Evaluation." EPRI CS-4087, Project 910-2, Final Report.

4. Yen, C. and Leu, R. (1992). "Treatability Study of Granular and Biological Activated Carbon for Groundwater Containing Fenac®, a Herbicide." To be presented in ASCE Water Forum '92.

25

CONCEPT AND APPLICATION OF FIXED-FILM TECHNOLOGY FOR BIODEGRADATION OF HAZARDOUS ORGANIC WASTE

Bruce E. Rittman
University of Illinois at Urbana-Champaign
Urbana, IL

INTRODUCTION

The premier issue of a newsletter on bioremediation, *The Bioremediation Report*, featured an article on biodegradation with biofilms (Hom, 1992). The organizers of this international conference specifically asked me to speak on the use of biofilms for biodegradation of hazardous organic wastes. Clearly, biofilms have great interest for those who must deal with biodegradation of waters and soils contaminated by industrial chemicals. Is this interest well founded, or is it just a fad? Specifically, do biofilms possess attributes that make them uniquely suited for biodegradation of hazardous organic chemicals?

Biofilms, being microbial aggregates attached to solid surfaces, share the two physical phenomena critical for all aggregated systems. First, aggregation allows cell retention significantly greater than the liquid detention time. Good cell retention allows for selection of slow-growing bacteria, stable performance, and economical designs. Although attachment as a biofilm is a particularly efficient means for achieving cell retention, flocculation can provide the same effect in suspended-growth systems, such as activated sludge.

The second phenomenon characterizing aggregates is the reduction of substrate concentrations inside the aggregate. The need for mass transport to supply substrates to bacteria within the aggregate sets up concentration gradients. Thus, bacteria inside a biofilm or a floc are exposed to lower substrate concentrations than are bacteria dispersed in the liquid.

At this point, a superficial conclusion might be that biofilms are not all that special. While good cell retention may be achieved somewhat more easily in a biofilm reactor, it is achievable in other systems, too. Furthermore, being in a biofilm can cause the bacteria to "see" lower substrate concentrations than they would if dispersed; normally, lower substrate concentrations slow growth and are disadvantageous.

Some postulate that the act of attachment to a surface brings about physiological changes that might provide a unique capability for biofilm bacteria. While attachment can alter membrane lipids in some cases (Valeur et al., 1988), the general consensus (van Loosdrecht et al., 1990; Stal et al., 1989; Palenick et al., 1989; Rittmann et al., 1991) is that bacteria immobilized in biofilms are no different from the bacteria that would be present in suspended growth, if the same environmental conditions were applied. Of course, biofilms are active participants in creating their internal environment through concentration gradients; however, suspended aggregates also create their own internal concentration gradients.

Although biofilms are not completely unique in comparison to other microbial aggregates, certain aspects of attachment give them advantages for treatment of hazardous organic compounds. Those advantages all stem from the *protection* afforded by being part of a biofilm. Three types of protection are described here:

- protection from shocks by the bacteria being attached to adsorbent media,
- protection of bacteria deep within the biofilm from toxic substrates, and
- protection of sensitive populations from detachment and washout.

Knowing when biofilms are especially useful in biodegradation of hazardous organic wastes requires an understanding of when the protection mechanisms are important. The remainder of this paper illustrates the protection mechanisms and when they are important.

PROTECTION FROM SHOCKS

The most commonly used adsorbent material is activated carbon, which can be incorporated into biological processes in two ways. First, granular activated carbon (GAC; 0.5-1.0 mm size) can be used effectively in a fluidized bed configuration (Wang et al., 1985; Rittmann, 1987), which combines a high specific surface area for bacterial attachment with large pore sizes that preclude clogging. The surface chemistry of carbon and its creviced texture make it ideal for biofilm colonization (Chang and Rittmann, 1988). Second, powdered activated carbon (PAC) can be added to a suspended-growth process (Schultz and Keinath, 1984; Rittmann, 1987). The PAC is quickly colonized and serves as nuclei for bacterial aggregation.

Because bacteria naturally form biofilms on exterior surfaces of activated carbon, biofilm processes occur whenever it is used. For fluidized beds, the processes usually are exclusively of the biofilm type, while systems with PAC addition become hybrids of suspended and biofilm growth.

When activated carbon is the attachment surface for biofilms, its adsorptive properties can protect the bacteria from two types of shocks:

- A shock from a sudden loading increase of a biodegradable substrate.
- A shock from a nonbiodegradable, but adsorbable toxicant.

The first type of protection is well illustrated by the work of Wang et al. (1986) in methanogenic systems and Chang and Rittmann (1987, 1988) for aerobic systems. Using phenol in both cases, these workers showed that a large loading shift-up did not result in immediate leakage of phenol in the effluent or to process upset. Instead, the unbiodegraded phenol initially was adsorbed in the GAC; later, the adsorbed substrate was biodegraded as the bacterial mass increased or the shock load ended. The course of events is illustrated in Figure 1.

This adsorption of a shock load improves process performance in three ways. One, it prevents immediate loss of substrate to the effluent. Two, by holding the substrate in the system for a longer time, adsorption increases the total mass of substrate degraded (Rittmann et al., 1992). Three, by keeping the substrate concentration low, it prevents process destabilization, such as by disaggregation due to high specific growth rates.

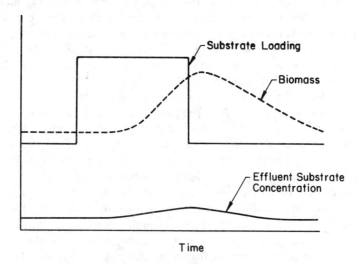

Figure 1 Schematic of how an adsorbent attachment medium keeps effluent quality stable during a shock loading.

Protection from deleterious inputs of a nonbiodegradable toxicant is well illustrated by the copious work of Suidan and co-workers (e.g., Suidan et al., 1983; Nakhla et al., 1990; Fox et al., 1990). When a nonbiodegradable, but toxic component, such as certain phenolics, can be adsorbed in the micropores of GAC, the bacteria are protected from toxicity. This protection only occurs when the toxicant is strongly adsorbable and when the activated carbon is replaced regularly in order to maintain adsorption capacity.

In summary, the protection from shocks afforded by activated carbon is associated with biofilms, because biofilms always proliferate when activated carbon is present. While suspended growths also could benefit from adsorption protection, they are not necessarily present when activated carbon is used.

PROTECTION FROM TOXIC SUBSTRATES

Many hazardous organic compounds are biodegradable, but also inhibitory to the bacteria that degrade them. Even when an adsorbing medium is absent, biofilms have means to protect themselves from the toxicity. The protecting mechanism is the concentration gradient, which can lower the inhibitory substrate's concentration to noninhibitory levels inside the biofilm. The key for having this type of protection, which can occur indefinitely, is that the toxic substrate be biodegradable by the bacteria in the biofilm.

To fully comprehend protection from toxic substrates, we must distinguish between cases in which the toxicant is the primary substrate or a secondary substrate. The primary substrate is the electron donor whose oxidation (or sometimes fermentation) provides electrons and energy for biomass synthesis and maintenance. On the other hand, utilization of a secondary substrate provides no or negligible electrons and energy for synthesis and maintenance (Rittmann, 1987). To have secondary utilization, a primary substrate must be utilized to grow and sustain the cells. The most interesting situation for a toxic secondary substrate is when it inhibits utilization of the primary substrate.

A good example of protection from a toxic primary substrate comes from the work of Saez et al. (1991), who degraded phenol in an anaerobic fluidized bed of anthracite coal. They demonstrated that steady-state phenol biodegradation was stable for liquid concentrations of 24 to 640 mg/ℓ in the biofilm process, even though a dispersed-growth process would have a stable range of only 24-76 mg/ℓ.

The reason for the biofilm's advantage is illustrated in Figure 2. As the primary-substrate concentration decreases in the biofilm—due to simultaneous diffusion and utilization via Haldane self-inhibition kinetics (Saez et al., 1991)--the specific growth rate for the bacteria is positive in the central zone. Near the outer surface, the specific growth rate is negative from inhibition, while it also is negative near the attachment surface due to the substrate concentration being less than S_{min} (Rittmann and McCarty, 1980), the minimum concentration giving a growth rate greater than the loss rate. Only by having simultaneous diffusion and utilization, one of the key attributes of biofilms being aggregates, can the concentration of the inhibitory substrate be lowered to noninhibitory levels in the biofilm, even though its bulk-liquid concentration remains inhibitory.

The situation is similar when a secondary substrate is the inhibitor. Figure 3 illustrates the situation when the secondary substrate (at concentration I) inhibits the utilization of the primary substrate (at concentration S) and the growth of the

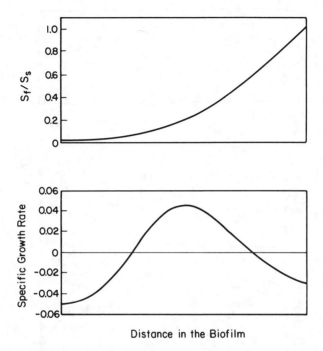

Figure 2 Example of how a substrate gradient allows a positive growth rate inside the biofilm for a self-inhibitory primary substrate. (Adapted from Saez et al., 1991.)

bacteria. The profiles are adapted from unpublished work by Dr. Pablo Saez. Good examples of primary and secondary substrates that fit this scenario are phenol and 4-chlorophenol (Saez and Rittmann, 1991). In Figure 3, the specific growth rate is positive in the interior of the biofilm, where the primary-substrate concentration is still high, while the secondary-substrate concentration has been reduced to a level low enough that its inhibition is mostly relieved. This kind of protection is most effective for extending the range of concentrations of inhibitor that the biofilm process can tolerate when S is high compared to I, the classic case of secondary utilization. Then, the biofilm process can stably treat substantially higher concentrations of an inhibitory secondary substrate than can a dispersed-growth process.

Because all aggregates can introduce substrate-concentration gradients, protection from inhibitory substrates is not completely unique to biofilms. However, the beneficial impact is accentuated for biofilms, compared to flocs. The reasons for the accentuation is connected to the different methods by which biomass is wasted from biofilm and suspended-growth systems, the subject of the next section.

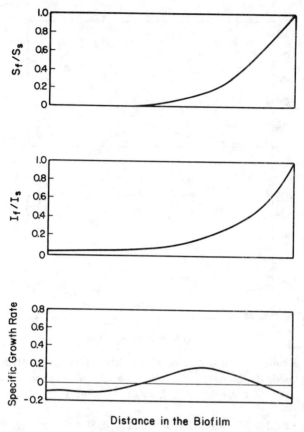

Figure 3 Example of how the simultaneous biodegradation of a primary substrate (S) and an inhibitory secondary substrate (I) can create a region of positive growth inside the biofilm. (Adapted from unpublished work by P.B. Saez.)

PROTECTION OF SENSITIVE POPULATIONS

The main way in which biomass is wasted from a biofilm process is through detachment and washout in the effluent. The most common cause of detachment is erosion, which is the continuous removal of small particles from the surface of the biofilm, primarily due to the shear stress of the water flowing past the biofilm surface (Rittmann, 1989). The most important feature of erosion is that it occurs from the outer surface of the biofilm.

Figure 4 illustrates a hypothesized distribution of biofilm loss relatively for thick biofilms (Furumai and Rittmann, manuscript in preparation). When the biofilm is thick enough (e.g., greater than about 30 μm in depth), the biomass located away from the outer surface is not subject to detachment loss. Thus, it is protected from

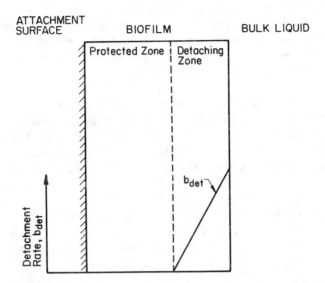

Figure 4 Schematic of how the thickness of biofilm over which detachment occurs extends only over a zone near the outer surface. Therefore, a zone protected from detachment loss develops deeper within the biofilm.

detachment loss and has a solids retention time greater than the average for the entire biofilm, which is approximated by $1/b_{det}$, where b_{det} is the specific detachment rate (Rittmann, 1989).

The situation is quite different for flocs in suspended-growth processes (Furumai and Rittmann, 1992; manuscript in preparation). Sludge wasting removes the entire floc, which means that bacteria inside the floc are not protected from loss, but have the average solids retention time.

When the process performance requires the stable presence of bacteria that are sensitive to washout, having them locate themselves inside a biofilm offers them protection from loss, but have the average solids retention time.

When process performance requires the stable presence of bacteria that are sensitive to washout, having them locate themselves inside a biofilm offers them protection from loss and increases their retention. Two situations are clear examples:

- The bacteria are exposed to a toxic substrate, which reduces their growth rate.
- The bacteria have very slow growth rates in all cases.

The first situation was described in the previous section. When an inhibitory substrate (primary or secondary) is present, a positive growth rate is achieved only in the interior of the biofilm. If that positive-growth zone is protected from detachment by an outer layer of bacteria, the positive growth rate is increased, making the biofilm more robust than a floc exposed to similar condition. Thus, biofilm accumulation accentuates protection from a toxic substrate.

The second situation occurs when fast-growing and slow-growing species must coexist. Two examples are evident:

- In anaerobic treatment, slow-growing methanogens must coexist with faster-growing eubacteria that ferment complex organic compounds to acetate and H_2, which are the primary substrates for methanogens. The presence of methanogens is required for stabilization of the total biochemical oxygen demand (BOD) in the wastewater and for several key detoxification reactions, including reductive dechlorination.
- In aerobic treatment, slow-growing nitrifiers must coexist with fast-growing aerobic heterotrophs. Because of high total Kjeldahl nitrogen contents of many industrial wastewaters and landfill leachates, nitrification must accompany general BOD removal and detoxification of hazardous organic chemicals.

Modeling work by several groups (Rittmann and Manem, 1992; Wanner and Gujer, 1986; Kissel et al., 1984) has demonstrated that slower-growing species tend to accumulate deeper in biofilms. Figure 5 offers a good example for nitrification. The nitrifiers mainly exist in a middle layer between faster growing heterotrophs and slowly accumulating inert biomass.

Furumai and Rittmann (manuscript in preparation) simplified the distribution shown in Figure 5 by assuming discreet layers. From that simplified concept, they compared mixed-culture nitrification in biofilm, floc, and dispersed growth systems. The conclusion was that protection from detachment allows nitrification to occur stably for shorter average solids retention times and higher BOD concentrations, as long as the dissolved oxygen is not depleted.

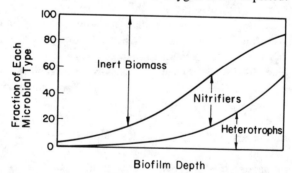

Figure 5 Example of species distributions in a multispecies biofilm. The slower-growing nitrifiers are more prevalent deeper in the biofilm and, therefore, are more protected from detachment than are the heterotrophs. (Adapted from Rittman and Manem, 1992.)

SUMMARY AND POSTSCRIPT

Biofilm processes possess no magical attribute that makes them a "silver bullet" for biological detoxification of hazardous organic wastes. Nevertheless, they have special advantages over suspended-growth processes when the following conditions apply:

1. The system is exposed to shocks of high substrate loading or nonbiodegradable toxicants. The use of adsorbent media, most usually activated carbon, is a natural complement to biofilm treatment, because the biofilms colonize the external surfaces of the medium.
2. The system must treat inhibitory substrates. Whether the inhibitor is a primary substrate or a secondary substrate, biofilm growth accentuates the benefits that substrate gradients inside the film have for extending the range of substrate concentration over which stable, operation is possible.
3. The system must sustain slow-growing species in a multispecies population. Biofilms offer an advantage of stability because the slow-growing species can accumulate in interior portions of the biofilm, where they are protected from detachment losses by the outer layer.

A final reflection on the importance of biofilms is that they are the dominant form of biological activity whenever surfaces are abundant. Therefore, situations in which biofilms will be critical, whether or not detoxification is a goal, include *in situ* bioremediation of aquifers and soils, fate and transport in groundwater and shallow streams, and any application in which a biofilm process is chosen for its cell retention, stability, and/or compact size.

REFERENCES

1. Chang, H.T. and Rittmann, B.E., "Verification of the Model of Biofilm on Activated Carbon," *Environ. Sci. Technol.*, Vol. 21, pp. 280-288, 1987.

2. Chang, H.T. and Rittmann, B.E., "Comparative Study of Biofilm Kinetics on Different Adsorptive Media," *J. Water Poll. Control Fedn.*, Vol. 60, pp. 362-368, 1988.

3. Fox, P., Suidan, M.T., and Pfeffer, J.T. "Hybrid Expanded-Bed GAC Reactor for Treating Inhibitory Wastewater," *J. Environ. Engr.*, Vol. 116, pp. 438-453, 1990.

4. Furumai, H. and Rittmann, B.E., "Advanced Modeling of Mixed Populations of Heterotrophs and Nitrifiers Considering the Formation and Exchange of Soluble Microbial Products," to appear in *Water Science & Technology*, 1992.

5. Hom, S., "Cleaning with Biofilms," *The Bioremediation Report*, Santa Rosa, CA, Vol. 1, No. 1, 1992.

6. Kissel, J.C., McCarty, P.L. and Street, R.L., "Numerical Simulation of a Mixed-Culture Biofilm," *J. Environ. Engr.*, Vol. 110, pp. 393-411, 1984.

7. Nakhla, G.F., Suidan, M.T., and Pfeffer, J.T., "Control of Anaerobic GAC Reactors Treating Inhibitory Wastewaters," *Res. J. Water Poll. Control Fedn.*, Vol. 62, pp. 65-72, 1990.

8. Palenik, B., Block, J.C., Burns, R.G., Characklis, W.G., Christensen, B.E., Ghiorse, W.G., Christina, A.G., Morel, F.M.M., Nicols, W.W., Tuovinen, O.H., Tuschewitzki, G.J., and Videla, H.A., "Biofilms: Properties and Processes," In W.G. Characklis and Wilderer, P.A., eds., *Structure and Function of Biofilms*, John Wiley & Sons, Inc., Chichester, England, pp. 351-368, 1989.

9. Rittmann, B.E. "Aerobic Biological Treatment," *Environ. Sci. Technol.*, Vol. 21, pp. 128-136, 1987.

10. Rittmann, B.E. "Detachment from Biofilms," In W.G. Characklis and P.A. Wilderer, eds., *Structure and Function of Biofilms*, John Wiley & Sons, Inc., Chichester, England, pp. 49-58, 1989.

11. Rittmann, B. E., Henry, B., Odencrantz, J. E., and Sutfin, J. A., "Biological Fate of Polydisperse Acrylate Polymer in Anaerobic Sand-Medium Transport," in press in *Biodegradation*, 1992.

12. Rittmann, B.E. and Manem, J.A., "Development and Experimental Evaluation of a Steady-State, Multi-Species Biofilm Model," *Biotechnology Bioengineering*, Vol. 37, pp. 914-922, 1992.

13. Rittmann, B.E. and McCarty, P.L., "Model of Steady-State-Biofilm Kinetics," *Biotechnology Bioengineering* Vol. 22, pp. 2343-2357, 1980.

14. Rittmann, B.E., Valocchi, A.J., Ray, C., Seagren, E., and Wrenn, B., *A Critical Review of In Situ Bioremediation*, North Dakota Energy and Mineral Resources Research Center, Grand Forks, North Dakota, 1991.

15. Saez, P.B., Rittmann, B.E., and Zhang, Q.B. Zhang, "Biodegradation Kinetics of a Self-Inhibitory Substrate by Steady-State Biofilms," *Proc. Purdue Industrial Waste Conf.*, May, 1990, Lewis Publishers, Inc., Ann Arbor, MI, pp. 273-279, 1991.

16. Saez, P.B. and Rittmann, B.E., "The Biodegradation Kinetics of 4-Chlorophenol, an Inhibitory Co-Metabolite," *Res. J. Water Poll. Control Fedn.*, Vol. 63, pp. 838-847, 1991.

17. Schultz, J.R. and Keinath, T.M., "Powdered Activated Carbon Treatment Process Mechanisms," *J. Water Poll. Control Fedn.*, Vol. 56, pp. 143-151, 1984.

18. Stal, L.J., Bock, E., Bouwer, E.J., Douglas, L.J., Gutnick, D.L., Heckmann, K.D., Hirsch, P., Kolbel-Boelke, J.M., Marshall, K.C., Prosser, J.I., Schutt, C. and Watanabe, Y., "Cellular Physiology and Interactions of Biofilms Organisms," in W.G. Characklis and P.A. Wilderer, *Structure and Function of Biofilms*, John Wiley & Sons, Inc., Chichester, England, pp. 269-288, 1989.

19. Suidan, M.T., Strubler, C.E., Kao, S.W., and Pfeffer, J.T., "Treatment of Coal Gasification Wastewater with Anaerobic Filter Technology," *J. Water Poll. Control Fedn.*, Vol. 55, pp. 1263-1270, 1983.

20. Valeur, A., Tunlid, A., and Odham, G., "Differences in Lipid Composition between Free-Living and Initially Adhering Cells of Gram-Negative Bacterium." *Arch. Microbiol.*, Vol. 149, pp. 521-526, 1988.

21. van Loosdrecht, M.C.M., Lyklema, J., Norde, W., and Zehnder, A.J.B., "Influences of Interfaces on Microbial Activity," *Microbial Rev.*, Vol. 54, pp. 7587, 1990.

22. Wang, Y.T., Suidan, M.T., and Rittmann, B.E., Performance of an Expanded-Bed Methanogenic Reactor," *J. Environ. Engr.*, Vol. 111, pp. 460-471, 1985.

23. Wang, Y.T., Suidan, M.T., and Rittmann, B.E., "Anaerobic Treatment of Phenol by an Expanded-Bed Reactor," *J. Water Poll. Control Fedn.*, Vol. 58, pp. 227-233, 1986.

24. Wanner, O. and Gujer, W., "A Multispecies Biofilm Model," *Biotechnol. Bioengr.*, Vol. 28, pp. 314-328, 1986.

26

EFFECTS OF INITIAL NITROGEN ADDITION ON DEEP-SOILS BIOVENTING AT A FUEL-CONTAMINATED SITE

John W. Ratz, Peter R. Guest, P.E.,
and Douglas C. Downey, P.E.
Engineering-Science, Inc.
Denver, CO

OVERVIEW

A ruptured pipe at a Burlington Northern Railroad (BNRR) fueling pump house resulted in over 60,000 gallons of No. 2 diesel fuel spilling onto the surrounding soil. An initial investigation of site conditions indicated that subsurface soils were contaminated with diesel fuel to ground water, which was observed approximately 70 feet below the ground surface. State regulatory agencies requested that BNRR develop and implement a remedial action plan to treat these diesel-contaminated soils and protect local ground waters. Engineering-Science, Inc. (ES) was retained for this work and, after evaluating a variety of remediation technologies, recommended using soil venting methods to enhance the immediate volatilization and long-term biodegradation of fuel residuals. ES designed and implemented a "bioventing" pilot test to determine soil properties such as air permeability, and to assess the potential for partial volatilization and long-term biodegradation of diesel fuel residuals at the site. Hydrocarbon concentrations, carbon dioxide, and oxygen levels were monitored at a vapor extraction well (VEW) and six vapor monitoring points (VMPs) to determine the rates of volatilization and biological degradation of fuel residuals. Pilot test results confirmed that full-scale bioventing was feasible for the remediation of this site. Following the installation of a full-scale bioventing system at the site, tests were conducted to determine the impact of nitrogen addition on the biological degradation of fuel residuals. The tests were deemed necessary because nitrogen was present in soils at concentrations that may limit future fuel biodegradation rates under enhanced oxygen conditions. Gaseous anhydrous ammonia was injected at one VMP to examine its mobility in the soil matrix, and an aqueous solution of ammonium nitrate fertilizer was added to four other VMPs. Ammonia concentrations, carbon dioxide, and oxygen levels were monitored during the tests to assess ammonia mobility and to determine if biological degradation rates increased as a result of nutrient addition. An evaluation of these monitoring results indicate that the moisture and nutrient addition had a positive effect on biological degradation, and these results are presented in this paper.

BACKGROUND

Fuel contamination in soil is often a long-term source of ground water contamination which must be addressed to ensure thorough site cleanup. Because fuel-contaminated soil is generally excluded as a hazardous waste under the Resource Conservation and Recovery Act (RCRA), excavation and placement of these soils in approved landfills has frequently been the most expedient option for site cleanup. Increased costs, increased restrictions on land disposal, and the risk of becoming a potentially responsible party in future landfill remediation have now made this option much less attractive.

A variety of *in situ* soil remediation options are now available for fuel-contaminated soils. The use of air as a medium to contact and remove volatile hydrocarbons has been extensively used in soil vapor extraction systems. More recently, soil vapor extraction has been implemented with the objective of supplying oxygen to subsurface bacteria for enhanced fuel biodegradation. Soil vapor extraction has been more successful than water flushing for delivering oxygen to fuel residuals trapped in soil micropores due to the greater diffusivity and lower viscosity of air. Wilson and Ward (1987) suggested that air was potentially 1,000 times more efficient than water at transferring oxygen for the bioremediation of fuels in deep, unsaturated soils. Researchers at the Texas Research Institute (1984), Chevron (Ely and Heffner, 1988), and others have observed or utilized *in situ* soil venting for supplying oxygen to subsurface microorganisms. Recent U.S. Air Force field tests using *in situ* soil venting to supply oxygen to the subsurface have documented this technique of stimulating aerobic biodegradation of fuel residuals.

Oxygen utilization rates measured during bioventing pilot tests have been used to estimate the rates at which fuel can be biodegraded. Complete biological mineralization of fuel hydrocarbons (e.g., n-decane) can be described by the equation:

$$C_{10}H_{22} + 15.5O_2 \rightarrow 10CO_2 + 11H_2O$$

Approximately 3.5 grams of oxygen is required to mineralize 1 gram of fuel hydrocarbons to carbon dioxide. This ratio will actually underestimate fuel biodegradation because a third or more of the fuel hydrocarbons may be used in cell production rather than carbon dioxide production (Atlas, 1981). Prior field research has shown that oxygen consumption provides the most accurate estimate of hydrocarbon degradation (Miller et al., 1990; Hinchee and Miller, 1991). Although carbon dioxide should provide an equivalent estimate of fuel biodegradation, natural sinks and sources of carbon dioxide in the soil carbonate cycle can mask carbon dioxide production associated with bacterial respiration.

A full-scale soil venting project to remediate a 27,000 gallon jet fuel spill at Hill Air Force Base (AFB), Utah was recently completed. During this 18-month project, jet fuel residuals in soils were reduced from an average total recoverable petroleum hydrocarbon (TRPH) concentration of approximately 900 milligrams per kilogram (mg/kg) to less than 10 mg/kg (Hinchee and Miller, 1991). Monitoring of vented soil gas indicated that volatilization accounted for 60 percent of the removal, and biodegradation accounted for the remaining 40 percent. Polycyclic aromatic hydrocarbon (PAH) compounds found in diesel fuels are more difficult to biodegrade than straight-chain hydrocarbons; however, successful PAH biodegradation using native microorganisms has been reported by several investigators (Lee et al., 1988; Srivastava, 1989). Because diesel fuel is less volatile than gasoline or jet fuel, long-term biodegradation will exceed volatilization as the primary removal mechanism at the BNRR site.

In addition to oxygen supply, the distribution of petroleum-degrading bacteria and soil moisture conditions influence the success of *in situ* biodegradation. Microbial populations were established at depths of 50 feet prior to initiation of the Hill AFB test. Respiration tests at the BNRR site indicated active microbial populations at depths exceeding 60 feet (Downey and Guest, 1991). Adequate soil moisture must also be available to sustain microbial populations. A column test using soils from the Hill AFB site showed increasing fuel biodegradation as soil moisture was increased from 6 to 18 percent by weight (Hinchee and Arthur, 1990). However, higher moisture levels can cause a reduction in air permeability which can limit oxygen supply.

Soil bacteria also require a variety of nutrients to sustain accelerated rates of hydrocarbon degradation. These nutrients, which include nitrogen, phosphorous, sulfur, and metals such as calcium and iron, are used by bacteria to synthesize new biomass and to manufacture enzymes. Soil bacteria must be able to obtain these nutrients from the subsurface environment to manufacture materials for cell maintenance and sustain a growing population. If nutrients are not available to the bacteria in adequate quantities, then the growth of the bacterial population and the rate of hydrocarbon degradation may be limited.

In fuel-contaminated soils, most nutrients are available in adequate concentrations to support fuel biodegradation at nonaccelerated rates. Attempts have been made to accelerate the biodegradation of petroleum hydrocarbons through the addition of nitrogen. Nutrients added at the Hill AFB site and during a recent bioventing demonstration at Tyndall AFB, Florida, produced little or no increase in hydrocarbon degradation (Miller et al., 1990). The demonstration at Tyndall AFB indicated that soil bacteria are able to recycle essential nutrients, and may also rely on nitrogenase bacteria to fix atmospheric nitrogen and to introduce useful forms of nitrogen for fuel degrading microbes.

SITE CHARACTERIZATION

The BNRR site is located south of a diesel fuel pump house, where the pipe rupture occurred (Figure 1). A bioventing pilot test was conducted in April 1991 to determine the potential for *in situ* biodegradation of fuel residuals at this site. A previous paper (Downey and Guest, 1991) presents the results of the pilot test, its implication for full-scale design, and the methods used to conduct the test. Based upon the successful results of the pilot test, a continuously operating full-scale vapor extraction system was installed at the VEW in September 1991.

Ground Water Conditions

Ground water was observed at depths ranging from 68 to 77 feet below the surface. Intermittent lenses of perched water were also observed on the top of an interbedded sand, silt, and clay zone. Dilute levels of benzene, toluene, ethylbenzene, and xylenes (BTEX) and PAH compounds have been detected beneath the site; however, only benzene and total petroleum hydrocarbon (TPH) concentrations exceed their respective cleanup goals of 5 micrograms per liter

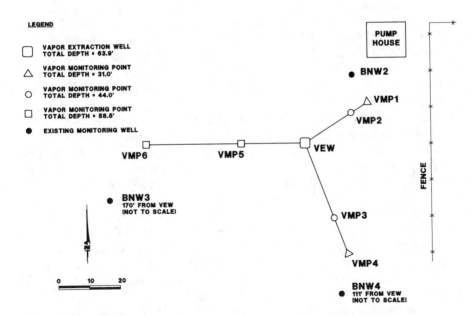

Figure 1 Site plan and monitoring point locations.

(µg/L) and 2 milligrams per liter (mg/L), respectively. Due to the thick column of soil above the water table, the vast majority of the diesel fuel spill is adsorbed and occluded in the unsaturated-zone soils and has not yet seriously impacted ground water quality. However, without soil treatment, soluble BTEX and PAH compounds could continue to migrate toward the ground water, creating a larger and more concentrated plume of contamination.

Soil Conditions

Soils at the BNRR site were characterized during the construction of the VMPs and the VEW in April 1991. During drilling, soil samples were collected and analyzed for physical properties, TRPH, total organic carbon (TOC), and nutrients.

Physical Soil Characteristics

A geologic section was constructed using the boring logs from the six VMPs and the central VEW in the pilot test area (Figure 2). The general lithology in this area consists of fine- to medium-grained, silty sands from the ground surface to approximately 30 to 35 feet below ground surface (upper sand zone), interbedded sand and silt/clay lenses that extend 45 to 50 feet below ground surface (interbedded zone), and another layer of fine- to medium-grained silty sand that extends to a depth of 70 to 75 feet (intermediate sand zone).

Sieve and moisture analyses were conducted to further characterize the soils. In the upper sand zone, the soil consisted of 86.1 percent sand and 13.9 percent silt/clay, and a moisture content of 5.9 percent was observed. The interbedded zone was 55.2 percent sand and 44.8 percent silt/clay, and the moisture content was 11.1 percent. The intermediate sand zone consisted almost entirely of sand (98.8 percent) with 1.2 percent silt/clay. The moisture content of the intermediate sand zone was 2.1 percent.

Air permeability was quantified through vacuum response tests conducted as part of the bioventing pilot test in April 1991. Vacuum response was measured at the VMPs and monitoring wells (Figure 1) while soil gas was extracted from the VEW using a vacuum blower powered by a 10-horsepower electric motor. Three extraction tests were performed at air flow rates ranging from 130- to 144- standard cubic feet per minute (scfm). The soil responded rapidly to the vapor extraction system, with a measurable vacuum response occurring at a distance of more than 170 feet from the VEW within the initial minutes of system operation. Because the contaminated soil zone was entirely contained within the radius of influence of the vapor extraction system, only one vapor extraction unit was specified for full-scale remediation. The highest vacuum response was measured in the interbedded zone. The upper and intermediate sand zones responded with lower vacuums, evidence of less restricted air flow through these more uniform and permeable soils.

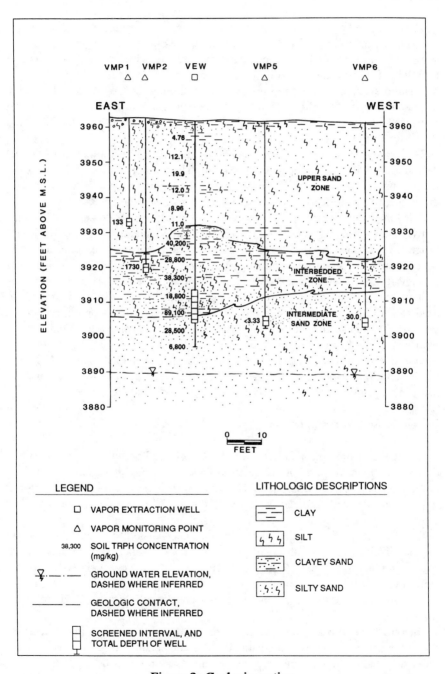

Figure 2 Geologic section.

Chemical Soil Characteristics

The diesel fuel contamination in the soil appears to be localized within a 60-foot radius of the pumphouse. TRPH concentrations are plotted on the geologic section shown in Figure 2, and are summarized in Table 1. Based on soil analysis and observations made during drilling, it appears that diesel fuel migrated rapidly downward at the spill site until it encountered the interbedded sand and silt/clay zone at approximately 30 to 35 feet. At this depth, the fuel spread laterally and continued its downward movement through more permeable sand lenses in the interbedded layer.

TOC analyses performed on noncontaminated site soils resulted in one TOC detection of only 600 parts per million (ppm). Three other samples did not exceed the 500 ppm detection limit. Thus, there is not a significant potential for nonfuel or competitive hydrocarbon consumption by bacteria at the BNRR site.

Soil samples collected from the screened intervals of VMP4 and VMP5 were analyzed for ammonia- and nitrate-nitrogen, total Kjeldahl nitrogen (TKN), and water-soluble phosphates. These analyses were performed to determine the concentrations of naturally occurring nutrients available in the soils to maintain biological activity. Ammonia-nitrogen was found at concentrations of 204 parts per million (ppm) in the upper sand zone (VMP4) and 4.2 ppm in the intermediate sand zone (VMP5), while nitrate-nitrogen levels ranged from 4 ppm in the upper sands to 11 ppm in the intermediate sand zone. TKN levels were found to be 4 ppm in both sand zones. Water-soluble phosphate concentrations ranged from 177 ppm in the upper sands to nearly 6,000 ppm in the intermediate sands. Although water soluble phosphates are present at elevated concentrations in the soil, the lower nitrogen concentrations may limit future biodegradation rates.

Biological Soil Characteristics

The fuel-consuming capability of the native bacterial population was examined during the *in situ* respiration test performed as part of the bioventing pilot test in April 1991. Oxygen consumption and carbon dioxide production were monitored in soil gas extracted from the VMPs after the vapor extraction system was shut down.

Based on the 3.5:1 oxygen to fuel mass ratio (Atlas, 1981) and the average oxygen utilization rate observed in contaminated soils, soil bacteria were consuming approximately 6 mg/kg TRPH per day under enhanced oxygen conditions. This rate is comparable to other bioventing tests in which fuel biodegradation rates of 1 to 20 mg/kg per day have been recorded (Hoeppel et al., 1991).

Uncontaminated background wells were monitored throughout the respiration test. Oxygen levels remained relatively constant at approximately 18 percent, indicating that biological oxygen consumption of natural (nonfuel) organic carbon and abiotic oxygen consumption were not significant in these soils.

Table 1. TRPH Concentrations in the Vadose Zone.

Sample Location	Sample Depth (feet)	Initial TRPH Concentration (mg/kg)
Upper Sand Zone		
BNW2	15	11
VEW	4.0-5.5	4.76
VEW	9.0-10.5	12.1
VEW	14.0-15.5	19.9
VEW	19.0-20.5	12
VEW	24.0-25.5	8.96
VEW	29.0-30.5	11
Interbedded Zone		
BNW2	35	64,000
VEW	33.0-34.5	40,200
VEW	33.0-34.5	41,900
VEW	39.0-40.5	28,800
VEW	44.0-45.5	38,300
VEW	49.0-50.5	18,800
VEW	54.0-55.5	89,100
VMP2	41.5-43.0	1,730
VMPH3	42.5-44.0	194
Intermediate Sand Zone		
BNW2	55	1,100
VEW	59.0-60.5	28,500
VEW	64.5-66.0	6,800
VMP5	58.0-59.5	<3.33
VMP6	56.8-58.3	30

Because the effects of nitrogen addition were observed through the use of *in situ* respiration tests, test procedures are discussed in greater detail in the Nutrient Testing Procedures section.

NITROGEN SUPPLEMENTATION

The potential benefit of nutrient addition for accelerating the biodegradation of fuel residuals is of particular interest at the BNRR site due to the high levels of TRPH found in some layers of the soil (Table 1). The current carbon:total nitrogen (C:N) ratio is approximately 300:1, much lower than ratios suggested for optimum microbial growth. One investigator found that oil sludge biodegradation in a land farm was optimal at a C:N ratio of 60:1 (Atlas, 1981). For *in situ* nitrogen application, a C:N ratio of approximately 25:1 may be appropriate due to matrix interference in the delivery of nitrogen. As initial soil analyses indicated that low nitrogen concentrations may limit future biodegradation rates, tests were conducted to determine if the addition of nitrogen to fuel-contaminated soils would significantly accelerate the biological degradation of fuel residuals. If so, the length of time required to achieve site remediation could be shortened substantially, translating into lower operating and maintenance costs. Two supplemental nitrogen sources were selected for testing, an aqueous solution of ammonium nitrate fertilizer and gaseous anhydrous ammonia. Both nitrogen sources are inexpensive, available in bulk quantities, and widely used as agricultural fertilizers.

In solution, ammonium nitrate (NH_4NO_3) disassociates into ammonium (NH_4^+) and nitrate (NO_3^-). Nitrogen in either of these forms can be easily utilized by most soil bacteria. Additionally, soil bacteria is not inhibited by either form of nitrogen even at concentrations of 600 ppm or higher (Kiehl and Netto, 1974). However, nutrients in aqueous solution are not highly mobile or easily dispersed in soils, particularly in fine-grained soils containing free petroleum product, such as those found in the interbedded zone at the BNRR site. Aqueous nutrient solutions may collect preferentially in coarse-grained soils with available void space. Furthermore, aqueous nutrient solutions may tend to collect in soils that contain negligible amounts of diesel fuel, rather than the contaminated regions containing free product where the need for nutrients is the greatest. Another disadvantage of an aqueous nutrient solution is interference with oxygen flow. The introduction of large volumes of liquid into the subsurface means that less pore space will be available for soil gas flow, and the oxygen supply to contaminated soils will be reduced.

Gaseous anhydrous ammonia has a potential advantage over liquid nutrients in soil mobility. Anhydrous ammonia may be more mobile and more easily dispersed through fuel-contaminated, fine-grained soils due to advantages in viscosity and diffusivity. However, the mobility of gaseous anhydrous ammonia is difficult to estimate. Mobility is dependent upon a large number of site-specific factors, including the clay content of the soil, permeability, soil moisture content, pH, organic content, microbial population present, and the strength of the subsurface vacuum. Fates of anhydrous ammonia in soil include dissolution into soil moisture, conversion to NO_3^- by nitrifying bacteria, and sorption to clay and organic material (Stevenson, 1982).

An additional potential problem with the use of gaseous anhydrous ammonia is that free ammonia (NH_3) can be inhibitory or toxic to soil bacteria at relatively low concentrations. One investigator found that 210 ppm of free ammonia as nitrogen (NH_3-N) inhibited the metabolism of nitrifying bacteria in soil samples, while fixed ammonium (NH_4^+) did not inhibit nitrification even at concentrations three times as high (Kiehl and Netto, 1974). In another study, NH_3-N applied at a concentration of 430 ppm to a fine, sandy loam with a background NH_3-N concentration of 7.0 ppm reduced the bacterial population from 7.22 to 1.79 million bacteria per gram of oven-dried soil. Additionally, nitrification was completely stopped at an NH_3-N concentration of 632 ppm in a fine sand (Eno et al., 1955). Gaseous ammonia, if added to contaminated soil at excessive concentrations, could result in localized sterilization of the soil.

NUTRIENT TESTING PROCEDURES

In general, the ammonium nitrate fertilizer solution offers the advantage of high toxicity and inhibitory thresholds, but is not highly mobile in soil. Conversely, anhydrous ammonia may possess mobility advantages in soil, but is inhibitory or toxic to bacteria at relatively low concentrations. Therefore the mobility of anhydrous ammonia mobility was tested in site soils, and the ammonium nitrate solution was used to evaluate the effect of nitrogen supplementation on the soil bacteria. A summary of the test schedule is provided (Table 2).

Table 2. Nutrient Testing Schedule.

Date	Event
April 1991	Bioventing Pilot Test (Respiration Test 1)
September 1991	Full-Scale System Installation
November 5–6, 1991	Respiration Test 2
November 6–7, 1991	Anhydrous Ammonia Applied at VMP6
November 7, 1991	Ammonium Nitrate Solution Applied to VMP1, VMP2, VMP3, and VMP5
December 1–2, 1991	Respiration Test 3

Nutrient Addition and Sampling Procedures

The application of gaseous anhydrous ammonia was field-tested to examine its mobility and dispersivity in site soils. Anhydrous ammonia was applied to VMP6 (Figure 1) on November 6 and 7, 1991, while the vapor extraction system was in operation. Approximately 75 pounds of anhydrous ammonia gas was allowed to flow into VMP6 over a 42 hour period. In order to track the potential movement of the gaseous ammonia, soil gas was sampled at VMPS and at the VEW.

The screened interval of VMP6 lies in the intermediate sand zone (Figure 2), which is the most critical soil zone in terms of ground water protection. TRPH concentrations in VMP6 and VMPS, also screened in the intermediate sand zone, are negligible (10 mg/kg and < 3.33 mg/kg, respectively). Therefore, overall site remediation will not be disrupted by potentially toxic ammonia concentrations at VMP6. The moisture content of this layer was measured at only 2.1 percent, and the silt/clay content of the intermediate sand zone is 1.2 percent. Therefore, gaseous ammonia should be more mobile in intermediate sands than in either the interbedded zone or the upper sands.

Three hundred gallons of ammonium nitrate fertilizer solution was applied to VMP1, VMP2, VMP3, and VMP5 on November 6 and 7, 1991 (Figure 1). The solution contained a concentration of 2,000 mg/L ammonium nitrate as nitrogen. Nutrients were not applied to the VEW or VMP4. VMP4 was used to provide background oxygen and carbon dioxide levels.

Background ammonia samples were collected from VMP5 and the VEW prior to the addition of either anhydrous ammonia at VMP6 or the ammonium nitrate solution at other VMPs. After anhydrous ammonia injection was initiated, samples were collected every 30 to 60 minutes at VMP5 and the VEW for the first 6 hours of ammonia injection. Samples were collected and analyzed every 6 to 8 hours later in the test. Immediately prior to Respiration Test 3 on December 1, 1991, another set of soil gas samples were collected and analyzed for ammonia content.

Sampling for ammonia in soil gas was conducted using Drager tubes in two different concentration detection ranges. Drager tubes with a detection range of 2 to 30 ppm were used when low ammonia concentrations were anticipated, and tubes with a detection range of 5 to 700 ppm were used when higher concentrations of ammonia were expected in the soil gas. These ranges include literature reported toxic and inhibitory thresholds for soil bacteria (Eno et al., 1955; Kiehl and Netto, 1974). Ammonia Drager tubes are relatively nonspecific, detecting anhydrous ammonia, ammonium, and naturally occurring ammonia in the soil gas. Oxygen and carbon dioxide were monitored using a portable gas analyzer which was calibrated daily with carbon dioxide and oxygen standards. Prior to soil gas sampling at each VMP, a small vacuum pump was used to draw at least three casing volumes of soil gas into the well. Extracted vapor samples were retained in sealed Tedlar® bags.

Respiration Test Procedures

Because oxygen utilization rates are proportional to rates of biodegradation, they were used to observe the effects of nitrogen addition. Three *in situ* respiration tests were performed and the results compared to evaluate the impact of nitrogen addition at the BNRR site. The pilot test in April 1991 (Respiration Test 1) established initial oxygen consumption rates. In November 1991, Respiration Test 2 was performed to evaluate the continued performance of soil bacteria under enhanced oxygen conditions and to establish background performance prior to nutrient supplementation. The ammonium nitrate solution was added to selected VMPs immediately following Respiration Test 2. Respiration Test 3 was performed in December 1991. A comparison of Respiration Tests 2 and 3 indicates the impact of nutrients and moisture at the site.

Prior to the start of each respiration test, the contaminated soil was oxygenated by running the vapor extraction system for an extended period of time. To start the respiration test, the vapor extraction system was shut off, thus ending oxygen enhancement in contaminated soils. Soil gas samples from the VMPs were collected and analyzed at 30 to 120 minute intervals to establish rates of oxygen consumption and carbon dioxide production.

RESULTS

Anhydrous Ammonia Mobility

Anhydrous ammonia was found to be slightly mobile in soil at the BNRR site. During anhydrous ammonia application at VMP6 in November, ammonia was not detected in any of the soil gas samples collected from the VEW and VMP5. However, when the soil gas was resampled for ammonia on December 1 and 2, results suggested that some of the anhydrous ammonia applied at VMP6 had dispersed to VMP5.

Average ammonia concentrations at sampling points are illustrated in Figure 3. In VMP6, ammonia was detected at concentrations ranging from 650 to 800 ppm. This result was expected, as VMP6 was the point of anhydrous ammonia injection. The VEW and VMP4, neither of which were supplemented with nitrogen, yielded no ammonia detections. Soil gas from VMP1, VMP2, and VMP3, supplemented with ammonium nitrate fertilizer, contained ammonia ranging from less than 1 ppm to 2 ppm. All ammonia detections from these three VMPs were under the 2 ppm practical quantitation limit, and these concentrations are estimated. Soil gas collected from VMP5, which also received ammonium nitrate fertilizer, contained ammonia ranging from 2 to 5 ppm. Ammonia concentrations in soil gas samples from VMP5 were substantially higher than those from VMP1, VMP2, and VMP3. This suggests that anhydrous ammonia applied at VMP6 has dispersed into soils adjacent to VMP5, but adsorption or other processes significantly limit subsurface transport of anhydrous ammonia.

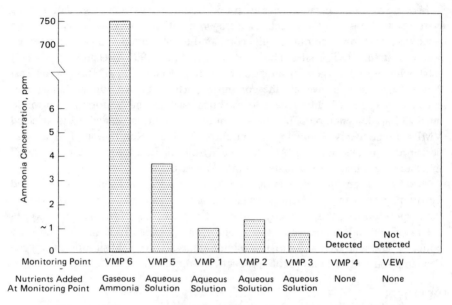

Figure 3 Average ammonia concentrations measured 24 days after nutrient addition.

Effects of Nutrient Addition on Biodegradation Rates

Oxygen utilization rates were calculated based on the first 5 to 7 hours of data from each *in situ* respiration test. The estimated oxygen utilization rates are summarized in Table 3, and are illustrated in Figure 4. These rates are within the range of .001 to .01 percent oxygen per minute ($\%O_2$/min) observed at other bioventing sites (Hinchee and Miller, 1991). Rates were assumed to be zero order; R^2 values ranged from 0.77 to 0.98, indicating the validity of the assumption.

The oxygen utilization rates at VMP1 and VMP2 increased between Respiration Tests 2 and 3, indicating that the addition of the ammonium nitrate fertilizer solution may have had a positive impact on fuel degradation rates. At VMP3 and VMP5, nutrient addition was found to have little effect on oxygen consumption. It was not possible to perform a respiration test at VMP6 following nutrient addition, because the high ammonia concentrations in the soil gas interfered with the operation of the gas analyzer.

Oxygen consumption rates at VMP5, where TRPH was not initially detected, were measured at approximately .002 $\%O_2$/min in all three respiration tests. Oxygen consumption rates at VMP5 may reflect bacterial activity in surrounding soils with less than 100 mg/kg TRPH. At VMP1 and VMP3, initial (April 1991)

Table 3 Biological Oxygen Uptake Rates.

Vapor Monitoring Point	Biological Oxygen Uptake Rate, (percent oxygen/minute)		
	Respiration Test 1 April 1991	Respiration Test 2 November 1991	Respiration Test 3 December 1991
1	0.0071	0.0023	0.0035
2	0.0041	0.0047	0.0087
3	0.0094	0.0043	0.0018
5	0.0031	0.0020	0.0021

TRPH concentrations in soil were found to be between 100 and 200 mg/kg, and oxygen consumption rates were measured at .0071 and .0094 %O_2/min, respectively. Since then, oxygen consumption rates in VMP1 and VMP3 have decreased to rates similar to those observed in VMP5 (Figure 4), suggesting that the majority of the biodegradable petroleum hydrocarbons have been consumed.

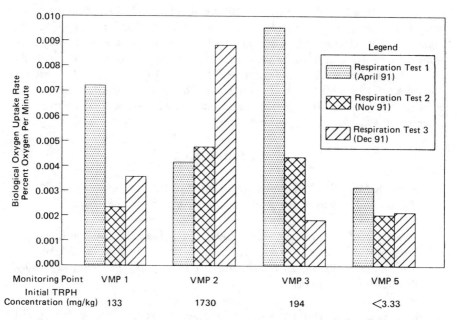

Figure 4 Changes in biological oxygen uptake during respiration tests.

Bacteria in soils with higher initial TRPH levels such as VMP2 (1730 mg/kg) and the VEW (89,000 mg/kg) continue to consume oxygen at a higher rate than less contaminated soils, and oxygen utilization rates at VMP2 appear to have increased as a result of moisture and nutrient addition.

Because moisture addition alone has also been shown to increase oxygen utilization rates by as much as 30 percent, it cannot be concluded that nitrogen alone was limiting biodegradation rates (Hinchee and Arthur, 1990). Additional respiration tests are planned at the site to determine if moisture additions alone can increase biodegradation rates at the BNRR site.

CONCLUSIONS

The bioventing pilot test clearly demonstrated the ability of a relatively small vapor extraction system to supply oxygen throughout 25,000 cubic yards of contaminated soil to a depth of at least 60 feet. The initial *in situ* respiration test indicated that native soil bacteria are capable of consuming fuel residuals at a rate of approximately 6 mg/kg TRPH per day when supplied with supplemental oxygen. Anhydrous ammonia was found to be slightly mobile in site soils. However, anhydrous ammonia does not disperse far enough or quickly enough to be used as a full-scale nutrient option at this site.

The addition of moisture and ammonium nitrate fertilizer increased biodegradation rates at two points on the site. Additional studies are required to determine if moisture, nitrogen, or both are limiting fuel biodegradation at the site. In the future, moisture will be added to the VEW. The VEW contained the highest TRPH detections at the site and biodegradation rates are most likely to be limited by moisture or nitrogen depletion at this point. To further investigate the cause of increased oxygen utilization at VMP2, 300 gallons of water, without nutrients, will be added at this point. Results of future *in situ* respiration tests conducted before and after moisture and nutrient addition will be compared to examine effects.

REFERENCES

1. Atlas, R.M. 1981, "Microbial Degradation of Petroleum Hydrocarbons: An Environmental Perspective," *Microbiological Reviews*. 45:180-209.

2. Downey, D.C., and Guest, P.R., 1991. "Combined Physical and Biological Treatment of Diesel-Contaminated Soils," *Proceedings of the NWWA/API Conference on Petroleum Hydrocarbons and Organic Chemicals in Ground Water*, Houston, Texas.

3. Ely, D.L., and Heffner, D.A., 23 August 1988, "Process for In-Situ Bioremediation of Hydrocarbon Contaminated Soil," U.S. Patent 4,765,902.

4. Eno, C.F., Blue, W.G., and Good, Jr., J.M., 1955, "The Effect of Anhydrous Ammonia on Nematodes, Fungi, Bacteria, and Nitrification in Some Florida Soils," Soil Science Society of America Proceedings, 19:55-58.

5. Hinchee, R.E., and Arthur, M.F., 1990, "Benchscale Studies on the Soil Aeration Process for Bioremediation of Petroleum Contaminated Soils," *Journal of Applied Biochemistry and Biotechnology.* Vol.28/29:901-906.

6. Hinchee, R.E., and Miller, R.N., 1991, "Bioventing for In-Situ Remediation of Jet Fuel," Proceedings of Air Force Environmental Restoration Technology Symposium. San Antonio, Texas.

7. Hoeppel, R.E., Hinchee, R.E., Arthur, M.F., 1991, "Bioventing Soils Contaminated With Petroleum Hydrocarbons," *Journal of Industrial Microbiology.* Vol 8:141-146.

8. Johnson, P.C., Kemblowski, M.W., and Colthart, J.D., 1990, "Quantitative Analysis for the Cleanup of Hydrocarbon-Contaminated Soils by In Situ Soil Venting," *Ground Water* Vol. 28, No. 3, May-June 1990.

9. Kiehl, J. de C., and Netto, A.C., 1974, "The Inhibitory Effect of Anhydrous Ammonia on Nitrification," Solo. 66:7-13.

10. Lee, M.D. et al. 1988, "Biorestoration of Aquifers Contaminated with Organic Compounds," *CRC Critical Reviews in Environmental Control.* Vol. 18:29-89.

11. Miller, R.N., et al. 1990, A Field Scale Investigation of Enhanced Petroleum Hydrocarbon Biodegradation in the Vadose Zone - Tyndall AFB, FL. Proceedings of the NWWA/API Conference on Petroleum Hydrocarbons and Organic Chemicals in Ground Water, Houston, Texas.

12. Srivastava, V.J. 1989, "Bioremediation of Former Manufactured Gas Plant Soils," Proceedings of 44th Industrial Waste Conference. Purdue University.

13. Stevenson, FJ., ed. 1982, "Nitrogen in Agricultural Soils," American Society of Agronomy. Madison, Wisconsin.

14. Texas Research Institute, 1984, "Forced Venting to Remove Gasoline Vapor From a Large-Scale Model Aquifer," Report 82101-F:TAV, American Petroleum Institute.

15. Wilson, J.T., and Ward, C.H., 1987. "Opportunities for Bioremediation of Aquifers Contaminated with Petroleum Hydrocarbons," *Journal of Industrial Microbiology.* Vol.27:109-116.

27

DEMONSTRATION OF ENHANCED SOIL BIODEGRADATION OF TPH-CONTAINING SOILS AT A NEW JERSEY ECRA SITE

Paul E. Levine and Donald R. Smallbeck
Harding Lawson Associates
Princeton, NJ

INTRODUCTION

Approximately 3,000 cubic yards of soil at an industrial facility in northern New Jersey were contaminated by an unknown quantity of No. 2 Fuel Oil. Soil contamination resulted from transfer operations adjacent to a 60,000-gallon above-ground storage tank (AST). The feasibility of several remedial alternatives was evaluated on the basis of contamination, the volume of soils, and a comparison of remedial costs. *Ex-situ* bioremediation was selected as the preferred alternative for on-site treatment. The objective of the treatment program was to reduce the total petroleum hydrocarbons (TPH) to below 1,000 milligrams per kilogram (mg/kg) as approved by the State of New Jersey.

Enhanced soil biodegradation (ESB) was an effective method for degrading petroleum hydrocarbons at this site. Pre-remediation studies indicated that indigenous microorganisms capable of degrading the petroleum hydrocarbons were present and, under enhanced growth conditions, would effectively reduce TPH concentrations in the soil. HLA supervised the construction of, and placement of the contaminated soil into a treatment unit. During a 12-week treatment period, initial average TPH concentrations decreased from 3,500 mg/kg to 185 mg/kg, well below the 1,000 mg/kg cleanup level. In addition, during the treatment period, hydrocarbon-utilizing microbial populations increased two to three orders of magnitude, indicating an environment conducive for enhanced microbial growth.

Because of successful treatment of TPH-containing soil by the ESB method, an in-situ biodegradation process is being considered for treatment of toluene-containing soils and groundwater in another location at the site.

This report presents the results of the application of enhanced soil biodegradation (ESB) for the remediation of approximately 2,750 cubic yards of soil containing total petroleum hydrocarbons (TPH) at an ECRA site in northern New Jersey. ESB uses indigenous bacterial populations to degrade and stabilize hydrocarbons in a soil environment.

A 60,000-gallon, above-ground storage tank (AST) was used at the site to store Bunker C (No. 6) and No. 2 fuel oils from 1970 to 1985 (Figure 1). The tank was converted from Bunker C to No. 2 fuel oil in 1972. A spill of No. 2 oil occurred at the time of the conversion. As a result of the spill, a soils investigation was implemented to determine the extent of the total petroleum hydrocarbon (TPH) contamination in the vicinity of the AST. The investigation was conducted by a previous consultant in 1989 and 1990, and was ultimately expanded to include sampling near the AST, an associated oil pipeline, and a 6,000-gallon day tank adjacent to the main plant.

The New Jersey Department of Environmental Protection and Energy (NJDEPE) approved remedial activities in the vicinity of the AST, pipeline, and day tank, including dismantling and disposing of the AST, excavation of soils containing TPH above the cleanup level of 1,000 ppm, and removal and disposal of the pipeline.

Approximately 2,750 cubic yards of soil were incrementally removed. Soil samples were collected at each stage of the excavation until TPH was below the 1,000 ppm cleanup level. Interim and final post-excavation samples were collected from the excavation walls according to NJDEPE protocol. Each sample was analyzed for TPH. The four samples with the highest TPH concentrations were analyzed for Base Neutral Extractables (by EPA Method 8270). A summary of TPH analytical results is presented in Table 1.

All final post-excavation samples were below the TPH cleanup level of 1,000 ppm. There were no base-neutral compounds detected in any of the samples submitted for analysis. On this basis, the excavation of TPH-contaminated soils was deemed complete. Excavated soils were stockpiled on-site at locations indicated on Figure 1. The soils were placed on heavy plastic and covered. Hay bales were placed around each stockpile to anchor the plastic cover.

An initial microbial evaluation concluded that, given proper conditions for microbial metabolism, the indigenous micro-organisms present in the soil could be stimulated to degrade petroleum hydrocarbons (Table 2).

Based on this microbial evaluation, HLA prepared an ESB Work Plan that was approved by the NJDEPE. The objective of the ESB program was to biodegrade TPH present in soil stockpiled from the excavation to below 1,000 milligrams per kilogram (mg/kg). This cleanup objective had been negotiated during earlier phases of the project.

350 Soil Biodegradation

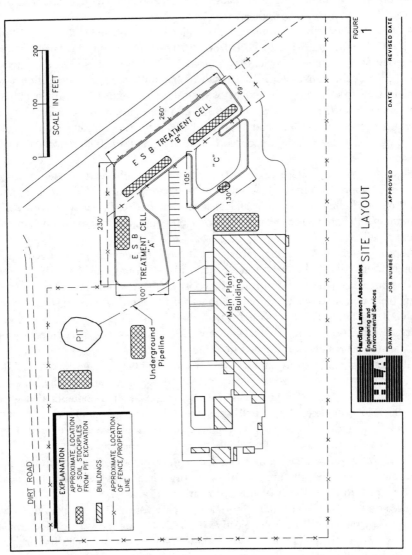

Figure 1 Site layout.

Table 1. Post-Excavation Soil Sampling Results

SAMPLE NUMBER	TPH (ppm)
PE-1	28
PE-2	40
PE-3	27
PE-4	31
PE-5	33
PE-6	35
PE-7	300
PE-8	42
PE-9	<25
PE-10	170
PE-11	290
PE-12	<25

Notes: PE-1   Post-excavation sample location
TPH   Total petroleum hydrocarbons (EPA Method 418.1)

Table 2. Enumeration of Total Petroleum Hydrocarbon Utilizing Microorganisms in Soils

Sample Designation[1]	Total Microorganisms	Petroleum Hydrocarbon-Utilizing Microorganisms (percent of total)
SA-1	5.9×10^6 cfu/gram*	4.9×10^4 cfu/gram* (0.8)
SA-2	2.1×10^6 cfu/gram*	2.7×10^4 cfu/gram* (1.3)
SA-3	6.0×10^6 cfu/gram*	4.9×10^4 cfu/gram* (0.8)
SA-4	3.5×10^6 cfu/gram*	7.9×10^4 cfu/gram* (2.3)
SA-5	3.7×10^6 cfu/gram*	1.3×10^4 cfu/gram* (0.4)
SA-6	8.0×10^5 cfu/gram*	2.2×10^4 cfu/gram* (0.3)

*cfu/gram. Colony-forming units per gram of soil wet weight.

ESB Unit Location and Construction

The ESB unit was constructed at the location shown on Figure 1. This location was selected because it was:

- Level to gently sloping;
- Easily accessible;
- Close to excavated soil stockpiles;
- Removed from areas where additional site cleanup may have impacted operations; and
- Contained sufficient area for the construction of the treatment bed.

Before construction of the ESB treatment cells, the area was leveled and rolled smooth. All sharp objects and large rocks that may have punctured the treatment bed liner were removed.

As part of the treatment cell construction, perimeter containment berms were constructed, using clean compacted on-site soil, to contain the hydrocarbon-containing soil (Figure 2). The minimum height of the berms was 2 feet.

A continuous, 20-mil PVC liner was used to line unpaved areas and keyed into the perimeter berm; paved areas were not lined, since the pavement acted as the barrier. A 2-inch to 6-inch layer of clean imported sand was placed over the paved or lined surface to serve as a buffer zone during tilling operations. The hydrocarbon-containing soil was moved by dump truck to the treatment unit and distributed evenly in lifts of 12 inches, to total depths of approximately 12 to 18 inches.

ESB UNIT OPERATIONS

Operations involved application of an aqueous nutrient solution and periodic aeration of the soil. The nutrient formulation was derived from the initial chemical and microbial evaluations of the stockpiled soils (Table 3).

The soil chemistry results indicated that low concentrations of the inorganic nutrients nitrogen and phosphorous, and a low pH, could limit the metabolism of the existing microorganisms capable of degrading hydrocarbons in the soil environment. Initially the soil pH was adjusted with lime (approximately 2.5 tons total) to a pH range of 6.5 to 7.5, conducive for enhanced microbial growth. The nutrient formulation was an aqueous mixture of ammonium sulfate and sodium and/or potassium phosphates. Application rates per 100 cubic yards of soil were approximately 400 gallons of water mixed with 165 pounds of nitrogen and 50 pounds of phosphate. Process monitoring data collected during the treatment operations was used to determine if adjustments to the content or application rate of the formulation were necessary. The nutrient solution was distributed over the soil twice: once at the beginning of operation, the second application at four weeks.

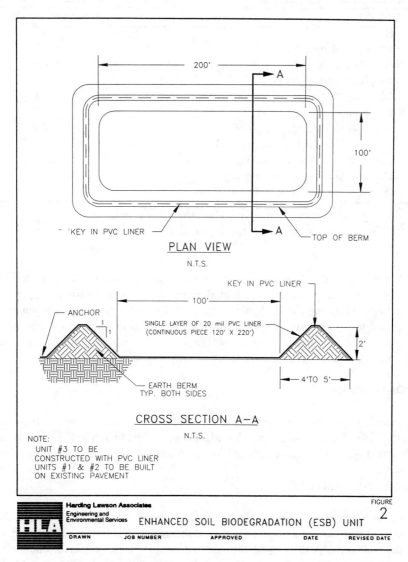

Figure 2 Enhanced soil biodegradation (ESB) unit.

Table 3. Initial Soil Chemistry Results of Stockpiled Soils

Parameter	Sample Results
pH	3.9
Ammonia-N (mg/kg)[1]	2.0
Nitrate-N (mg/kg)	0.1
Orthophosphate (mg/kg)	3.3
Sulfate (mg/kg)	10.0
Water Soluble Iron (mg/lg)	27.7
Water Soluble Manganese (mg/kg)	2.6
Water Soluble Magnesium (mg/kg)	9.7
Water Soluble Calcium (mg/kg)	27.2
Water Soluble Potassium (mg/kg)	3.2
Cation Exchange Capacity (meq/100 gm)[2]	2.5

[1] Milligrams per kilograms-equivalent to parts per million.
[2] Milliequivalents per one hundred grams soil.

The soil was tilled to a depth of 12 to 18 inches twice a week to promote aeration and biodegradation. The underlying sand, by virtue of its color, provided visual control during tilling. Tractor-mounted rotor-tilling and disking equipment was used to till the soil. A sprinkler system was used to water the treatment cell as needed to maintain moisture in the soil. The sampling and operational chronology is presented on Figure 3.

MONITORING

Monitoring events included: baseline sampling, process monitoring, and verification/clearance sampling. A grid was established to divide the ESB unit into 43 cells (Figure 4). Within each cell, samples were collected at random locations.

BASELINE SAMPLING

Baseline sampling followed placement and initial tilling of the soils. Data generated was viewed as representative of the initial distribution of hydrocarbons in the ESB cells. A total of 43 discrete soil samples were collected, one random sample from each cell. Samples were analyzed for TPH using EPA Test Method 418.1. Sample collection and analysis for this and later sampling was performed in accordance with NJDEPE protocol for sample collection and analysis. In addition, 14 composite soil samples were analyzed for heterotrophic and hydrocarbon-utilizing micro-organisms to establish baseline conditions. Soil moisture content and pH were similarly measured.

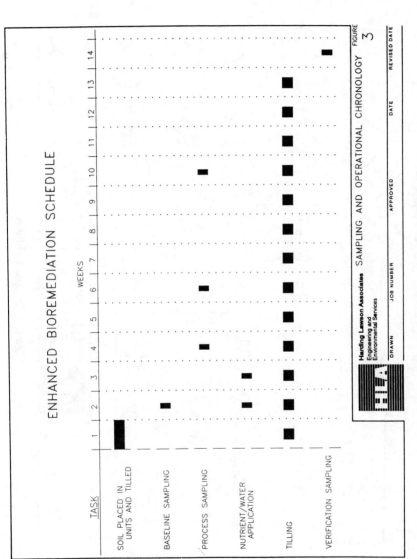

Figure 3 Sampling and operational chronology.

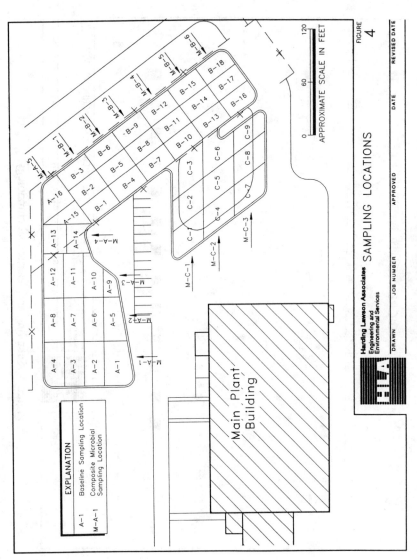

Figure 4 Sample locations.

PROCESS SAMPLING

Process sampling was conducted periodically to assess remediation progress (Figure 3). Two types of process monitoring samples were obtained: samples for TPH analyses; and samples for microbial and nutrient evaluation. Composite soil samples were collected and analyzed for TPH using EPA Method 418.1 during the four process sampling events. In addition, 39 composite soil samples were collected for heterotrophic and hydrocarbon-utilizing micro-organisms.

VERIFICATION/CLEARANCE SAMPLING

Final verification samples were collected from each of the 43 cells to verify that bioremediation reduced soil TPH to less than 1,000 mg/kg. Two additional samples were collected for QA/QC purposes. Field-prepared soil duplicates were collected as part of the verification/clearance sampling. Analytical data are summarized in Tables 4 and 5.

CONCLUSIONS

Significant conclusions are as follow:

- The microbial evaluation indicated that the initial microbial population contained a subpopulation of hydrocarbon-utilizing microorganisms.
- The soil chemistry results indicated that low concentrations of the inorganic nutrients nitrogen and phosphorous (and a low pH) could limit the metabolism of the existing microorganisms capable of degrading hydrocarbons in the soil environment. This limitation was overcome by addition of a nutrient formula.
- Consistent with the observed decreases in TPH, microbial data indicated that the addition of the nutrients, periodic aeration, and soil moisture maintenance enhanced the growth of hydrocarbon-utilizing microorganisms producing a decrease in TPH concentrations.
- Verification analysis confirmed that TPH in soil samples collected from each of the 43 cells is below the cleanup level of 1,000 ppm. Treatment of the hydrocarbon containing soils took 11 weeks to achieve concentration of less than 1,000 ppm of TPH. Average concentrations decreased from an initial 3,537 ppm to 185 ppm.

Based on these observations, it is proposed that the soil contained in the treatment unit be used to backfill the original AST excavation. Backfilling would include placing and compacting the soil in no greater than 12-inch lifts to 90 percent of its maximum dry density, as determined by the Standard Proctor Test. The excavation should be restored to original grade and seeded for erosion control.

358 Soil Biodegradation

Table 4. Enhanced Soil Degradation Data Total Petroleum Hydrocarbons Analysis

Date Sampled Sample Location	Base Line Samples 07/26/91 (ppm)	Date Sampled Sample Location	Process Samples (1) 08/15/91 (ppm)	08/27/91 (ppm)	09/23/91 (ppm)	10/09/91 (ppm)	Verification Samples 10/22/91 (ppm)
A1	3,200	A1	1,100	280	250	NS	110
A2	840						48
A3	3,400						130
A4	1,200						370
A5	1,200	A2	900	620	450	NS	290
A6	2,600						220
A7	3,000						530
A8	2,700						360
A9	3,100	A3	870	890	340	NS	220
A10	1,000						160
A11	860						72
A12	1,100						38
A13	1,100	A4	820	170	120	NS	180
A14	3,200						60
A15	590						170
A16	1,100	A5	720	91	180	NS	290

Notes:
ppm = parts per million
NS = not sampled
ND = not detected
(1) Process samples are composite samples as follows:

A1 = Composites of A1 – A4
A2 = Composites of A5 – A8
A3 = Composites of A9 – A12
A4 = Composites of A13 – A14
A5 = Composites of A15 – A16

B1 = Composites of B1 – B3
B2 = Composites of B4 – B6
B3 = Composites of B7 – B9
B4 = Composites of B10 – B12
B5 = Composites of B13 – B15
B6 = Composites of B16 – B18

C1 = Composites of C1 – C3
C2 = Composites of C4 – C6
C3 = Composites of C7 – C9
FB = Field Blank

Table 4. Enhanced Soil Degradation Data Total Petroleum Hydrocarbons Analysis (Continued)

Date Sampled Sample Location	Base Line Samples 07/26/91 (ppm)	Date Sampled Sample Location	Process Samples (1) 08/15/91 (ppm)	08/27/91 (ppm)	09/23/91 (ppm)	10/09/91 (ppm)	Verification Samples 10/22/91 (ppm)
B1	1,800	B1	1,800	63	300	NS	260
B2	760						93
B3	2,600						150
B4	3,300	B2	3,000	220	220	NS	250
B5	4,100						100
B6	8,400						130
B7	3,900	B3	3,300	1,200	2,200	390	88
B8	4,100						87
B9	3,400						50
B10	2,400	B4	2,300	620	340	300	120
B11	5,300						130
B12	5,300						140
B13	17,000	B5	1,600	470	200	220	75
B14	15,000						87
B15	15,000						130

Notes:
ppm = parts per million
NS = not sampled
ND = not detected
(1) Process samples are composite samples as follows:

A1 = Composites of A1 – A4
A2 = Composites of A5 – A8
A3 = Composites of A9 – A12
A4 = Composites of A13 – A14
A5 = Composites of A15 – A16

B1 = Composites of B1 – B3
B2 = Composites of B4 – B6
B3 = Composites of B7 – B9
B4 = Composites of B10 – B12
B5 = Composites of B13 – B15
B6 = Composites of B16 – B18

C1 = Composites of C1 – C3
C2 = Composites of C4 – C6
C3 = Composites of C7 – C9
FB = Field Blank

360 Soil Biodegradation

Table 4. Enhanced Soil Degradation Data Total Petroleum Hydrocarbons Analysis (Continued)

Date Sampled Sample Location	Base Line Samples 07/26/91 (ppm)	Process Samples (1)					Verification Samples 10/22/91 (ppm)
		Date Sampled Sample Location	08/15/91 (ppm)	08/27/91 (ppm)	09/23/91 (ppm)	10/09/91 (ppm)	
B16	2,500						230
B17	4,800	B6	2,600	280	390	NS	74
B18	1,500						250
C1	2,100						540
C2	2,600	C1	1,400	460	160	NS	160
C3	2,200						160
C4	1,300						260
C5	820	C2	820	500	260	NS	560
C6	2,300						260
C7	4,200						48
C8	3,800	C3	820	450	160	NS	170
C9	1,400						320
FB	NS	NS	NS	NS	NS	NS	ND

Notes:
ppm = parts per million
NS = not sampled
ND = not detected
(1) Process samples are composite samples as follows:

A1 = Composites of A1 – A4
A2 = Composites of A5 – A8
A3 = Composites of A9 – A12
A4 = Composites of A13 – A14
A5 = Composites of A15 – A16

B1 = Composites of B1 – B3
B2 = Composites of B4 – B6
B3 = Composites of B7 – B9
B4 = Composites of B10 – B12
B5 = Composites of B13 – B15
B6 = Composites of B16 – B18

C1 = Composites of C1 – C3
C2 = Composites of C4 – C6
C3 = Composites of C7 – C9
FB = Field Blank

Table 5. Microbial Results Enhanced Soil Biodegradation

Sample Designations	July 29, 1991 Results			August 15, 1991 Results			August 27, 1991 Results	
	Total Count CFU/Grams*	Hydrocarbon Utilizers CFU/Grams*	pH	Total Count CFU/Grams	Hydrocarbon Utilizers CFU/Grams	pH	Total Count CFU/Grams	Hydrocarbon Utilizers CFU/Grams
A-1	5.0×10^6	1.1×10^5	5.5	6.9×10^7	1.6×10^7	6.6	2.8×10^8	2.2×10^8
A-2	1.2×10^7	7.9×10^5	5.8	1.6×10^8	2.4×10^7	6.6	5.4×10^8	1.7×10^8
A-3	2.3×10^7	4.9×10^5	6.2	1.1×10^8	9.2×10^6	6.6	2.3×10^8	2.3×10^8
A-4	1.2×10^7	7.9×10^5	6.8	1.1×10^8	1.6×10^7	6.5	1.7×10^8	1.7×10^8
A-5	6.3×10^6	1.4×10^4	5.7	9.0×10^7	3.5×10^6	6.4	4.5×10^8	4.5×10^8
B-1	2.9×10^7	1.4×10^5	6.0	1.3×10^8	9.2×10^6	6.6	5.8×10^8	2.4×10^8
B-2	5.0×10^6	1.1×10^5	5.4	1.4×10^8	2.4×10^7	6.7	4.8×10^8	4.8×10^8
B-3	8.7×10^6	7.9×10^4	5.9	1.8×10^8	2.4×10^7	6.7	5.8×10^8	4.3×10^8
B-4	2.0×10^7	2.8×10^5	5.4	1.4×10^8	1.6×10^7	6.8	4.8×10^8	2.3×10^7
B-5	7.8×10^6	1.7×10^5	5.6	1.7×10^8	1.6×10^7	6.8	4.5×10^8	4.9×10^7
B-6	3.2×10^6	7.9×10^4	6.4	7.6×10^7	5.4×10^6	6.7	2.3×10^8	2.2×10^7
C-1	1.5×10^7	3.5×10^5	6.7	1.8×10^8	2.4×10^7	7.1	3.5×10^8	7.0×10^6
C-2	5.9×10^6	2.3×10^4	6.1	1.0×10^8	2.4×10^7	6.9	3.4×10^8	3.3×10^7
C-3	5.8×10^6	3.3×10^4	6.8	7.1×10^7	1.6×10^7	6.6	2.6×10^8	3.3×10^6

Matrix: Soil

Analysis	Method
Total Plate Count	Spread plate method 30°C Plate Count Adar Methods of Soil Analysis. 1965. Agronomy No. 9 pp. 1460–1466
Hydrocarbon Utilizer Count	Most Probable Number Method 30° C/2 weeks. Basal Media and diesel. Canadian Journal of Microbiology 1978: 24 pp 552–557

* Colony forming unit/gram of soil wet weight

REFERENCES

1. NJDEPE, *Field Sampling Procedure Manual*, February 1988.

2. "Spread Plate Method 30° hours. Plate Count Adar. Methods of Soil Analysis," 1965. *Agronomy* No. 9, pp. 1460-1466.

3. "Most Probable Number Method 30° C/2 weeks." Basal Media and Diesel. *Canadian Journal of Microbiology* 1978, 24, pp. 552-557.

28

SUCCESSFUL IMPLEMENTATION OF CONTROLLED AEROBIC BIOREMEDIATION TECHNOLOGY AT HYDROCARBON CONTAMINATED SITES IN THE STATE OF DELAWARE

C. Darrel Harmon
WIK Associates, Inc.
New Castle, DE

Anne V. Hiller
Delaware Department of Natural Resources
and Environmental Control
New Castle, DE

Judith B. Carberry, Ph.D.
Department of Civil Engineering
University of Delaware
Newark, DE

INTRODUCTION

WIK Associates, Inc. of New Castle, Delaware, has been working over the last two years to improve and advance a cost effective method of treating hydrocarbon contaminated soils which avoids common problems such as contaminated leachate collection, uncontrolled air emissions and projects postponed due to low winter temperatures. This process, known as controlled aerobic bioremediation, is cost competitive, performed on site and allows the soil to be reused as clean fill by the client.

The first section of this paper describes treatment methods and associated benefits such as increased control over environmental parameters. The second part of this paper describes work performed in attempting to predict degradation rates for varying types of hydrocarbon contamination under varying conditions. This research is based on data gathered in performing on-site bioremediation as described.

A third section included in this paper describes the unique perspective of a State regulator responsible for overseeing remediation efforts evolving from leaking underground storage tanks. This section describes regulatory issues and procedures in Delaware and how the Department handles the submission and implementation of corrective action work plans, through project closure with thorough documentation of the remediation.

THE CONTROLLED AEROBIC BIOREMEDIATION PROCESS

Sources of Contamination

Hydrocarbon contaminated soils are a common problem most often associated with leaking underground storage tanks. While some contaminated soils are generated by spills and accidents, the majority are the result of tanks leaking over a long period of time. These tanks are most commonly used to store four types of hydrocarbons: gasoline, heating oil, fuel oil and waste oil. Upon the removal of these tanks, varying amounts of contaminated soil are often found.

Construction

In preparation for treatment, a site is chosen, marked out and covered with a sheet of polyethylene. This poly sheet forms an impermeable liner used as the bottom of the containment structure. The contaminated soil is loaded onto the plastic and covered to prevent emissions into the atmosphere. Over the soil pile, a portable greenhouse structure is constructed and covered with ultraviolet stabilized poly sheeting. The ends of the structure are fully enclosed, incorporating a large door for access with tilling equipment and provisions for a large fan which is used for ventilation. A carbon drum filter can be attached directly to the ventilation fan if required. Once fully enclosed, the soil is uncovered and the edges of the bottom liner lifted and attached to the inside of the structure. This forms a complete containment system which avoids a number of problems, as discussed later. Being fully enclosed and contained, the soil is now ready for treatment to begin.

Initiating Treatment

Once the structure is enclosed, the soil is spread to an even depth, rocks and debris removed and treatment is initiated. The first step in bioremediation is to loosen and aerate the contaminated soil while adding a solution of nutrients. The solution is primarily water with nitrogen and phosphate, sometimes supplemented with cultured bacteria proven able to degrade hydrocarbons. The specific mixture of the solution is determined by soil tests performed on composite samples collected earlier. If necessary, pH is also adjusted through the addition and incorporation of lime. This allows balancing of the nutrients added with those already present to provide ideal conditions for bioremediation to occur.

Maintenance and Monitoring

As treatment continues, a number of parameters must be monitored to maintain ideal conditions for continuous bioremediation. Regular inspections are carried out to monitor the important factors of air emissions, aeration, moisture and temperature.

Air emissions of benzene, toluene, ethyl benzene and xylenes (total BTEX) from gasoline contaminants are often a concern. A hand held photo-ionization detector (PID) is calibrated to isobutylene and used to measure volatile organic compounds, reading in parts per million equivalents. The readings are reported to State regulators on a regular basis. At their request, a carbon drum filter can be attached to the ventilation fan to provide total filtering of air from inside the structure, resulting in zero levels of air emissions.

Aeration is performed on a regular basis to avoid the development of anaerobic conditions which would impede the degradation of hydrocarbon compounds. This aeration is performed with an agricultural tractor fitted with a scarifier. This implement is dragged through the soil, turning and loosening it, allowing oxygen to come in contact with the pore spaces within the soil. Depending on the level of contamination and environmental factors such as moisture level, the soil may be aerated at intervals from twice per week to once every second week.

Moisture control is also greatly enhanced through use of this approach. The enclosed structure eliminates wind effects and allows improved control over the soils' moisture level. Monitoring is carried out with a tensiometer for measuring soil moisture. The meter is read on a scale of 0 to 100 indicating a scale between no moisture and saturation. Soil moisture is brought up to 60 to 70% of saturation and maintained there throughout the treatment. Due to the minimized evaporation the soil dries less quickly than if outside. Because the soil is isolated and enclosed, precipitation is completely controlled and saturation is not a possibility. The need to collect and recycle or dispose of leachate is thereby eliminated. Another benefit of precise moisture control is that it forestalls the development of anaerobic conditions, which allows the project to be carried out in a shorter time frame.

Temperature is also of significance, particularly in winter. The enclosed transparent structure allows maximum thermal loading to occur from the sun. The soil serves as an excellent heat sink, storing the energy absorbed. As a result, the soil temperature is maintained well above the ambient temperature. In marginal climates, this difference allows treatment to continue uninterrupted throughout the winter.

RESULTS

Controlled aerobic bioremediation results from seven sites with differing petroleum contamination are reported below.

The seven sites chosen for illustration included three classifications of molecular weight contaminants. Light weight contaminants measured as total BTEX indicate gasoline contamination. Medium weight contaminants consisting of fuel oil, diesel, etc., are measured as total petroleum hydrocarbons (TPH). Heavy asphaltic tar, oil or coal tar contamination represents a high molecular weight petroleum, and is also measured as TPH.

The sites were remediated during both winter and summer weather conditions in order to determine if the controlled aerobic bioremediation process might be affected by the following seasonal temperature conditions: retardation of microbial degradation rates under cold ambient air or soil temperatures near the freezing point of water, and enhanced volatilization of light weight fractions requiring controlled hydrocarbon vapor removal from the greenhouse structure during hot weather. A summary of site conditions is presented in Table 1.

Table 1. Summary of Remediation Conditions

Site	Contaminant	Molecular Weight Fraction	Contaminated Soil Volume
1	fuel oil/gasoline	medium/light	200 yd^3
2	fuel oil	medium	450 yd^3
3	asphaltic oil	heavy	70 yd^3
4	diesel fuel/gasoline	medium/light	250 yd^3
5	gasoline	light	250 yd^3
6	gasoline	light	50 yd^3
7	fuel oil	medium	300 yd^3

For each site, an initial TPH or total BTEX reading was determined by a certified, independent laboratory. Appropriate ambient air temperatures were maintained during hot weather by periodic venting with the fan at the closed end of the greenhouse. Adequate moisture was provided by intermittent sprinkling and re-scarifying with a tractor when measurements indicated the need. Periodic integrated soil samples were collected from each pile and analyzed for TPH and/or total BTEX concentrations. The frequency of hydrocarbon monitoring was determined by the estimated molecular weight fraction of the known contaminant. At least three soil samples were collected at appropriate intervals from the pile and analyzed. From these results, a first order plot was made and a biodegradation rate constant determined from the slope of that line. An example from Site 2 is illustrated in Figure 1. Comparable degradation rate constants from three sites are presented in Table 2.

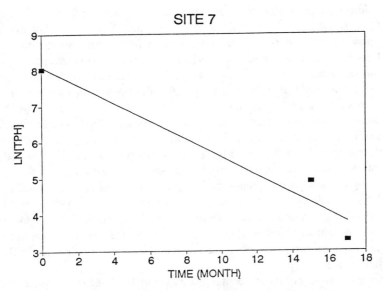

Figure 1 Plot of log total TPH vs. time for first order rate constant determination.

Table 2. Biodegradation Rate Constants

Site	Rate Constant Week^{-1}	Molecular Weight Range	Regression Coefficient
1 (BTEX)	0.20	light	0.51
(TPH)	0.04	medium	0.90
2 (TPH)	0.07	medium	0.87
3 (TPH)	0.007	heavy	0.93
4 (TPH)	0.08	medium	0.70
5 (TPH)	0.05	medium	0.93
6 (TPH)	0.04	medium	0.92
7 (TPH)	0.06	medium	0.95

Regression analyses on data from early sites indicated that no satisfactory correlation existed between the biodegradation rate constants presented in Table 2 and the average soil temperature or the soil particle grain size. For example, medium fuel oil contaminated both Sites 1 and 2. Despite the fact that the average temperature at Site 1 was more than double that of Site 2, the biodegradation rate constant for Site 1 was approximately half that of Site 2.

In addition, analyses demonstrated that approximately 65% of each soil was composed of particles with a diameter of less than 425 μm. As all grain size determinations were very similar, any variation in biodegradation rate constant could not be attributed to that characteristic of the site. Grain size analysis of soil particles at subsequent sites was eliminated. Attempts to further correlate biodegradation rate constants with temperatures in the soil or ambient air inside the greenhouses were not carried out.

The variation in molecular weight, however, demonstrated the best correlation with variation in biodegradation rate constants. If average weights are assigned to the different petroleum fraction, Figure 2 then illustrates the results obtained from correlation of molecular weight with biodegradation rate constant.

Although the semi-log regression appears to provide a favorable linear correlation, resulting in a correlation coefficient of 0.97, such a result may prove to be invalid when more cases are evaluated. Until then, this correlation of molecular weight and biodegradation rate will be utilized in order to estimate the time required to carry out controlled aerobic landfarming and to verify this correlation, if possible.

No vapor phase petroleum contaminants could be detected at concentrations greater than 10 ppm in the enclosed greenhouse atmosphere at any time under any ambient air temperature. Under the conditions provided by the controlled remediation process described, therefore, the volatilization of petroleum components must have been minimized. Since the downward percolation of petroleum components has been prohibited by the impervious liner, any measured decrease in the petroleum contamination concentration must be due to the optimized natural microbial biodegradation of these contaminants.

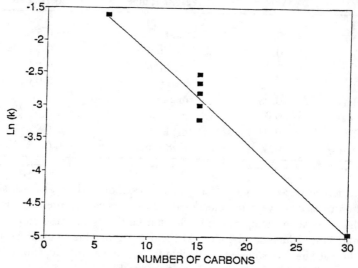

Figure 2 Correlation of log K vs. molecular weight of hydrocarbon contaminant.

Although the correlation of aerobic biodegradation rate constants with estimated molecular weight ranges is quite acceptable, deviations from the predicted values would be expected within any molecular weight range. Such deviation from predicted rates could be due to the type or number of indigenous bacteria present in the soil at any contaminated site. The authors have not investigated possible microbial variations at each site.

REGULATORY VIEWS ON BIOREMEDIATION AT L.U.S.T. SITES

Introduction

For more than five years the Delaware Department of Natural Resources and Environmental Control Underground Storage Tank Branch has approved on-site bioremediation of petroleum contaminated soils resulting from leaking underground storage tanks. Interest in and use of this remediation technique significantly increased in 1990 when restrictions were placed upon the disposal of petroleum contaminated soils at landfills within the State. Independently owned service stations and other small businesses began to find off-site treatment or disposal methods prohibitively expensive. However, these generators of contaminated soil often have the space and the time needed for on-site surface bioremediation.

Most of the on-site bioremediation projects approved by the Department have involved remediation of excavated piles of petroleum contaminated soil on the surface. Overexcavation of small areas of petroleum contaminated soil and subsequent bioremediation of the soil is a popular low cost remediation technique in Delaware. Often the initial work on this type of project can be combined with underground storage tank removal activities at a site. However, responsible parties who do not have the needed space on-site, who cannot disturb the site, or who have too much soil to economically excavate are increasingly interested in *in-situ* bioremediation of petroleum contaminated soil.

The Department encourages the use of innovative remediation techniques for leaking underground storage tank sites and is continually working to enable the use of these techniques while at the same time ensuring that effective remediation of the site occurs. Bioremediation fits well into the Department's remediation approach by providing an effective remedy for petroleum contamination in soils that is economically attractive to responsible parties, especially those with limited funds.

Regulatory Perspectives

The Department views the use of bioremediation in corrective action at leaking underground storage tank sites in the same way as it views the use of soil vapor extraction or other techniques. The technique may be combined with other corrective action options at a site.

370 Aerobic Bioremediation

All types of bioremediation corrective action at leaking underground storage tank sites are approved and overseen by the Underground Storage Tank Branch of the Department. The most common type of bioremediation currently being used in Delaware is on-site surficial treatment of excavated petroleum contaminated soil. Off-site surficial bioremediation is also possible. If the responsible party owns the off-site location then the only additional requirement is that the soil be moved by a transporter with a State waste hauler's permit. If the off-site location is owned by another party, then the second party must provide documentation indicating its acceptance of responsibility for the remediation of the soil and the soil must be moved by a transporter with a State waste hauler's permit.

The decision to use bioremediation for corrective action at leaking underground storage tank sites becomes a balance between time and money for most responsible parties.

The advantages of bioremediation of small quantities of petroleum contaminated soil often include less expensive costs than those for available off-site treatment or disposal methods. Also, risks due to transportation of the material to the off-site facility are eliminated. Often the soil can be reused at the site where it was contaminated. However, there are some disadvantages associated with using bioremediation. The responsible party must have the necessary time and space available for the project. Regular maintenance and monitoring is required. As with many innovative techniques, regulatory concerns may exist regarding the proposed methodology. But in Delaware, especially for responsible parties with small quantities of petroleum contaminated soils, the advantages frequently outweigh the disadvantages.

Responsible parties with large amounts of excavated petroleum contaminated soil have also used bioremediation as the least cost alternative. If large amounts of petroleum contaminated soils are present at the site, however, the most prudent course of action is often to treat it in place. *In-situ* bioremediation then becomes the preferred alternative provided that certain treatability parameters are amenable to successful treatment.

Interest in *in-situ* bioremediation is growing throughout Delaware. However, due to Delaware's widespread reliance on groundwater for water supply, regulatory concerns about *in-situ* bioremediation do exist. The Department is developing generic specifications for infiltration galleries which will help to alleviate some of these concerns.

Bioremediation Corrective Action Work Plans

In all cases a Corrective Action Work Plan must be prepared and approved by the Department before work begins at the site. Work Plans must provide detailed information including a description of the site, proposed methodology for bioremediation, schedules for soil sampling and monitoring, compliance with applicable local, municipal, county, state, and federal laws and regulations, methods to be employed to avoid potentially harmful effects on human health, safety and the environment, proposed goals for soil clean up, and proposed soil reuse options. In 1990, the Department developed a Technical Guidance Manual for underground storage tank sites which provides guidelines on the content of such work plans.

Additional site-specific guidance on the preparation of Corrective Action Work Plans for bioremediation is available from the Department. The Department works with the responsible party and its consultants and contractors to develop an appropriate and approvable Corrective Action Work Plan. Assistance coordinating the proposed work with other Branches in the Department, such as the Hazardous Waste Management Branch or the CERCLA Section, is available if necessary.

Many successful bioremediation work plans are technologically simple. Complex pilot tests are not usually needed because of the small size of most leaking underground storage tank sites in Delaware. At a minimum, some concrete indication that the project will be successful must be present in the workplan. Previous successful use of the proposed methodology is important.

Implementation of Bioremediation Corrective Action

Once the bioremediation corrective action workplan is approved, work may begin at the site. Department representatives perform regulatory oversight activities while the work is being performed.

The Department sets clean-up goals for the petroleum contaminated soil and the site based upon individual site characteristics and soil reuse plans. Individual site characteristics include depth to groundwater, land use, distance to surface water bodies, distance to water supply wells, and other criteria. Soil reuse plans must be determined before bioremediation begins since cleanup goals for the soil may vary from those for the site if the soil is to be reused on another piece of property the responsible party owns.

Site sampling and monitoring schedules may vary depending upon the bioremediation methodology used, the season of the year, and other factors. However, the Department requires enough data to demonstrate that bioremediation of the soil did occur and was successfully completed.

After successful completion of the bioremediation of the petroleum contaminated soil, the soil is often used as clean fill on the site where it was contaminated. Occasionally the soil may be used as clean fill on another piece of property the responsible party owns or for some other purpose, but only with advance Department approval.

The Department will inactivate the leaking underground storage tank site when a closure request with all supporting documentation is submitted and approved. If bioremediation of excavated soils was combined with other *in-situ* remediation techniques at the site, the site will not be inactivated until all remediation is complete.

Summary

Bioremediation of petroleum contaminated soil is an effective remediation technique for remediation of leaking underground storage tank sites. Bioremediation in Delaware is a relatively new technology, but one with a high rate of success.

CONCLUSIONS

WIK Associates, Inc. has developed this technique of controlled aerobic bioremediation at a number of sites in the Delaware Valley. The process has proven to be both effective and economic, meeting the needs of all concerned parties. Laboratory testing and careful monitoring of the discussed environmental parameters have shown that the objectives are being achieved. Levels of contamination measured as total petroleum hydrocarbons and total BTEX are being reduced by orders of magnitude in treatment periods on the order of months.

On site controlled aerobic bioremediation of petroleum contaminated soil can be carried out safely, efficiently and economically in enclosed greenhouses.

Downward percolation of contaminants was prevented by providing an impervious liner under the soil pile. No evidence of significant volatilization of contaminants could be detected, under hot weather conditions.

Proper conditions of soil moisture and nutrient content must be provided under aerobic conditions, but neither fine soil grain size nor low ambient or soil temperatures appear to predictably affect the biodegradation rate under the observed conditions.

The correlation of biodegradation rate appears to be significant with molecular weight of the petroleum contaminants. Deviation from this correlation would most probably be due to variations in concentrations or species types of indigenous soil microbes present at each site.

Based on the interests they represent, the State of Delaware has found bioremediation to be an effective method for remediation of hydrocarbon contaminated soils. By remaining open to innovative technologies, the State has encouraged the development of successful remediation strategies, providing benefits to both the generators and the overflowing landfills in the state. Careful monitoring of projects and work plans have allowed this technology to improve and enjoy a high success rate in Delaware.

29

FULL SCALE UPFLOW ANAEROBIC SLUDGE BED PROCESS FOR INDUSTRIAL WASTEWATER TREATMENT

Gwo-Dong Roam
Bureau of Environmental Sanitation and
Toxic Substances Control, EPA
Taiwan, R.O.C.

Hsin Shao
Union Chemical Lab, Industrial Technology Research Institute
Taiwan, R.O.C.

INTRODUCTION

Great progress has been made in the last decade at the Union Chemical Laboratories, ITRI, in the development of anaerobic techniques for industrial wastewater pollution control. The thirty full scale Upflow Anaerobic Sludge Bed (UASB) reactors with the capacity of 3500 m^3 were employed to treat wastewater from fermentation, brewery and chemical industries. The start-up experience and long-term performance results obtained from these reactors will be presented.

During the last ten years, a series of research and development has been done for the modified UASB process. The performance of this process has been verified by bench as well as pilot scale studies with various kinds of wastewater. Table 1 is a summary of the performance of pilot scale UASB reactors obtained from 1982 to 1985. After considerable bench / pilot scale studies have been conducted, a 85 m^3 UASB was established to treat wastewater of citric acid fermentation process, in 1986 and more than 30 UASB reactors with total capacity of 3000 m^3 were constructed and operated during 1988-1989 at six winery plants of Taiwan Tobaces & Wine Monopoly Bureau. Another 600 m^3 UASB reactor was constructed for treating bottom residue of a solvent distillation unit from an adhesive tape manufacturing plant. Table 2 is a summary of the performance of

Table 1. Summary of Performance of Pilot Scale UASB Reactor with Various Kinds of Wastewater

Wastewater Origin	Period of Experiment	Scale of Reactor (m^3)	COD (inf) (mg/l)	COD Loading (kg COD/m$^3 \cdot$ day)	COD Removal (%)	HRT (hr)	Inhibitory Compound	Reference
Brewery:								
Rice Wine	1982.7-1983.7	0.1-0.2	3000-24000 <20000>	10-30 <20>	65-85 <80>	18-24 <24>		Roam, et al., 1983
Shaw Hsing	1982.7-1983.7	0.1-0.2	2680-6960 <5700>	3.6-10 <5700>	85-94 <90>	12-24 <18>		Roam, et al., 1983
Kao Liang	1985.8-1986.3	1.0	8000-20000 <12000>	8-30 <14>	70-90 <80>	12-24 <20>		Roam, et al., 1986
Fruit Wine	1985.1-1985.6	1.0	6000-15000 <8000>	6-15 <12>	70-90 <80>	12-30 <24>		Roam, et., 1986
Antibiotics	1985.1-1985.6	3.0	5000-8000 <8000>	5-10.5 <8>	55-70 <60>	12-24 <24>	CTC OTC	Tzou, et al., 1985
Mosodium glutamate	1984.4-1985.2	0.1-0.2	8000-30000 <15000>	10-22 <15>	55-70 <60>	12-24 <24>	NH$_3$	Yang, et al., 1985

(a) The data in < > refers to the recommended balue for reactor design
(b) CTC: Chlortetracycline
(c) OTC: Oxytetracycline

Table 2. Summary of Performance of Commercialized Full Scale UASB Process

Wastewater Origin	Date of Start-Up	Scale of Reactor (m^3)	COD (inf) (mg/l)	COD Loading (kg COD/m$^3 \cdot$ day)	COD Removal (%)	HRT (hr)	Inhibitory Compound	Reference
Citric Acid	1986	85	18000-27000	20	90-96	27		Roam, 1986
Winery Plant:								
Grape and Brandy wine	1988	440 (88 x 5 set)	11000	10	82	26		Cheng, et al., 1990
Sorghum Wine	1988	889 (127 x 7 set)	1500-2500	2	85	24		Cheng, et al., 1990
Rice Wine	1988	1397 (127 x 11 set)	1000-35000	15	95	40		Cheng, et al., 1990
Shaw Hsing	1988	508 (127 x 4 set)	2000-4000	3	95	16		
Adhesive Tapes Manufacturing Plant	1988	600	5000-12000	2	99	90	Toluene Benzene	

commercialized full scale UASB process has obtained since 1986 to present. Table 1. Summary of the performance of the pilot scale UASB reactor with various kinds of wastewater.

The objectives of this paper are to investigate the process performance in short-term (start-up period) and long-term (1300 days operation) periods, and to compare the operation loading of the UASB process in different kinds of wastewater.

MATERIAL AND METHODS

Characteristics of Citric Acid, Winery & Adhesive Tapes Wastewater

The wastewater from citric acid plant contain several volatile organic acids with high concentration, such as acetic acid, propionic acid and butyric acid. There is no special toxic compound in the stream. It has been found, however, that citric acid at a concentration of 50m mol/l will cause inhibition (Koepp, et al., 1983.)

Four winery plants are located north, center and south of Taiwan. Lon-Ten plant manufactures Kao-Liang wine through fermentation and distillation of sorghum. Nan-Tou plant produces grape wine and brandy through fermentation and distillation, whereas both Su-Lin plant and PingTon plant ferment and distill rice to make rice wine. All winery plants produce wastewater with high concentrations of organics and suspended solids. After centrifugal separation, the COD, BOD and SS in the distilled wastewater are still higher than 30,000, 25,000 and 3,000 mg/l respectively. The pH of wastewater is relatively low and in the range of 4 to 6. More than three times dilution with treated water is necessary before the wastewaters are discharged into the UASB process. A big equalization tank provides both functions of dilution and acidification. However, the acidified wastewater is neutralized again with $NaHCO_3$ before being fed into the UASB reactors.

A chemical plant manufacturing adhesive tape is located north of Taiwan. Process wastewater mainly from solvent recovery distillation unit contains high concentration of residual organic solvent, such as ethyl acetate, acetic acid, ethyl alcohol, toluene and benzene. The characteristics of wastewater are as follows. According to the results analyzed by UV spectrophotometer, the highest concentration of toluene is 943 mg/l, 94% of the sample analyzed is below 180 mg/l, 60% of the sample is between 80~120 mg/l.

```
COD (mg/l)   : 5000~10000 (7745±1965, n=534)
pH           : 3.0
Temp.(°C)    : 70
toluene(mg/l): 80943 (146+201, n=17)
```

Upflow Anaerobic Sludge Bed (UASB) Reactors

Three to seven UASB reactors of identical size are employed in the previously description four winery plants. The steel-plate reactors are constructed with diameters of six meters and depth of six meters. The effective volume of liquid in each reactor is 127m^3. Nine sampling ports with 0.5 meter intervals are installed along the depth of each reactor. About one-third of the depth contains high concentrations of anaerobic sludge. The upflow stream passes through a plastic filter to separate gas and solid from liquid. A side armed settler is installed to provide secondary filtration and sedimentation of the effluent. The settled sludge will return to the sludge bed by gravity. The UASB bioreactors are operated with hydraulic retention time (HRT) of 20 to 30 hours depending on different loadings. The designed loading is 8.3 kg BOD/m^3 · day, equivalent to 15 kg COD/m^3 · day. So the gas production rate will be 500 m^3/day. Figure 1 illustrates the wastewater treatment plant including a 600 m^3 UASB reactor and aerobic polishing unit. The reinforced concrete reactor is constructed with a diameter of ten meters and depth of eight meters.

RESULTS AND DISCUSSION

Case I. Wastewater from manufacture of citric acid

Influent: COD : 18,000~27,000 mg/l
BOD$_5$: 15,00020,000 mg/l
volatile acid: > 10,000 mg/l
pH(adjusted): 6.8~7.5
citric acid: 2~25 mmol/l
acetic acid: 110~160 mmol/l
propionic acid: 30~70 mmol/l
butyric acid: 20~50 mmol/l
glucose: 0.3~23 mmol/l
fluctose: 0.2~3 mmol/l

Design basis: flow rate: 80~100 CMD
UASB reactor: 85 m^3 (5m φ × 6mH)
loading rate: 20kg COD/m^3 · day
0.4~0.7 kgCH$_4$ -COD/kg VSS.day

Performance: effluent COD: 1000~3000 mg/l
COD removal: 90~96 %
sludge yield: 0.06 kg VSS/kg COD removal
biogas yield: 0.5 m^3 /kg COD removal
CH$_4$/CO$_2$: 0.60~0.65/0.3~0.25

Figure 1. A view of wastewater treatment facilities in an adhesive tape manufacturing plant. The biggest cylinder-type reactor is UASB unit with effective volume of 600 m^3.

Start-up Considerations

A reactor start-up is defined as the initial operating procedure of a reactor seeded with the unadapted seed culture. A practical definition of the UASB start-up time would be the time required to meet the design criteria. For more empirical or experimental meaning, the start-up period is defined as the time required for development of the macroscopic sludge granules to about 1mm in diameter. For comparison with the start-up period in different studies, the time required to reach a methane production rate of 10 kg CH_4-COD/m^3. day is recommended and used as a definition of the startup period by W.J. de Zeeuw, 1984.

In this study, the start-up period is discussed on the basis of de Zeeuw's definition. Figure 2 shows that the start-up period of the commercialized full-scale UASB system at San Fu Chemical Co. is about 50 days, while the biological activity is controlled in the range of 0.4~0.7 kg CH_4-COD/kg VSS. day to prevent the reactor from overloading during the start-up period.

As shown in Figure 3, a very close relationship between methane production and volatile acid loading is revealed. After 35 days of operation, the volatile acid loading rate starts to increase sharply from 1.7 kg VA/m^3. day to 9.4 kg VA/m^3. day, and the system performance responds well. This sharp increase after 35 days operation can be explained by the acclimation of the microbes in the reactor. As shown in Figure 4, a breakpoint of the curve of the amount of accumulated biosludge vs. acclimation time occurred at the 35th operation day. Before that .

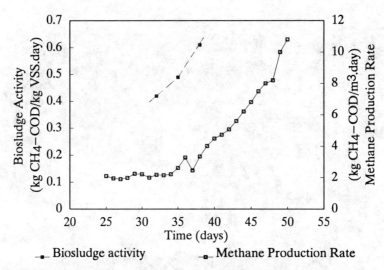

Figure 2 The relationship between biosludge activity and methane production rate during start up period of citric acid wastewater treatment.

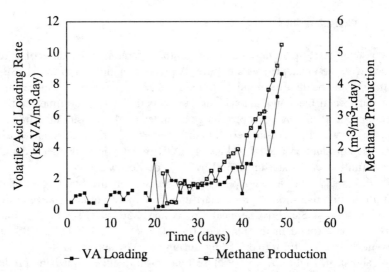

Figure 3 The relationship between VA loading rate and methane production rate in treatment of citric acid wastewater at San fu Chemical Company.

period, so called selecting phase, the amount of biomass decreased due to washout of unadapted biosludge and the low substrate input (about 3 kg COD/m^3. day). After this selecting phase, the adaptable biomass growth rate would be greater than the rate of washout and, thus, the amount of adaptable biosludge in the reactor increases sharply. Therefore, the loading rate increases sharply after the 35th day of operation. Samples were taken from the reactor at the intervals between the inlet and outlet along reactor height, and the volatile suspended solid (VSS) content were measured. Figure 5 shows the COD and VSS distribution along the height of the reactor. The portion sludge bed with VSS excessing 6000 mg/l is defined as high density sludge bed, and almost 90% of the COD is removed at this portion. Curve 1a in Figure 5 shows that the high density sludge bed is about 2 meter high. After one week operation without withdrawing excess sludge, the height of the high density sludge bed extended to 3.5 meter (curve 2a). Sludge yield can be estimated by calculating the amounts of drained sludge and total amounts of COD removal. The value is about 6% which is consistent with our previous results (Roam, et al., 1983).

Restart-up Considerations

Owing to the annual maintenance of the whole plant, the operation of 85m^3 UASB reactor at San Fu Chemical Co. had been stopped for about 3 weeks, During that 3 weeks. the ambient temperature was about 10 to 20°C. The sludge in the reactor settled down becoming quite compact; some sludge particles broke down and the VS/TS ratio decreased to 0.2 before the restart-up. From microscopic observation, the bio-sludge was dominated by the rod type methanogenic population with a few sarcina type organisms. It took about 10 days to restart up the reactor to the loading rate of 20 kg COD/m^3. day under low temperatures (10-20°C) as shown in Figure 6. It is believed, however, 5 to 7 days would be enough to restart up the reactor in warm season.

For the restart-up of UASB reactor, it is suggested that a microscopic observation of microorganisms in the sludge is necessary, and the VS/TS ratio must be measured. If VS/TS ratio is smaller then 0.2, reseeding with supplemental sludge into the reactor is recommended.

Case II. Waste from Grape wine and Brandy Winery Plant

Influent: COD: 160,000 mg/l (non-diluted)
 BOD: 0.78 of COD concentration
Design basis: flow rate: 400 CMD
 UASB reactor: 88 m^3/reactor × 5 reactors
 loading rate: 15 kg COD/m^3day
Performance: effluent COD: <3000 mg/l
 COD removal: 82%
 gas yield: 0.625 m^3/kg VSS.day
 granule diameter: (0.57~0.67)±0.16 mm

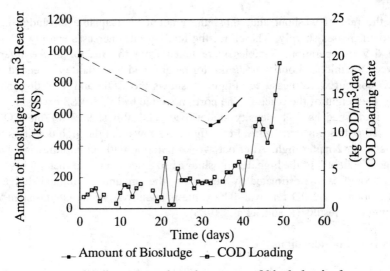

Figure 4 COD loading rate and total amounts of biosludge in the reactor of citric acid wastewater treatment.

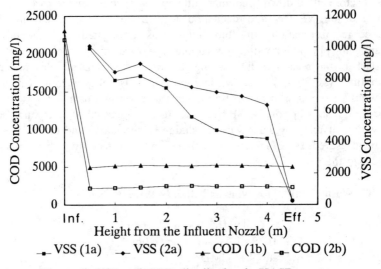

Figure 5 VSS and COD distribution in UASB reactor.

Wastewaters of grape wine and brandy in Nan-Tou plant were discharged from the fermentor bottom residue and the distillate. A very high COD concentration of 160,000 mg/l and high BOD/COD ratio of 0.78 indicated that biodegradability of this wastewater was extremely high for a biological treatment process. The performance of the UASB process which treated 400 m/day of wastewater is shown in Figure 7. Significant fluctuation of the influent COD concentration attained relatively stable leveling off of the effluent COD below 3,000 mg/L which would be feasibly treated by the subsequent stage of activated sludge process. The average COD removal efficiency of UASB was 82% while the volumetric loading was increased up to 15 kg COD/m^3 day.

Case III. Waste from Sorghum Wine Manufacture Plant

Influent: COD: 2000 mg/l with tannin and ploysaccharides.
Design basis: flow rate: 900 CMD
                UASB reactor: 127 m /reactor × 7 reactors
                Loading rate: 2 kg COD/m·day

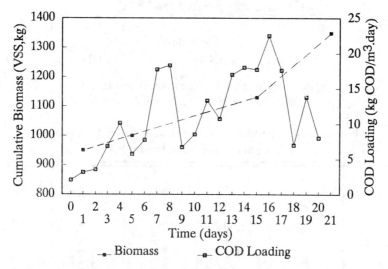

Figure 6 The COD loading and the cumulative amount of VSS in reactor during the periods of restart-up operation after 3 weeks starving.

Performance: effluent COD: 300 mg/l
                COD removal: 85%

Wastewater characteristics of sorghum wine in Lon-Ten Plant is different from that of grape wine. More refractory constituents are presented in sorghum grain, such as tannin and polysaccharides. So the UASB process was acclimated with relatively low loading at 2 kg COD/m^3 · day and low concentration of feeding COD at 2,000 mg/l. More than six months was taken to start up the UASB with low COD concentration of wastewater.

During the start-up period, five sludge samples were taken from the UASB bottom and tested by means of Biochemical Methane Potential (BMP) incubation series. The UASB operating conditions and the BMP results are compared in Table 3.

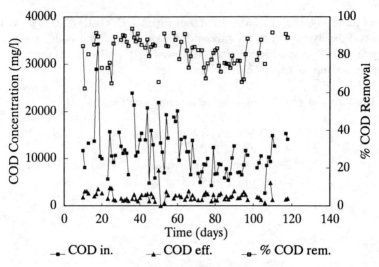

Figure 7 Influent and effluent COD concentration of UASB in Nan-Tou Winery Plant.

Table 3. UASB Operating Conditions and Sludge Characteristics in Sorghum Wine Manufacture Plant

Run	Days	Influent (COD mg/L)	Volumetric Loading (kg COD/m^3 · day)	Max Rate (mL gas/g VSS · day)	Granule Diameter (mm)
1	5	Swine Waste	–	1.9	–
2	16	1,600	1.9	2,500	0.18
3	31	1,500	2.3	2,500	0.29
4	94	1,500	1.0	1,217	0.55
5	201	2,500	2.8	1,700	0.33

Case IV. Waste from Rice Wine Manufacture Plant

Influent: COD: 25,000 mg/l
 BOD$_5$: 15,000 mg/l
Design basis: UASB reactor: 127 m^3/reactor × 9 reactors
 organic loading: 15 kg COD/m^3 · day

Performance: effluent COD: 1500 mg/l
              COD removal: 95%

High-strength wastewater was produced through fermentation and distillation during the rice winery process in the Su-Lin plant and Ping-Ton plant. Both plants employed two-stage UASB processes to treat the influent COD from 10,000 to 35,000 mg/l. Without effective solid separation, the UASB system of the Ping-Ton plant received high concentrations of suspended solids above 10,000 mg/l and a certain amount of lime-soda from the chemical coagulation system. So different sludge characteristics of these two UASB systems were evident even with similar winery process. In Su-Lin winery plant, the first stage UASB unit was operated with a long HRT of 40 hours and the influent COD of 25,000 mg/l during start-up period. Applied with high loading of 15 kg COD/m^3 · day, the first-stage effluent discharged COD 3,000 mg/l into the second-stage UASB with HRT 13 hours. COD removal efficiencies of two stages could be maintained at 88% and 50% respectively. After four mouths of start-up operation, the UASB systems were very stable in effluent quality: pH: 6.5, COD <1,500 mg/l, BOD <600 mg/l and SS <1,000 mg/l. A subsequent aerobic activated sludge process was employed to improve the effluent quality. Meantime, the bioactivity of anaerobic sludge was measured as the highest gas production rate of 751 ml Gas/g VSS-day compared with the previous data.

Case V. Waste from Adhesive Tape Manufacturing Plant

Influent: COD: 5,000~10,000 mg/l
                 pH: 3
                 toluene: 80~943 mg/l
Design basis: flow rate: 160 CMD
                 UASB reactor: 600 m^3
                 COD loading: 2kg COD/m^3 · day
Performance: COD removal: > 98%
                 effluent of COD: < 100 mg/l

A full scale UASB process was established in July 1988, after a feasibility study; engineering design and construction began in April, 1988. During the last four years, after 3 months of start- up operation started in August, 1988, the treatment plant established excellent performance. According to the results of 1300 days of operation, the COD concentration of effluent with a total of 80% was below 100 mg/l, and with a total of 53% of effluent below 60 mg/l. The long-term average of COD in effluent was 76 m,g/l (σ = 59 m/l, n = 388). Table 4 shows this result.

Table 4. The COD Removal Efficiency of UASB Reactor in
Adhesive Tape Manufacturing Plant

Parameter	Average	Standard Deviation	C.V.%*	Remark
COD-in (mg/l)	7745	1965	25	n = 534
COD-out (mg/l)	76	59	78	n = 388
Rem.%	98.99			
C.V.% = Coefficient of Variation				

A long-term record based on 1300 days operation of the UASB process was shown in Figure 14. The system indicates strong buffer capability and steady COD removal capacity.

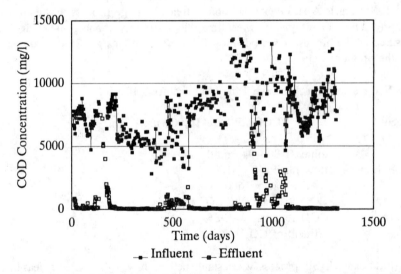

Figure 8 The influent and the effluent COD concentration of the UASB process which treated wastewater from solvent recovery unit in an adhesive tape manufacturing plant (COD-in 7745 mg/l ± 1965 mg/l, n = 534; COD-out 76 mg/l ± 59 mg/l, n = 388).

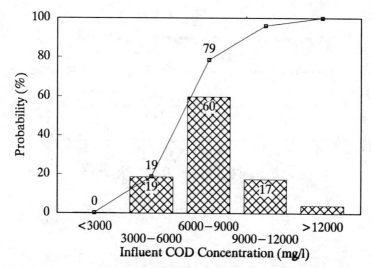

Figure 9 A distribution probability of the influent COD concentration based on 1300 days operation.

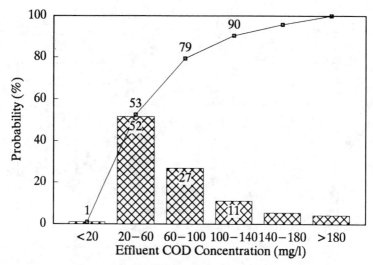

Figure 10 A distribution probability of the effluent COD concentration based on 1300 days operation.

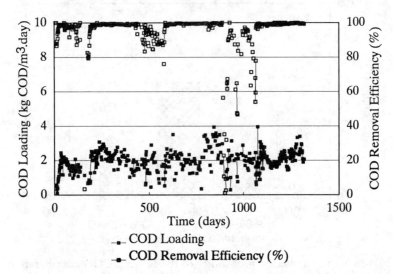

Figure 11 A long-term record of the UASB performance in the adhesive tape plant based on COD loading and removal efficiency.

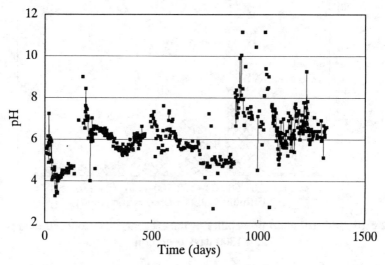

Figure 12 The fluctuation of influent pH of the UASB process in the adhesive tape plant (pH = 6.05 ± 1.14, n = 549).

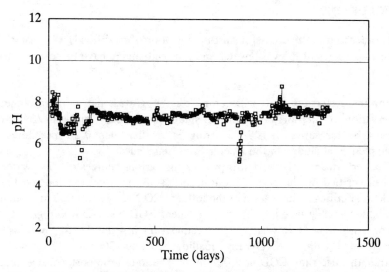

Figure 13 The effluent pH of the UASB process in the adhesive tape plant (pH = 7.39 ± 0.39, n = 556).

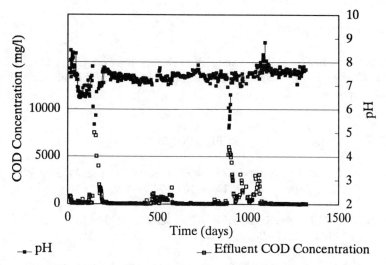

Figure 14 The correspondence of pH and COD concentration of the effluent of the UASB process in the adhesive tape plant.

CONCLUSIONS

1. In order to make a successful start-up operation of modified UASB reactor, the characteristics of the wastewater and the seed sludge have to be evaluated in detail.
2. For the restart-up of UASB retor, it is suggested that a microscopic observation of microorganisms in the sludge is necessary, and the VS/TS ratio must be measured. If VS/TS ration is smaller then 0.2, reseeding with supplemental sludge in the reactor is recommended.
3. The performance of the UASB process for various sources of wastewater has been verified on the basis of long-term results. In a grape wine and brandy plant, significant fluctuation of the influent COD above 160,000 mg/l attained relatively stable leveling-off of the effluent COD below 3,000 mg/l.
4. For sorghum wine wastewater treatment plant, the UASB process was acclimated with relatively low loading at 2 kg COD/m^3 · day and low strength of feeding COD at 2,000 mg/l. The long-term performance attained the effluent COD below 300 mg/l.
5. For adhesive tape manufacturing plant, the UASB process established excellent performance. According to the results of 1,300 days operation, the COD removal efficiency is greater than 98% and the effluent COD is below 100 mg/l.

ACKNOWLEDGEMENT

We thank our colleagues, Wen-Yaung Tzou, Ming-Jing Perng, Yee Chung Chen and Iow-Ching Lin for their excellent technical support to the full scale UASB process operation.

REFERENCES

1. Cheng, S.S., et al., "A Modified UASB Process Treating Winery Wastewater", *Wat, Sci, Tech.*, Vol.22, No.9, 167-174 (1990).

2. de Zeeuw, W.J. "Acclimatization of Anaerobic Sludge for UASB Reactor Start-up," PhD. Thesis (Sept. 1984).

3. Koepp, H.J., Schoberth, S.M., and Sahm, H., "Evaluation of the Anaerobic Digestion of an Effluent from Citric Acid Fermentation" Proc. of Anaerobic Wastewater Treatment in Netherlands, 13-29 (Nov. 1983).

4. Roam, G.D., Yeh, Y.Y., Chang, T.C., Tzou, W.Y. and Perng, M.J., "Treatment of Rice Liquor Stillage by Anaerobic Process" proc. 8th Conf. on Wastewater Treatment Technology in R.O.C., 395-412 (Sept. 1983) (in Chinese).

5. Roam, G.D. and Cheng, S.S., et .al. "Anaerobic Process for Kao Liang and Fruit Wine Wastewater Treatment," UCL Report, unpublished (1986).

6. Tzou, W.Y., Chang, T.C., Roam, G.D., and Huang, A.L., "Anaerobic Process for Antibiotics Wastewater Treatment Results of Bench Test and Pilot Plant," Proc. 10th Conf. on Wastewater Treatment Technology in R.O.C., 147-161 (Dec. 1985)(in Chinese).

7. Yang, Y.Y., Roam, G.D., et al., "Comparison of Two Anaerobic Processes for Treating MSG Waste Liquor," Proc. 10th Conf. on Wastewater Treatment Technology in R.O.C., 551-560 (Dec. 1985)(in Chinese).

30

KINETIC MEASUREMENT OF BIODECOMPOSITION OF ANILINE, 2,4-DICHLOROPHENOL AND 1,2,4-TRICHLOROBENZENE USING A CALORIMETRIC METHOD

Kuanchun Lee, Ming-Chin Chang, Yeun C. Wu
Department of Civil and Environmental Engineering
New Jersey Institute of Technology
Newark, NJ

Lun-Wen Yao
Powel Duffryn Terminals, Inc.
Bayonne, NJ

INTRODUCTION

Biological treatment techniques for detoxification of toxic organic compounds have been investigated and employed for field applications for many years. In fact, it has been reported that the biotreatment is often found to be more economic and efficient than other treatment methods.

The traditional substrate and biomass measurements are time-consuming tasks. So, many researchers have intended to develop a simple method that can quickly and precisely determine the metabolic activity of organisms during the course of biodegradation of toxic substances. This study has planned to use the microcalorimetric method to monitor the cell energy production resulting from the biodecomposition of aniline, 2,4-dichlorophenol (2,4-DCP) and 1,2,4-trichlorobenzene (1,2,4-TCB). By utilizing these results obtained from the present study, it is possible to determine both kinetic data and biodegradability of all three target compounds. More importantly, the kinetic information presently reported can additionally serve as a database for designing the future hazardous and toxic wastewater treatment plant. The effect of toxic substance concentration on cell decay during the phase of endogenous respiration is also studied by the direct use of calorimetric method in this research work.

MODELING OF MICROBIAL GROWTH IN BATCH CULTURE

The cultivation of microorganisms has been observed to have several distinguished growth phases as shown in Figure 1. When a small amount of inoculum is placed into a substrate and nutrient rich medium, the bacteria population will remain relatively constant for a period of time which is called "lag phase." In this phase, the substrate degrading microorganisms first have to get accustomed to the new cultural condition and then generate enough extracellular enzymes for metabolizing food substrate. After the organisms have already adopted to the media, they begin to multiply or reproduce exponentially and enter the "log phase." During this phase, the food is rapidly and continuously consumed by organisms until its concentration becomes limiting. As a result of food restriction, the rate of cell growth is nearly equal to the rate of cell die-off. Eventually, the food supply becomes insufficient and causes the death rate to be greater than the cell synthesis rate. A decrease in the viable cell concentration occurs. During this phase of growth (called "endogenous respiration"), the cells actually utilize the stored ATP energy for maintaining their activities and motion until the ATP is depleted. Finally the cells die.[1,2]

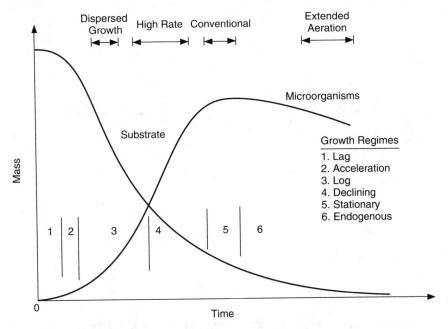

Figure 1 Typical growth curves for a batch reactor.

Two basic kinetic models including the traditional Monod equation and the substrate inhibition type equation are used for this investigation. These equations describe the actual relationship between the specific growth rate (μ) and the limiting substrate concentration (S). Monod equation proposed a functional relationship between μ and S and it can be expressed by Equation (1) and (2).[3]

$$dX/dt = \mu X \qquad (1)$$

$$\mu = \mu m\, S/(Ks + S) \qquad (2)$$

where X is the biomass concentration, μm is the maximum specific growth rate, S is the growth limiting substrate concentration and Ks is the saturation constant which is the substrate concentration with respect to $1/2\mu m$. A graphic presentation of this model is shown in Figure 2. It is apparent from the present study that the cell energy production measured directly by a calorimeter perfectly reflect the microbial growth and hence, a conceptual model similar to the Monod model can be established by the following Equations (3) and (4).

$$dE/dt = \mu_e E \qquad (3)$$

$$\mu_e = \mu_E\, S/(K_c + S) \qquad (4)$$

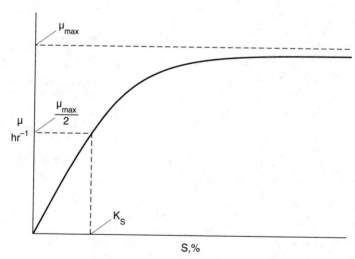

Figure 2 Specific growth rate for biomass at various substrate levels.

Where E is the cell energy production, μ_e is the cell energy production rate, μ_E is the maximum cell energy production rate and K_c is the substrate concentration at half of μ_E. Equation 3 can be rearranged to Equation 5 in which μ_E and K_c can be determined if μ_e at various substrate concentrations S are known.

$$1/\mu_e = (K_c/\mu_E)(1/S) + 1/\mu_E \tag{5}$$

By plotting $1/\mu_e$ versus $1/S$, the values of K_c and μ_E can then be estimated by the slope and intercept which equal to K_c/μ_E and $1/\mu_E$, respectively.

In 1980, Neufeld has illustrated that the specific growth rate of mixed culture in anaerobic degradation of phenol decreases significantly after the substrate concentration exceeds its maximum allowable level of 725 mg/l.[4] Also, he has used the following equation to state the function of specific cell growth rate against the inhibitory substrate concentration.

$$\mu = u_m/[1 + K_S/S + (S/K_i)^N] \tag{6}$$

or

$$\mu_e = \mu_E/[1 = K_c/S + (S/K_i)^N] \tag{7}$$

Where the value of N determines the order of inhibition and Ki is an inhibition constant.

The approach used in evaluating the parameters μ_E, K_c, K_i and N is one of nonlinear curve fitting. The solution for any set of experimental data can, however, be simplified by an understanding of the influence of these parameters on curve shape. Equation 6 may be converted as follows:

$$1/\mu_e = 1/\mu_E + (K_c/\mu_E)(1/S) + S^N/(K_i^N * \mu_E) \tag{8}$$

At low values of S, the term $(S/K_i)^N$ is negligible nd the equation is reduced to Monod type. For an illustration, the solid line shown in Figure 3 reflects the influence of the inhibition term $S^N/(K_iN^*\mu m)$ in Equation 6 or $S^N/(K_iN^*\mu_E)$ in Equation 7. Figure 4 shows a plot of Equation 7 with typical but assumed values of K_c, μ_E, K_i, and varied values of N. As can be seen by a comparison of the generated family of curves with the curve of non-inhibition, at levels of S in the neighborhood of or less than the observed maxima, small values of N cause relatively greater deviations from the curve of no inhibition term than do larger values of N. This effect is reversed at levels of S/K_i greater than 1.0, or at substrate levels beyond the intersection of the generated family of curves.

In the past, little study has been done to examine the effect of substrate type and concentration on cell decay during the phase of endogenous respiration. This maybe because most of the batch-type microbiological processes are terminated before the endogenous phase begins. Usually death of microbes is assumed to follow an exponential decay. The following equation can be used to express the cell decay numerically in which the biomass concentration is stated by the term of cell energy production (E):

$$dE/dt = -K_d *E \qquad (9)$$

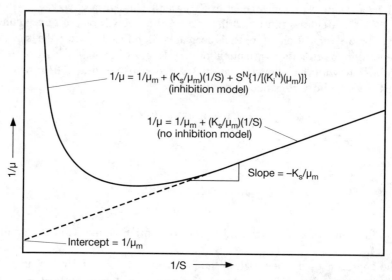

Figure 3 Representative plot of $1/\mu$ vs. $1/S$ for the substrate inhibition model.

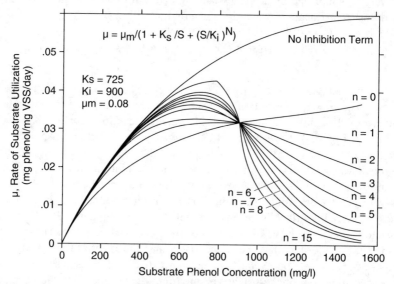

Figure 4 Calculate family of substrate inhibition curves showing influence of parameter N on curve shape.

where the K_d is the cell decay constant. Since three organic compounds presently investigated have been classified by the U.S. EPA as priority pollutants, it is interested in knowing their concentration effects on cell die-off rate.

EXPERIMENTAL PROCEDURE

The bench-scale microcalorimeter employed for this study was originally developed and designed by the Department of Biological and Agricultural Engineering at Rutgers University. This apparatus could successfully monitor the cell energy production generated from the result of substrate decomposition. This technique offers a convenient and relatively quick way to determine the interaction of the microbiological community and a contaminant.[5,6,7]

The microbial activity determined by cell energy production for a one gram sample is termed a "thermogram." A normalization test performed on a sample that is the organic medium sustain the microbiological community will reflect the metabolic activity of the microorganisms present. The energy output will be proportional to bacterial population density. For these tests, dewatered primary sewage sludge collected from the Camden City Waste Water Treatment Plant, New Jersey, was used as the organic medium. To obtain a baseline from which to measure the activity contributed by the microorganisms in the organic medium, a sample of the sludge is tested in the microcalorimeter. The thermogram obtained from the sample provides the information that shows the feasibility of using the microbiological community to degrade an organic contaminant. Measuring the area between the baseline and the normalization thermogram for the sewage sludge provides a quantitative measurement of the heat contributed by the sewage sludge.

In this study, the organic compounds are aniline, 2,4-dichlorophenol and 1,2,4-trichlorobenzene with high purity of 97% to 99%. The chemical structure of these compounds are shown in Figure 5.[8] All of the experimental work was performed at fifty degrees Celsius surrounding the microcalorimetric cell and under aerobic conditions. The liquid - liquid extraction method of analytical techniques was used to extract the organic compounds from aqueous leachate samples. The samples were analyzed with Gas Chromatography (Hewlett Packard Model 5730 with dual flame ionization detector).

A predetermined amount of each target compound was mixed with the sewage sludge in order to test for the biodegradability of the contaminants. These compounds are pre-prepared according to the concentrations shown in Table 1 to detect the effect of the toxicant concentration on the microbiological community. A blank sample test was carried out to account for the heat output due to the utilization of the organic medium employed to sustain the microbiological community. In all studies, one gram of this sewage sludge was placed in the microcalorimeter cell. After a period of time, the cell energy output was measured by the area of thermogram.

Aniline

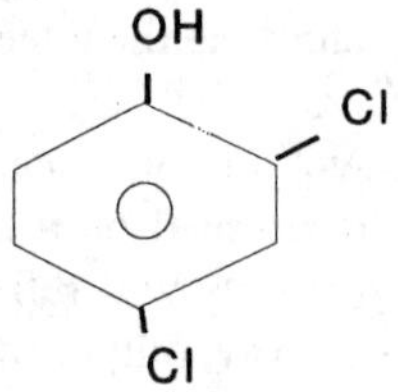

2,4-Dichlorophenol

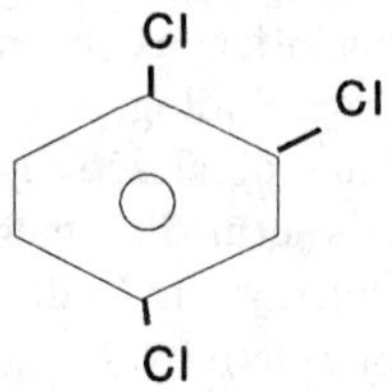

1,2,4-Trichlorobenzene

Figure 5 Molecular structure of target compounds.

Table 1. Target Compounds and Their Concentrations

Compound	Concentration
Aniline	0.1%, 0.3%, 0.5%, 0.7% and 1.1%
2,4-DCP	0.05%, 0.10%, 0.15% and 0.20%
1,2,4-TCB	0.1%, 1.5%, 3.0% and 3.5

RESULTS AND DISCUSSION

Figure 6 (a)-(d) showed several typical thermograms generated under various concentrations of 1,2,4-TCB with sludge. It is apparent that the heat energy production first increased and then decreased as the concentration of 1,2,4-TCB increased. Figure 7 reported the method of data transfered from the thermogram printout area into energy value for the above-mentioned chemical compound. The heat productions from various concentrations of all three target compounds are shown in Figure 8, 9, and 10, respectively. Apparently, the results have clearly indicated that if the added target compound becomes a toxicant or its dosage exceeds the maximum allowable concentration, the heat output will be less than that observed from the normalized thermogram. Conversely, if the target compound can be effectively metabolized, the heat energy will be greater. For comparison of the energy production between base line (no target compound added) and test line (various concentrations of target compound added) with sewage sludge. Figures 11, 12, and 13 showed the net energy gain or less (energy production from test run minus energy production from base control run) for aniline, 2,4-DCP and 1,2,4-TCB, respectively.

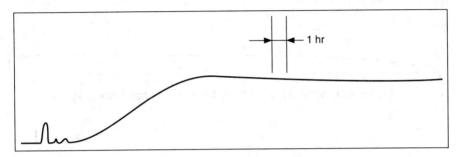

Figure 6(a). Sewage sludge thermogram.

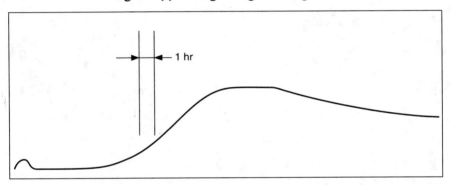

Figure 6(b). 1% 1,2,4-TCB and sewage sludge thermogram.

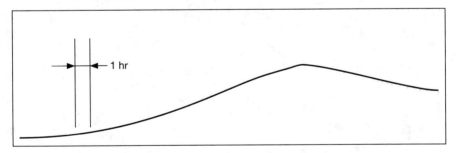

Figure 6(c) 3.0% 1,2,4-TCB and sewage sludge thermogram.

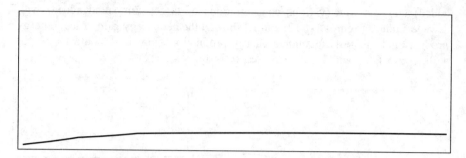

Figure 6(d) 5.0% 1,2,4-TCB and sewage sludge themogram.

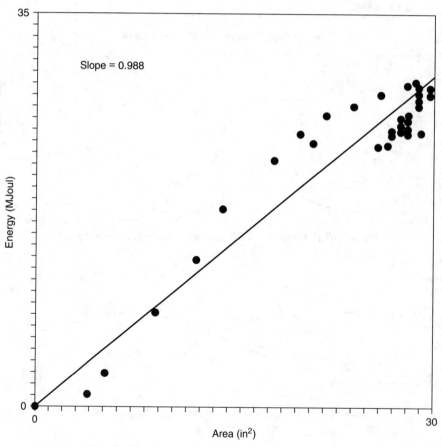

Figure 7 Ratio of energy production vs. thermogram area
(target compound 1,2,4-TCB).

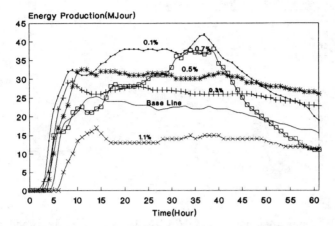

Figure 8 Biodegradation of aniline at 0, 0.3%, 0.5%, 0.7 and 1.1%.

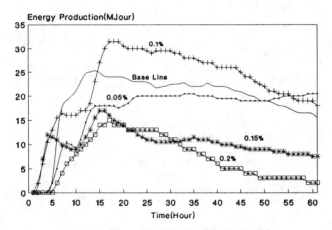

Figure 9 Biodegradation of 2,4-DCP at 0.05%, 0.1%, 0.15% and 0.2%.

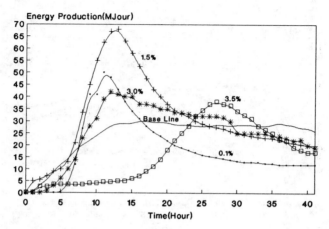

Figure 10 Biodegradation of 1,2,4-TCB at 0.1%, 1.5%, 3.0% and 3.5%.

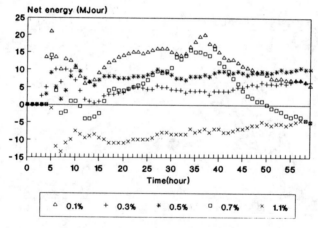

Figure 11 Net energy effect of aniline.

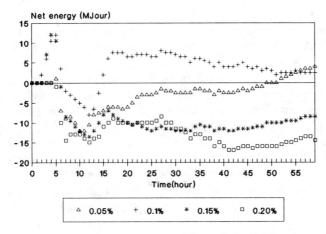

Figure 12 Net energy effect of 2,4-DCP.

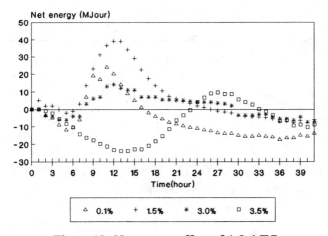

Figure 13 Net energy effect of 1,2,4-TC.

In general, it was found from the present study that the cell energy production versus time curve is almost identical to the results of cell growth in batch culture, which consists of lag phase, log phase, stationary phase, and endogenous phase. In addition, it has also found that the substrate degrading organisms evidently need a longer time for adopting to the new external growth living environment (i.e., longer lag time), particularly when the toxic substrate concentration is higher. Moreover, the cell energy output during a log growth phase seems to be highly dependent upon the initial concentration of toxicant involved. The higher the concentration of toxic contaminant, the greater the bioinhibition may take place.

All investigations presently performed revealed that the addition of toxic target compound to the bacterial culture could possibly result in a negative heat output during each growth phase. When the excessive amount of toxic chemical is added, the total heat output will drop markedly to below the standard heat energy value solely produced by sewage sludge due to its concentrated effect on cell metabolism and growth.

The kinetic data was collected from all experiments presently conducted to understand how the substrate degrading bacterial reacted to the different types, and different concentrations, of target compounds applied. By using Equation 5, plot $1/\mu_e$ versus $1/S$ to calculate the value of K_c and μ_E. After getting μ_E and K_c from Monod Equation, substitute these values into Equation 8 to compute K_i and N by using Rosenbrock method (nonlinear curve-fitting program). The method can obtain the most optimal K_i and N values with minimum error in standard deviations. The curve-fitting results are illustrated in Figures 14, 15, and 16, respectively, for aniline, 2,4-DCP and 1,2,4-TCB. The values of μ_E, K_c, K_i and N are summarized in Table 2 as follows:

Table 2. The Values of μ_E, K_c, K_i and N

Compounds	Kinetic Data			
	μ_E (hr-1)	K_c (%)	K_i (%)	N
Aniline	3.76	1.435	0.546	3.36
2,4-DCP	8.188	0.745	0.155	9.0
1,2,4-TCB	3.537	0.57	1.63	3.2

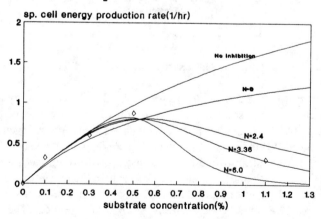

Figure 14 Substrate inhibition curve of aniline.

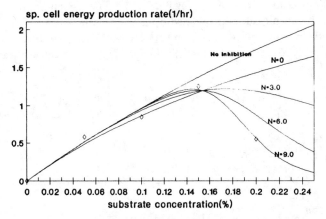

Figure 15 Substrate inhibition curve of 2,4-DCP.

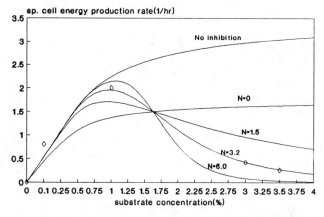

Figure 16 Substrate inhibition curve of 1,2,4-TCB.

The decay constants K_d in endogenous phase can be computed by using Equation 9 for all three different toxic compounds. The effect of substrate concentration on μ_e and K_d is shown in Figures 17, 18, and 19 (a)-(c). It is apparent from Figure 17 that the rate of cell energy production (μ_e) is directly proportional to the substrate concentration that is equal to or less than 0.2%, 0.5% and 1.1% for 2,4-DCP, aniline, and 1,2,4-TCB, respectively. However, when the concentration of any target has passed through the maximum tolerable level, the rate of cell energy production becomes inversely proportional to its dosage applied, i.e., the higher the concentration of toxic compound involved, the lower the μ_e. The present study has also indicated that the biodegradability of these three chemicals follows the order of 2,4-DCP > 1,2,4-TCB > aniline, according to Figure 17. This result can be explained by Eq. 7 as follows:

At non-substrate inhibition condition (2,4-DCP < ϕ.2%, aniline < 0.5% and 1,2,4-TCB < 1.1%), the term $(S/K_i)N$ in Eq. 7 can be neglected and then it becomes

$$\frac{\mu_e}{S} = \frac{\mu_E}{S + K_c}$$

(10)

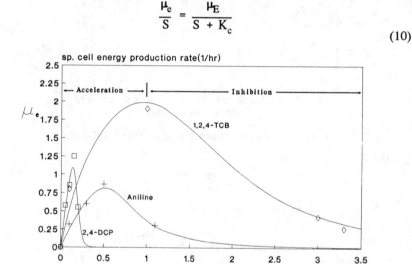

Figure 17 Comparison of inhibitory effect of target compounds.

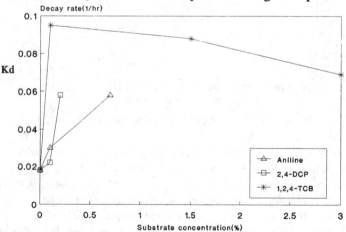

Figure 18 Comparison of endogenous decay rate vs. concentration.

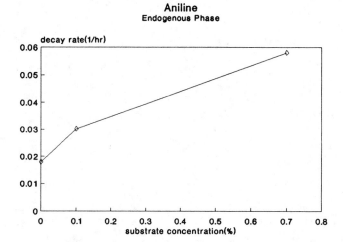

Figure 19 (a) Endogenous decay rate vs. substrate concentration of aniline.

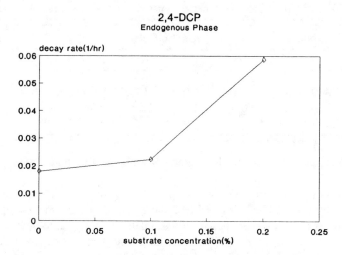

Figure 19 (b) Endogenous decay rate vs. substrate concentration of 2,4-DCP.

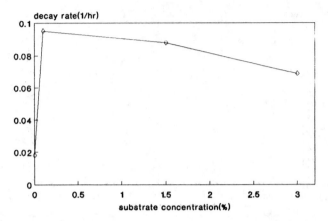

Figure 19 (c) Endogenous decay rate vs. substrate concentration of 1,2,4-TCB.

Table 3. Comparison of Kinetic Constants μ_E, K_c, K_i and N Values

Compounds	Kinetic Data			
	μ_E	K_c	K_i	N
2,4-DCP	1.0*	1.00*	1.00*	1.00*
1,2,4-TCB	0.43	0.77	10.51	0.35
Aniline	0.45	1.92	3.52	0.37

* Assume the values of μ_E, K_c, K_i and N are unity or equal to one.

The term of μ_e/S is the highest if μ_E and K_c are, respectively, the largest and smallest values. In Table 2, it is evident that 2,4-DCP has the highest μ_E and the second lowest K_c. However, based on these data given in Table 3, it is realized that the influence of μ_E on μ_e/S is more significant than that of K_c. This is why the target compound 2,4-DCP produced the highest value of μ_e/S among these three curves shown in Figure 17 and 1,2,4-TCB has the second highest μ_e/S value.

At high substrate concentration, the term $\mu_E/(S/K_i)^N$ in Equation 7 becomes important and the equation can be rewritten as

$$\mu_e/S = \mu_E/[1/S + K_c/S^2 + S^{N-1}/K_i^N] \tag{11}$$

It is clear in Figure 17 that the negative slope of all three curves expressed by the μ_e/S term is greatly affected by the important value so-called "inhibitory factor" or $1/(S^{N-1}/K_i^N)$ term. The greater the inhibitory factor value, the higher the μ_e/S and K_i. It is apparent from Table 3 and Figure 17 that the substrate inhibitory effect on μ_e/S between 1,2,4-TCB and aniline can be easily defined. The former compound has a higher value in K_i and almost equal values in N and μ_E that makes 1,2,4-TCB more toxic than aniline. However, the present study has indicated that the substrate degrading bacterial can metabolize a wider concentration range of 1,2,4-TCB than aniline. To compare the toxicity of 2,4-DCP with two other compounds, it seems that the 2,4-DCP has an extremely high value in N and μ_E that somehow discount the effect of the other factors such as K_i and K_c. It is important to point out that the compound 2,4-DCP is more biodegradable than 1,2,4-TCB and aniline but it becomes very toxic to microbes when it exceeds the maximum tolerable level.

The influence of toxic substrate concentration on K_d can be seen in Figure 18. It is obviously shown that the constant K_d is significantly affected by the added substrate concentration. The K_d constant becomes greater as a result of adding more toxic target compound to the bacterial culture. However, in the study of 1,2,4-TCB, the decay constant K_d decreased slightly after the substrate concentration exceeded 0.1%. No study has been done to understand whether the same result will also exist with the other two compounds 2,4-DCP and aniline. The dosage of these two compounds mentioned above did not reach a level as high as 1,2,4-TCB. The reason that it causes a decrease in K_d with increasing the concentration of toxic chemical is unknown.

CONCLUSIONS

The present study evidently shows that the metabolic activities of microorganisms in degrading the toxic contaminants such as aniline, 2,4-DCP and 1,2,4-TCB can be precisely determined by the direct measurement of cell energy heat output. A thermogram which describes the heat production as a function of time during the bioassimilation of toxic organics can be easily established and the resulting curve resembles a normal cell growth of batch culture consisting of lag phase, log phase, stationary phase, and endogenous phase.

The rate of heat energy production during log phase could be mathematically expressed by Monod first-order reaction model. However, when the concentration of any toxicant mentioned above exceeded the maximum tolerable level, the substrate inhibitory effect on its biodegradation occurred.

The constants of μ_E (the maximum rate of heat energy production), K_c (saturation constant), K_i (constant of inhibition) and N (degree of inhibition) were determined to be of 3.76 hr^{-1}, 1.435%, 0.546% and 3.36 for aniline; 8.188 hr^{-1}, 0.745%, 0.155%, and 9.0 for 2,4-DCP; and 3.537 hr^{-1} 0.57%, 1.63% and 3.2 for 1,2,4-TCB.

The biodegradability of toxic target compounds presently investigated followed the order of 2,4-DCP > 1,2,4-TCB > aniline at the concentration equal to or less than 0.2% DCP, 0.5% aniline and 1.1% TCB. This study has successfully demonstrated that the chemical nature and property of toxic compound can significantly influence its biodegradability due to high toxicity. For example, 2,4-DCP is more biodegradable than other compounds 1,2,4-TCB and aniline but it becomes highly toxic to microbes after the concentration of 0.2% it exceeds.

No research has been done to study the result of toxic substrate concentration effect on cell growth in the endogenous phase. This study has reported that the endogenous decay rate in the 1,2,4-TCB case was not concentration-dependent at high substrate level (>0.1%). However, for all cases presently investigated, it was found that at the substrate concentration (2,4-DCP < 0.2% and aniline < 1.1%) the cell decay rate increased with an increase in chemical dosage.

REFERENCES

1. Sundstrom, D.W. and Klei, H.E., *Wastewater Treatment*, Prentice-Hall, Inc., NJ 1979.

2. Gaudy, A.F. Jr., and Gaudy, E.T., *Microbiology for Environmental Scientists and Engineers*, McGraw-Hill, NY, 1980.

3. Shamat, N.A. and Maier, W.J., "Kinetics of Biodegradation of Chlorinated Organics," J. Wat. Pollut. Control Fed., Vol. 52, 1980, pp. 2158-2165.

4. Neufeld, R.D., Mack, J.D. and Strakey, J.P., "Anaerobic Phenol Biokinetics," J. WPCF, Vol. 52, No. 9, 1980, pp. 2367-2377.

5. Yao, L.W., "Measurement of Biodegradation Aniline, 2,4,-Dichlorophenol, 1,2,4-Trichlorobenzene Using Microcalorimetric Method," Master Thesis, NJIT 1989.

6. Singley, M.E., Higgins, A.J., Rajput, V.S., Pilapitiya, S., Mukherjee, R. and Mercade, V., "Biotechnology for the Treatment of Hazardous Waste Contaminated Soils and Residues," Symposium of Int'l Conf. on Innovative Biological Treatment of Toxic Wastewaters, Virginia, 1986, pp. 362 - 378.

7. Rahman, M.S., "Microcalorimetric Measurement of Heat Production and the Thermophysical Properties of Compost," PhD. Dissertation, Rutgers University 1984.

8. Marshall, S., *Handbook of Toxic and Hazardous Chemicals and Carcinogens*, 2nd edition, Noyes Publications, 1985.

31

SUCCESSFUL LEACHATE TREATMENT IN SBR-ADSORPTION SYSTEM

Wei-chi Ying , John Wnukowski and David Wilde
Occidental Chemical Corporation
Grand Island, NY

Donald McLeod
TreaTek-CRA Co.
Niagara Falls, NY

INTRODUCTION

Based on results of extensive bench- and pilot-scale treatability studies employing sequencing batch reactor (SBR) biotreatment and post-treatment by carbon adsorption, a full scale SBR-adsorption system was built to treat leachate from the Hyde Park chemical landfill in Niagara Falls, New York. Since April 1990, up to 100,000 gal/day of leachate, containing a wide variety of organic contaminants, has been treated in three 130,000 gal SBRs for removal of biodegradable constituents and then in carbon adsorbers (two to three 20,000-lb beds in eries) for removal of persistent constituents. Treatment performance in two SBRs started when biomass grown on the leachate was found to be better than in the third SBR started with activated sludge from a local POTW. Excellent treatment results were maintained during build-up of biomass in the SBR, allowing a very high MLSS (up to 15,000 mg/L) before wasting. Overall, the treatment system has consistently removed more than 95% TOC (from up to 2000 mg/L) and virtually all phenol (from up to 800 mg/L) in the leachate. Because of the very effective biotreatment in the SBR, the carbon exhaustion rate has been very low, less than 10 of the carbon requirement of the former leachate treatment employing carbon adsorption alone. The treatment system is also equipped with vapor phase carbon adsorption and regeneration as well as sludge thickening and dewatering units to provide complete treatment of leachate in a safe, clean and pleasant environment.

The Hyde Park Landfill site is located in an industrial complex in the extreme northwest corner of the Town of Niagara, New York (Figure 1). The site is roughly triangular in shape and occupies approximately 6.1 hectares. The Hyde Park Landfill was used from 1953 to 1975 as a disposal site for an estimated 73000 metric tons of chemical waste, including halogenated organics. A compacted clay cover was placed over the landfill in 1978, and a tile leachate collection system was installed around the perimeter in 1979. Until the start-up of the on-site treatment plant in April 1990, the leachate was collected and then trucked to the nearby Occidental Chemical's Niagara Plant for treatment. After mixing with plant wastewaters, pH adjustment, and filtration for removal of metals and suspended solids, the combined wastewater was treated first in two small sacrificial carbon adsorbers (2000 lb of granular activated carbon each) in series for removal of PCBs, dioxins and carry-over organic liquids and then by three larger serial adsorbers (20000 lb of carbon each) for removal of the rest of organic contaminants.

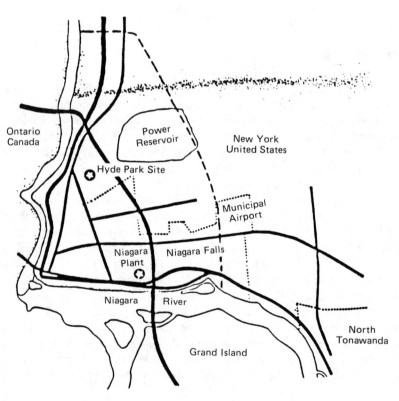

Figure 1 Location of the Hyde Park Landfill site.

The carbon adsorption system was installed to ensure the treated leachate meet the effluent limits (Table 1). The adsorption treatment was inefficient because carbon capacities for most of target organic contaminants were significantly reduced due to the high total organic carbon (TOC) level of leachate[1]. Table 2 shows that cumulative loadings of five major organic leachate constituents at the end of an adsorption cycle were far less than those estimated from the adsorption isotherms for the same compounds in pure water. The fact that the TOC and o-chlorobenzoic acid loadings were almost the same as the capacities predicted by the leachate isotherms indicates that the low bed utilization rates were primarily the result of competitive adsorption rather than poor adsorber design and/or

Table 1. Discharge Limits of Final Effluent

Parameter	Maximum Concentration*
pH	5–10
Phenol	1 mg/L
TOC (excluding methanol) or	300 mg/L
TOC (total)	1,000 mg/L
Trichloroethylene	10 µg/L
Tetrachloroethylene	10 µg/L
Monochlorobenzene	10 µg/L
Monochlorotoluene	10 µg/L
Benzene	10 µg/L
Trichlorogenzenes	10 µg/L
Tetrachlorobenzenes	10 µg/L
Monochlorobenzotrifluoride	10 µg/L
Hexachlorocyclobutadiene (C-46)	10 µg/L
Hexachlorocyclopentadiene (C-56)	10 µg/L
Hexachlorocyclohexanes (C-66)	10 µg/L
2,4,5-trichlorophenol	10 µg/L
Endosulfan	10 µg/L
Mirex	1 µg/L
2,3,7,8-tetrachlorodibenzo-p-dioxin (2,3,7,8-TCDD)	not detectable
* Except pH	

operational problems. Further, the adsorption system would have to be substantially expanded to handle the increased leachate volume as a result of an additional remediation project planned for the site.

A comprehensive experimental program was undertaken over a six year period from 1980 to 1986 to develop a treatment technology capable of reducing the treatment cost while meeting the discharge limits. Since carbon treatment was found necessary for complete removal of many chlorinate organic contaminants, the search was then focused on a compatible pretreatment technology which could significantly reduce the carbon consumption rate. A variety of wastewater treatment processes, including ion-exchange, chemical oxidation, solvent extraction, ultrafiltration, reverse osmosis, aerobic and anaerobic biodegradation were evaluated. Based on considerations of treatment cost and organic removal efficiency, biodegradation in sequencing batch reactors (SBRs) was selected as the best pretreatment process to reduce the future carbon consumption rate of an expanded adsorption system.[2,3] Since the start-up two years ago, the treatment performance of the full size on-site treatment system, which consists of pH adjustment, precipitation and sedimentation, SBR biotreatment, and carbon adsorption, has been excellent, and the carbon saving potential of this integrated system has been fully demonstrated. This paper presents a brief recount of the bench- and pilot-scale SBR treatability studies, including powdered activated carbon (PAC) enhanced SBR treatment, as well as the design and operation of the full size treatment system.

TREATABILITY STUDIES

Leachate Characteristics

Soon after the carbon adsorption treatment began, large populations of bacteria were found in the adsorber effluent. The TOC, chemical oxygen demand (COD), and concentration for some major organic constituents of a refrigerated raw leachate sample were found to decrease over time. The rate of concentration reduction increased when the sample was stored at room temperature. The biochemical oxygen demand (BOD) to TOC ratios for several leachate and the combined wastewater samples were all greater than 2, indicating that most organic compound in these samples were readily biodegradable. Many persistent leachate constituents were found to be biodegraded by bacteria isolated from soil samples of the Hyde Park landfill site.[4,5,6] Concentration of pollutants in raw leachate and the combined wastewater fluctuated widely over the study period; TOC had a range from 850 to 10000 mg/L and suspended solids (SS) from 200 to 2000 mg/L. Table 2 presents the composition, in terms of TOC and total organic halogen (TOX) and their major components, of various samples - raw leachate, combined wastewater feed, and adsorber effluent taken during a carbon adsorber treatment performance evaluation study. Characteristics of typical raw and pretreated (neutralization, aeration, and sedimentation) leachate samples are shown in Table 3.

Table 2. Adsorptive Capacities of Carbon for Major Leachate Constituents

Concentration Parameter	Concentration (mg/L)			Adsorptive Capacity (mg Adsorbed/g Carbon)		
	Raw Leachate	Combined* Waste Feed	Adsorber† Effluent	Carbon** Loading	Leachate†† Isotherm	Pure‡ Compound Isotherm
pH	5.3	5.5–6.4	5.5–6.4	5.5–6.4	4.9–5.5	5.0–6.0
Phenol	981	780	$ND_{0.1}$‡	41.0	74.9	174
Benzoic acid	830	910	0.8	48.0	74.1	171
o-chlorobenzoic acid	562	372	7.4	19.6	22.9	109
m-chlorobenzoic acid	61	120	$ND_{0.5}$	6.4	23.0	160
p-chlorobenzoic acid	40	80	$ND_{0.1}$	4.2	15.7	171
TOC	3,080	2,618	318	137	143	
TOX	264	299	2.7	15.8	11.7	

* Average concentrations for the adsorber feed during an adsorption service cycle.
† Concentrations were measured at the end of an adsorption cycle.
** Total removal of the compound at the end of an adsorption cycle.
†† Capacities were estimated at the feed concentration from the raw leachate isotherms.
: Capacities were estimated at the feed concentration from the pure compound isotherms.
‡ ND = not detected at a detection limit of x mg/L.

SBR Biotreatment

The SBR biotreatment process is a fill-and-draw activated sludge process. Biodegradation of organics and separation of biomass are accomplished in one tank under a cyclic schedule of operation consisting of five sequential steps: FILL, REACT, SETTLE, DRAW, and IDLE. The wastewater is fed, during FILL, to a tank which contains acclimated activated sludge from the previous cycle. Aeration and mechanical mixing are provided while feeding, or during REACT, to enhance the rate of aerobic biodegradation. After REACT, the mixed liquor is biologically stabilized. Air and mixing are stopped, and clarification takes place in the SETTLE step. During DRAW, the clear supernatant is withdrawn from the reactor for direct discharge or additional treatment. The IDLE period finally completes the SBR cycle. The five SBR steps are often overlapped, and one or two steps may be omitted in a particular treatment cycle. The withdrawal of effluent may start as soon as a clear zone of supernatant is formed, and the wastewater feeding may begin immediately after the completion of the DRAW step of the last SBR cycle. Figure 2 depicts the SBR biotreatment process. Many combinations of feeding, aeration, and mixing strategies are possible. The required nutrients are either supplemented to the feed or added directly to the bioreactor. The sludge wasting is accomplished by removing a portion of the settled sludge in the DRAW or IDLE step. The optimum SBR operating and cycle schedules

Table 3. Characteristics of Typical Raw and Pretreated Hyde Park Leachate Samples

Parameter*	Raw Leachate	Pretreated Leachate[†]
pH	4.3	7.5
TOC	3,500	3,200
COD	10,040	9,200
BOD	7,500	7,200
SS	900	80
VSS	300	40
TDS	25,700	22,400
PO_4-P	<1	<1
Acid-P	3	3
Total-P	131	92
NH_4-N	150	130
TKN	180	160
NO_3-N	20	20
NO_2-N	<5	<5

* All values, except pH, are given in mg/L.
[†] Pretreatment consisted of neutralization with NaOH to a pH of 7.5, two hours of aeration, and two hours or longer of settling.

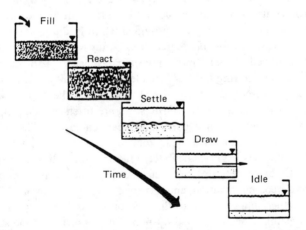

Figure 2 Illustration of SBR biotreatment technology.

must be experimentally established for wastewater to achieve the specific treatment objectives. [7] The operation and control for bench-scale SBR biotreatment experiments are simple, and the requirements for laboratory space and wastewater volumes are small. It is therefore often the process of choice for study of wastewater treatability.[8,9,10] The advantages of more complete treatment, greater operational flexibility to accommodate changing feed characteristics, intermittent treatment, and single tank for biodegradation, nitrification, denitrification and sludge separation make the SBR process an attractive technology for treating municipal and industrial wastewaters.[11,12,13]

SBR Treatability Study

A feasibility study on SBR treatment of a leachate sample was first performed at University of Notre Dame; 90% TOC reduction was achieved under a 24-hr cycle/10-day hydraulic retention time (HRT) operating schedule. A comprehensive SBR biotreatment study wa then conducted at Occidental Chemical's research laboratories to: 1) establish the best start-up procedure, 2) identify the nutrient requirements, 3) treat the combined wastewater, 4) define the optimum strategies for achieving the maximum TOC reduction under various SBR operating and cycle schedules, 5) develop a procedure for improving the performance of an upset SBR, 6) quantify the sludge production rate and select a method for it disposal, 7) verify the carbon saving due to biotreatment of wastewater, and 8) study the benefits of bacterial supplementation. Long term biotreatment performance data under a variety of conditions necessary for design of a full-size SBR system were collected during the 18-months study, utilizing three sets of SBRs: four 1-L, four 12-L, and three 500-L. Table 4 presents the treatability study programs and the associated specific objectives. Examples of the operating schedules for the four 12-L and the three 500-L units are given in Table 5, and the routine maintenance, sampling, and monitoring schedule are presented in Table 6.

PAC-SBR Treatment

A feasibility study on PAC-SBR treatment of leachate was conducted to determine whether the improved treatment by simultaneous adsorption and biodegradation in a SBR would produce an acceptable effluent without the pot-treatment in a granular activated carbon adsorber. This was indeed accomplished with the simple addition of PAC, near the end of REACT, to a SBR operated at a HRT of 5 days or less. The PAC dosage was just below the expected carbon requirement of the two-stage SBR-adsorption process. The PAC-SBR effluent quality was still much better than that of the control unit three week after the PAC addition wa stopped. Furthermore, the removal of the persistent HET acid in the PAC-SBR wa consistently more than that predicted by the adsorption isotherm for the SBR-treated leachate (Figure 3). A comprehensive PAC-SBR treatability study program wa therefore undertaken to define the PAC requirement and to optimize the treatment process. The operating and cycle schedule for the eight 1-L PAC-SBR are presented in Table 7.

Table 4 Experimental Programs for SBR Treatability Study

Program	Objectives	Tank/Adsorber Used
A	Determine the nutrient (N and P) requirements, compare methods for wastewater feed preparation, obtain performance data for replicate units, and study the effect of feeding period	four 1 L and four 12 L SBRs
B	Compare performance in treating full strength leachate with the combined waste, simulate full size SBR operation, and develop procedure for improving the performance of upset SBR	three 500 L and four 12 L SBRs
C	Verify carbon savings, correlate carbon saving to biotreatment performance, select the best carbon, and perform bioassay on raw and treated leachates	three 500 L SBRs, two 2.5 cm and two 7.5 cm adsorbers
D	Determine the fate of volatile constituents, establish aeration requirement for air-stripping, and test air pollution control device	two feed tanks, three 500 L SBRs, one vapor adsorber
E	Quantify sludge production rate, characterize the precipitation and biological sludges, test sludge digestion, compaction, and dewatering	two feed tanks, three 500 L SBRs, one 500 L digester
F	Obtain performance data at 1 d, 1.7 d, 2 d, 5 d, and 10 d HRT; operate the SBRs without weekend feedings	three 500 L, four 12 L, and four 1 L SBRs
G	Obtain performance data under 12 h and 24 h cycles; establish the optimum SBR schedule	four 12 L SBRs
H	Obtain performance data at temperatures of 9, 12, 15, and 20° C; determine the maximum organic loading for winter operation	four 1 L SBRs
I	Study effects of bacterial supplementation	four 12L and four 1 L SBRs

Table 5. Examples of SBR Operating and Cycle Schedule

Operating Schedule	Sequencing Batch Reactor						
	I	II	III	IV	A	B	C
Leachate feed	pretreated[†]				pretreated[†]		
Sterilization of feed	no				no		
Bacterial supplementation	no				no		
SBR cycle time, h	24				24		
Working volume, L	10				300		
Feeding, % working volume	10				20	20	50
HRT, d	10				5	5	2
MLSS, mg/L	not controlled				5,000	10,000	10,000
Time per SBR cycle, h							
FILL (air & mixing)	10				6		
REACT (air & mixing)	12				10		
SETTLE	1				2		
DRAW	0.5				5		
IDLE	0.5				1		

[†] The raw leachate was neutralized to pH of 7.5, aerated for two hours, and then settled for at least two hours. No dilution was made. Ammonia and phosphate were supplemented to a TOC/NH$_5$N/PO$_5$P ratio of 150/10/2.

SBR Start-up Procedure

The return activated sludge from a POTW (Wheatfield, NY) was utilized to seed the bioreactors for treating Hyde Park leachate and the combined wastewater. The SBRs, with more than half of the desired working volume filled with diluted POTW sludge (to MLSS = 6000 mg/L), were fed, over four days, at an increasing per cycle volume of wastewater to 10 of the reactor working volume (after FILL). The amount of effluent discharged was about 50 of the feeding until the full SBR working volume was attained. When treating a 3000 mg TOC/L leachate using a 10-d HRT and 24-h cycle schedule, the effluent TOC was found to increase gradually to about 180 mg/L. No mass die-off of the seed was observed, and the effluent SS was consistently less than 100 mg/L. Successful start-ups were accomplished without the use of any supplementary sources of organic carbon. The feed preparation procedure was established in a series of experiments to compare the performance in treating two types of leachate feed - one prepared by neutralization (to pH = 7.5) and filtration, a practiced in the feasibility study, and the other by neutralization, aeration (2 h), and sedimentation (2 h). More complete reductions in TOC and its components, as well as greater

Table 6. Routine Maintenance, Sampling, and Monitoring Schedule for SBRs

	Monday	Tuesday	Wednesday	Thrusday	Friday
pH*	x	x	x	x	x
Turbidity	x		x		x
TOC	x	x	x	x	x
Settled sludge volume	x			x	
Mixed liquor SS/VSS	x/x			x	
Sludge wasting†		x			x
Effluent SS/VSS		x/x			x
Nh$_4$-N		x			x
NO$_2$-N/NO$_3$-N		x			x
PO$_4$-P		x			x

* Acid or base was used to maintain pH within 7.0–7.5 after REACT
† The volume of the settled sludge to be wasted each time, VW (L):
$$VW = VT \times (MLSS2 / MLSS2)/(MSLL1 \times TMB/SV) \text{ where:}$$
VT = working volume (L), after FILL
TMV = the sample volume used in measuring the settled sludge volume
SV = settled sludge volume after two hours of settling
MLSS1 = mixed liquor suspended solids (mg/L) before wasting
MLSS2 = the MLSS to be maintained in the reactor
Tapwater was used for making up the settled sludge volume wasted

MLVSS/MLSS ratios were achieved in the 1-L SBRs receiving the aerated feed. The pretreatment procedure of neutralization, aeration, and sedimentation wa employed to provide feed for the SBR in all treatability programs.

PAC-SBR Start-up Procedure

About 5 liters of the POTW return activated sludge (MLSS = 6000 mg/L) were utilized to seed a 10-L SBR bioreator. The SBR was then fed, over a four-day period, at an increasing daily feed volume of Hyde Park leachate to 2 L/d. The amount of effluent discharged was about 50 of the daily feed until the full SBR working volume (8 L) was attained. Within three weeks, the effluent TOC was stabilized at about 250 mg/L for a leachate feed having a TOC of 3000 mg/L. The biotreatment was accomplished at the room temperature of 20 °C in the SBR which was operated under a 24-hr cycle (Table 5). No mass die-off of the seed sludge was observed, and the effluent SS was consistently less than 100 mg/L. When the MLSS increased to about 10000 mg/L, the liquid content of the 10-L SBR was divided evenly to eight 1-L units. A quantity of PAC which had been pre-saturated with compounds remaining in the leachate after the biotreatment (by contacting with a large volume of effluent from the 10-L SBR) was then introduced to each PAC-SBR to provide a desired mixed liquor PAC concentration (Table 7).

Table 7. Examples of PAC-/sbr Operating and Cycle Schedule

	PAC-SBR Units 600 ml working volume, 24 hour cycle, 20°C 4 day hydraulic retention time (25% daily feeding)							
Operating Schedule	1C	3A	3B	4A	4B	6A	6B	6C
Wastewater feed	pretreated leachate							
Sterilization of feed	no							
Bacterial supplementation	no							
Mixed liquor biological suspended solids, mg/L	10,000	10,000*						
Mixed liquor PAC, mg/L	0	3,000	3,000	4,500	4,500	6,000	6,000	6,000
PAC inventory, g	0	1.8	1.8	2.7	2.7	3.6	3.6	3.6
PAC dose, g/day	0	0.9	0.18	0.135	0.27	0.18	0.18	0.36
Mixed liquor wasting, ml/day	vw†	30	60	30	60	30	30	60
Time per SBR cycle, hour								
FILL (air & mixing)	6**							
REACT (air & mixing)	14							
SETTLE	3							
DRAW	0.25††							
IDLE	0.75							

* Initial value at the beginning of PAC_SBR studyj it gradually declined with the daily wasting of mixed liquor. The steady state mixed liquor biological solids concentration was dependent on the feed concentration and daily wasting volume.
† The mixed liquor wasting volume of the control unit was calculated (see † in table 6) to maintain a MLSS concentration of 10,000 mg/L.
** The PAC-SBR units were fed twice, 12.5% of working volume each time, at the beginning and the end of FILL.
†† Effluent discharge was accomplished using a 100-ml pipet.

RESULTS OF TREATABILITY STUDIES

SBR Biotreatment Results

The nutrient requirements were determined using the small 1-L and 12-L SBRs operating at a 24-h cycle schedule. The effects of nitrogen supplementation in the forms of NH_4-N and NO_3-N can be seen from the results given in Table 8. The leachate feeds to SBRs II and III were respectively supplemented with sodium nitrate and ammonium hydroxide to a TOC/NH_4-N or NO_3-N ratio of 15, as established in a separate series of experiments. Since Hyde Park leachate is deficient in orthophosphate (Table 3), phosphoric acid was added to all feeds to provide a 30 mg/L of PO4-P for bacterial growth. A 75-ml aliquot of pretreated leachate was fed to the SBRs four times in six hours; aeration and mixing were

Table 8. Results of Biotreatment of Leachate in 1 L SBRs

SBR† Sample	TOC (mg/L)	TOX (mg/L)	HET Acid (mg/L)	Phenol (mg/L)	Benzoic Acid (mg/L)	o-CBA (mg/L)	m-CBA (mg/L)	p-CBA (mg/L)
Feed	1,800	205	110	436	708	227	62	74
I Effluent (no nitrogen)	283	126	84	13	28	98	26	21
II Effluent (nitrate)	164	91	80	3	12	3	6	15
III Effluent (ammonia, 20°C)	152	95	76	4	12	1	5	17
IV Effluent (ammonia, 9°C)	155	88	69	3	11	1	<3	13

* Samples were taken at the end of the program.
† 24 h SBR cycle, HRT = 2.5 d (four 10% batch feedings in 6 h aerated FILL), MLSS = 10,000 mg/L, working volume = 750 ml.

provided between the batch feedings. The results obtained from the 2.5-d HRT SBRs showed that both forms of nitrogen addition had improved the treatment efficiency. Ammonia was used in the subsequent treatability programs since it also neutralized the naturally acidic leachate and the combined wastewater. Results for SBR IV, which was operated similarly as SBR III but at 9°C, showed that excellent treatment was also achieved at a lower temperature expected for the winter operation. The optimum nutrient supplementation for the leachate was validated by the result of the 12-L SBRs, as shown in Table 9. SBR IV, with a nitrogen and phosphate supplemented feed (TOC/NH_4-N/PO_4-P ratio of 150/10/2), accomplished the best treatment of the high TOC (8135 mg/L) leachate. Supplementing only N produced almost the same results as adding both N and P, indicating that the total-P in the leachate (Table 3) was available for bacterial synthesis. Aeration and mixing were provided during the 4-h FILL period to accelerate the biodegradation process. The high MLSS of 10000 mg/L was responsible for the excellent SBR performances in treating the high TOC leachate, since a reactor with a MLSS of 5000 mg/L failed early in this study program. All SBRs operated at a MLSS from 8000 to 13000 mg/L achieved at least 90 reduction in TOC and COD. Another important advantage of establishing a wide range of acceptable MLSS is that the excess biological sludge can be wasted periodically at the operator's convenience. Nitrification and denitrification were observed in a single SBR with no aeration during the last two hours of REACT. The second portion of Table 9 data show results obtained at much higher hydraulic loadings; SBR B was operated at a 1.7-d HRT (24-h cycle and 60% feed) and SBR III at a 1-d HRT (12-h cycle and 50 feed). Good treatment performances were achieved, using again the high MLSS for reducing the food to biomass ratio, the F/M, in the SBRs and a slower rate of feed rate with aeration and mixing. The 500-L SBRs were utilized for simulating full scale treatment. They provided data on on long-term treatment performance, under various operating conditions, fate of volatile compounds, and production and disposal of sludges. The feed was prepared by neutralization, aeration, nitrogen supplementation, and sedimentation in a 2000-L tank. Table 10 presents the results of the pilot-scale SBR biotreatment

Table 9. Results of Biotreatment of Leachate in 12 L SBRs

SBR[†] Sample	TOC (mg/L)	TOX (mg/L)	HET Acid (mg/L)	Phenol (mg/L)	Benzoic Acid (mg/L)	o-CBA (mg/L)	m-CBA (mg/L)	p-CBA (mg/L)
Feed	8,135	780	435	1,650	2,475	840	240	285
I Effluent (no N & P)	603	320	240	16	43	148	<9	23
II Effluent (ammonia)	393	270	239	<2	21	3	<4	33
III Effluent (phosphate)	595	330	273	16	47	127	<4	45
IV Effluent (N & P)	409	240	238	<1	7	3	<2	6
SBR Sample[**]								
Feed	1,784	210	135	390	590	220	60	70
B Effluent (24 h, 60% feed)	219	98	115	6	11	2	9	14
III Effluent (12 h, 50% feed)	256	98	90	7	8	34	7	12

* Samples were taken at the end of the program.
[†] 24 h SBR cycle, HRT = 10 d (10% feedings over 4 hFILL), MLSS = 10,000 mg/L, working volume = 10 ml.
[**] MLSS = 10,000 mg/L, FILL period = 6 h, working volume = 10 L.

with two HRTs (2 and 5 d) and two MLSSs (5000 and 10000 mg/L). The better performance of SBR B relative to SBR A demonstrated again the advantage of the high MLSS. The low quality of the effluent from SBR C indicated that it was overloaded at a HRT of 2 days. After increasing the HRT to 5 days, the SBR C performance gradually improved, and in two weeks its effluent was almost the same as SBR B. Table 10 also shows significant reduction in aquatic toxicity with well-treated effluent samples.

Table 10. Results of Biotreatment of Leachate in 500 L SBRs

SBR[†] Sample	TOC (mg/L)	COD (mg/L)	TOX (mg/L)	SS (mg/L)	HET Acid (mg/L)	Phenol (mg/L)	Benzoic Acid (mg/L)	o-CBA (mg/L)	m-CBA (mg/L)	p-CBA (mg/L)	Microtox[**] Toxity (EC20)
Feed	2,000	5,300	325		260	530	730	350	110	110	1.7%
A Effluent	140	510	110	114	170	6	6	12	25	3	41.9%
B Effluent	120	400	105	100	150	1	2	2	3	2	68.0%
C Effluent	536	1,700	235	400	175	12	6	20	25	3	9.3%

* Samples were taken at the end of the program.
[†] The experimental conditions are described in table 5.
[**] Average of three observations. EC20s (% dilution of test sample which would cause 20% corrected light reduction) are shown, since EC50s for the effluent samples (A, B and C) cannot be calculated as a result of SBR biotreatment.

PAC-SBR Treatment Results

Table 11 presents the treatment results in terms of reductions in TOC, TOX, HET acid, phenol, benzoic acid, and CBAs; Table 12 presents the reductions in TCDDs, PCBs, and five other halogenated compounds covered in the discharge permit (Table 1). The improvements in effluent quality due to the addition of PAC

Table 11. Results of PAC-SBR Treatment of Leachate

PAC–SBR† Sample	TOC (mg/L)	TOX (mg/L)	HET Acid (mg/L)	Phenol (mg/L)	Benzoic Acid (mg/L)	m-CBA (mg/L)	p-CBA (mg/L)
Feed	3,570	440	150	820	1,160	130	160
1C Effluent	286	196	102	3	6	20	16
3A Effluent	207	141	80	<1	4	5	9
3B Effluent	179	114	77	<1	2	4	7
4A Effluent	207	130	80	<1	2	10	8
4B Effluent	143	83	51	<1	2	5	5
6A & B Effluent**	179	106	71	<1	2	3	7
6C Effluent	121	55	63	<1	2	2	3

* Samples were taken at the end of the treatability study program.
† The experimental conditions are described in table 7.
** Average of the duplicate units.

Table 12. Removal of TCDDs, PCBs and Halogenated Organic Compounds in PAC SBRs

PAC–SBR† Sample	TCDDs* (ppt)	PCBs† (ppb)	Trichloro-benzenes (ppb)	C-56 (ppb)	2,4,5 Tri-chlorophenol (ppb)	Endosulfan (ppb)	Mirex (ppb)
Feed	1.5	9	68	37	39	51	26
1C Effluent	$ND_{0.8}$**	ND_2	ND_{10}	ND_{10}	ND_{10}	ND_{10}	ND_1
3A Effluent	$ND_{0.8}$	ND_2	ND_{10}	ND_{10}	ND_{10}	ND_{10}	ND_1
3B Effluent	$ND_{0.8}$	ND_2	ND_{10}	ND_{10}	ND_{10}	ND_{10}	ND_1
4A Effluent	$ND_{0.8}$	ND_2	ND_{10}	ND_{10}	ND_{10}	ND_{10}	ND_1
6C Effluent	$ND_{0.8}$	ND_2	ND_{10}	ND_{10}	ND_{10}	ND_{10}	ND_1

* 2,3,7,8-TCDD and co-eluting isomers.
† Arocolor-1248.
** ND = Not detected at a detection limit of x ppt or ppb.

were apparent. Most importantly, the effluent from PAC-SBR 3A, the least PAC-dosed unit, had consistently met the discharge limit during the entire study period. The PAC supplementation rate was less than 4% of the estimated granular activated carbon requirement for treating the raw Hyde Park leachate. The nearly complete removal of TCDDs, PCBs and other persistent halogenated organic compounds in the PAC-SBRs was due to the much higher adsorptive capacities of activated carbon relative to pretreatment precipitates and SBR biomass[3]. Because of the large inventory of PAC (3000 to 6000 mg/L), quality of effluents from the PAC-SBRs was much more stable than the control unit (No. 1C) when the organic loading was increased as a result of higher feed TOC and/or more daily feed volume. Nitrification and denitrification were also observed in all bioreactors with no aeration during the last two hours of REACT.

Observations of Treatment Performance

Up to 15% variations in the effluent TOC, COD and SS were observed for the replicated SBRs. Hyde Park leachate was well treated either alone or combined with other Niagara Plant wastewaters. The treatment performances were almost identical for the three sizes of SBRs when they were operated under the same conditions. Such results validated the use of the smaller (1-L and 12-L) units for the treatability study. Virtually the same performances were obtained for SBRs with several FILL periods - batch feed, 2, 4, and 6 h. The complex nature of the high TOC wastewater had diminished the effects of FILL period.

Insufficient dissolved oxygen in the mixed liquor was the major cause of low TOC reduction ($<85\%$). In treating a 2000 mgTOC/L leachate at a HRT of 1.5 d and a MLSS of 10000 mg/L, the maximum oxygen utilization rate (OUR) was about 150 mg DO/L·h near the end of the FILL, while the minimum OUR was 40 mg DO/L·h at the end of REACT. With at least 1 mg/L of DO during the REACT period, TOC and COD reduction were more than 90% for the SBR operated at a F/M as high as 0.2 mg TOC/mg MLSS·d (MLSS = 10000 mg/L).

Cloudy effluents (SS > 250 mg/L), due to large populations of dispersed and/or filamentous bacteria, were observed several times during this study. They were caused by excessive organic loading, short REACT period, low DO, nutrient deficiency, and accumulation of toxic compounds. After correcting these causes, the SBR treatment performance slowly improved. Effluent SS was less than 100 mg/L except when the feed TOC was higher than 3000 mg/L.

No significant improvement in effluent quality was observed in SBRs supplemented with doses of bacteria isolated from a landfill soil sample which were capable of degrading many persistent leachate constituents.[4] This was due to the fact that the leachate feed contained many of the same bacteria.

The treatment performance was nearly unchanged when the feeding was suspended on holidays, Saturday and/or Sunday. The REACT period for these units was extended over the weekend, with either continuous or periodic aeration and mixing. This resulted in slightly lower effluent TOC, more complete nitrification and/or denitrification. When normal feeding was resumed, the same performance as the 7 days-a-week units was then observed.

Many aromatic and straight-chain halogenated hydrocarbons were present in Hyde Park leachate, typically in less than 5 mg/L except for chlorotoluenes (as high as 40 mg/L). Air stripping was known to be the predominant mechanism for removal of volatile organics in the aerobic biological treatment system.[4] About 99% of the volatile constituents were air-stripped from the feed during pretreatment, and the rest were removed during aerated FILL and/or REACT. These compounds were removed from the air by carbon adsorption.

Depending on the wastewater characteristics, 200 to 2000 mg of pretreatment precipitates per liter of feed were produced by neutralization, aeration, and sedimentation. It was mainly ferric hydroxide which included adsorbed organic compounds, i.e., HET acid, PCBs and other chlorinated organics.[15] The sludge was well compacted in the feed tank, to about 7% solids, and easily dewatered by either vacuum or press filtration, to more than 30% solids. Biomass yield was estimated at 0.20 mg/mg feed TOC for the SBRs and was slightly lower in the PAC-SBRs.[16] The settled biomass sludge contained about 3.5% solids, and the dewatered sludge (by vacuum filtration and filter press) had about 30% solids.

Carbon Savings

Figure 3 shows the capacities of carbon for the persistent HET acid was significantly increased after the leachate was biologically treated in the SBRs. The improvement was actually higher than that indicated by the isotherms. The carbon capacity for HET acid ($pk_a s$ = 2.6 and 4.8) was higher at an acidic pH. In the unbuffered isotherm experiment, pH was found to increase with the carbon dose, resulting in reduced observed capacity at lower residual concentration.[17] Table 13 shows that the carbon capacity for TOC in the SBR-treated leachate sample was about the same as in the raw leachate and that the capacity for TOX was much greater after biotreatment. Similar results were obtained for five different types of activated carbon. Since the SBR biotreatment could easily accomplish 90% reduction of leachate TOC, the carbon exhaustion rate could be reduced by at least 90% with SBR pretreatment. Although the PAC-SBR treatment of leachate was slightly more cost effective than the two-stage SBR-granular activated carbon treatment because of the even lower carbon consumption rate and also because no need for post treatment by carbon adsorption, the latter process was selected for Hyde Park leachate treatment system to ensure complete removal of such recalcitrant compounds ag HET acid.[18]

FULL SIZE SBR-ADSORPTION LEACHATE TREATMENT SYSTEM

Because of the successful SBR treatability studies, an economic analysis was performed in 1984 for a proposed SBR biodegradation-carbon adsorption system for treating the Hyde Park leachate. With favorable economics established, the treatability data and estimated leachate volume and composition were utilized to formulate the design parameters (Table 14) for the proposed system [19]. During 1985-89, a decision was made to build a SBR-adsorption system at the landfill site, the leachate treatment process flow sheet was defined, the final design (Figure 4) was completed and subsequently approved by the regulatory agencies, process control and equipment were specified, and various contracts were awarded. In early 1990, construction was completed and the on-site treatment system started to receive leachate, and after several months of process calibration and start-up, the SBR-adsorption system began regular mode of treatment in July. The routine sampling schedule is shown in Table 15.

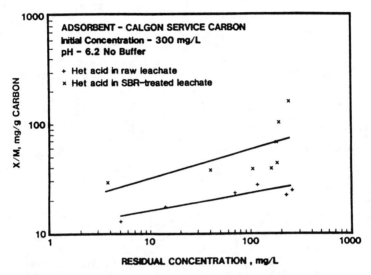

Figure 3 Freundlich adsorption isotherms for HET acid in raw and SBR-treated leachate samples.

Pretreatment

Raw leachate is received in two large storage tanks. The non-aqueous phase liquid (NAPL) fraction settled in the bottom and is periodically trucked to the Niagara Plant for disposal by incineration. The aqueous phase liquid (APL) fraction is pumped to the pH adjustment tank where caustic i continuously added under automatic pH control raising the pH to about 7.5. The pH adjustment tank is vented to one of the two vapor phase adsorbers (4000 lb carbon each) for removal of volatile organics. The vapor phase adsorber, which also process discharge air streams from the aeration tank, continuous sand filter, and SBRs which may also contain volatile organics, are steam regenerated. APL then flows into the aeration
tank by gravity; compressed air is introduced near the bottom of the tank through a ring sparger. Aeration IR provided to achieve both oxidation of ferrous ion to form insoluble iron hydroxide and stripping of volatile organics in the leachate. APL is then pumped to an inclined plate settler for removal of suspended solids. The solids move down along the plates into the thickener portion of the settler and, periodically, are pumped to a filter press for sludge dewatering. APL from the settler flows by gravity through a continuous sand filter and into a SBR feed tank (40,000 gal). In the sand filter, the APL flows upward while compressed air is introduced into the bottom to carry the solidcoated sand, along with a portion

Table 13. Adsorptive Capacities of Carbon for TOC and TOX in Raw and SBR-treated Leachate Samples

Activated Carbon Type	Raw Leachate		SBR-Treated Leachate**	
	TOC	TOX	TOC	TOX
	(mg adsorbed/g carbon)			
Calgon F-300	133	11.7	152	127
Calgon Service Carbon	97.9	8.8	113	75.9
Carborundum 30	173	19.6	268	172
ICI Hydrodarco 3000	103	11.5	87.8	83.8
Laboratory Reactivated Spent Calgon Service Carbon	148	18.3	115	91.6

* Adsorptive capacities were estimated from the Freundlich adsorption isotherms.
† Raw leachate: TOC = 3,080 mg/L, TOX = 264 mg/L, pH = 5.3. The TOC capacities were estimated at TOC = 1,500 mg/L, and the TOX capacities were estimated at TOX = 125 mg/L.
** SBR-treated leachate: TOC = 400 mg/L, TOX = 334 mg/L, pH = 6.8. (The raw leachate had a TOC of 8,100 mg/L and a TOX of 780 mg/L). The TOC capacities were estimated at TOC = 300 mg/L, and the TOX capacities were estimated at TOX = 125 mg/L.

of APL, up a center tube to a settling chamber where the sand is separated from the liquid and solid and then returned to the sand filter. The liquid stream is recycled back to the aeration tank.

SBR Treatment

The pretreated APL is pumped from the SBR feed tank through one of the two banks of two serial cartridge filters (100 and 5 micron) to remove remaining suspended solids. The standby bank of cartridge filters is automatically placed in the service mode when pressure dropped thru the running bank is increased to a pre-set limit. The APL then passes through two serial sacrificial adsorbers (2000 lb carbon each) for removal of dioxin and PCB. Ammonia and phosphoric acid are added to the APL, to a TOC/NH_4-N/PO_4-P ratio of 150/10/2 as determined in the treatability study, before it enters the SBR. Depending on the APL volume, up to 100,000 gal/day, one, two or all three SBR (40 ft dia. × 15 ft height, 80,000 gal working volume) are employed for biotreatment. The SBR are operated on a 24-hr cycle with a treatment schedule based on volume and TOC of the APL feed; for example, FEED - 6 h (120 gal/min, with aeration at 1400 SCFM), REACT - 10 h (aeration), SETTLE - 3 h, DRAW - 3 h, and IDLE - 2 h. The control of the SBR process is semi-automatic with the operator initiating each step of the process

Table 14. Design Parameters for the Hyde Park SBR-Adsorption Leachate Treatment System

Flowrate	75 gal/min; 10,800 gal/day
APL Leachate Composition	
pH	7.5 after neutralization of the raw leachate
COD	9,200 mg/L
BOD	7,200 mg/L
TOC	3,200 mg/L
TDS	22,400 mg/L
Total P	92 mg/L
P-PO$_4$	< 1 mg/L
TKN	160 mg/L
N-NH$_4$	130 mg/L
N-NO$_3$	20 mg/L
Major Organic Constituents	
Benzoic Acid	2,290 mg/L
Chlorobenzoic Acids	1,290 mg/L
Phenol	746 mg/L
Chlorendic Acid	450 mg/L
Biotreatment Specifications	
Treatment Objective	90% TOC Removal
TOC/N-NH$_4$/P-PO$_4$	150/10/2
Oxygen Transfer Rate	150 mg-DO/L-hr
Biomass Density in SBR	8,000 mg/L

and the valves are then automatically set in the proper position. To control the biomass density in the SBR, mixed liquid is periodically withdrawn to the sludge storage tank and subsequently dewatered in another filter press.

Carbon Adsorption

The SBR treated APL is chlorinated before entering the filter feed tank (40,000 gal) which is provided to enable continuous equalized feed to three parallel high rate sand filters. The filtrate is finally treated in the three-bed (20000 lb carbon each) adsorption systems. Two of the adsorbers are in the service mode receiving the feed in series, while the third adsorber is on standby. The standby adsorber is put in the polishing position when more than 1 mg/L of phenol is observed in the effluent from the leading adsorber which is then taken off from the treatment train for carbon replacement. The sand filters are backwashed, based on a pre-set time interval or pressure drop, using the final effluent from the effluent storage tank.

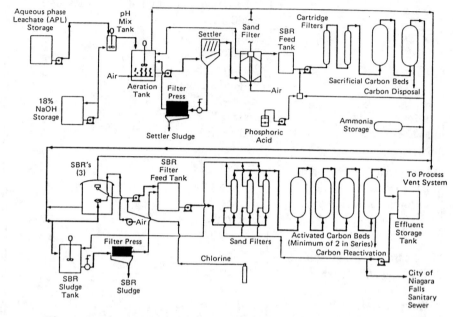

Figure 4 Process flow diagram for the hyde Park SBR-Adsorption leachate treatment system.

Process Control

The SBR-Adsorption treatment process employs a combination of continuous and batch processes. A Baily distributed control system, which allows both automatic operation and operator management of the process, is used as the primary control.

Treatment Performance

The overall SBR-adsorption treatment for the last two years has been excellent. Figures 5 and 6 presents the SBR treatment results for removal of TOC and phenol shortly after the start-up. Table 16 shows the treatment results were slightly better in 1991 compared to the same period in 1990. The two SBRs which were started with naturally grown bacteria had performed very well; in fact, they produced better quality effluent than the SBR started with imported activated sludge from the POTW which provided seed culture for the treatability study. The

Table 15. Hyde Park Leachate Treatment Sampling Schedule

Source	G-Grab C-Comp	Analysis	Frequency
Clarifier Feed	G	Visual – Floccing	1/day
Cont. Filter	G	Visual – Turbidity	1/day
SBR Feed	G	TOC, phenol, pH	1/day
SBR Feed	G	$N-NH_3$, $P-PO_4$	2/wk/reator
SBR Mixed Liquor	G	DO decay, phenol, 5 min/30 settle rates	1/batch
SBR Mixed Liquor	G	SS (Suspended solids)	2/wk/reator (VSS 1/wk)
SBR Effluent	G	SS	1/wk/reator (VSS 1/wk)
Carbon Bed Feed	G	TOC, phenol, pH	1/day
Carbon Bed Interstage	G	TOC, phenol, pH	1/day
Carbon Bed Interstage	G	Specific organics (See Effluent list)	1/week (Monday)
Effluent	C	TOC, phenol, pH	1/day
Effluent	C	Trichloroethylene Tetrachloroethylene Monochlorotoluenes Monochlorobenzenes Trichlorobenzenes Monochlorobenzotrifluorides Hexachlorobutadiene Hexachlorocyclohexanes 2,4,5-Trichlorophenol Octachlorocyclopentene	

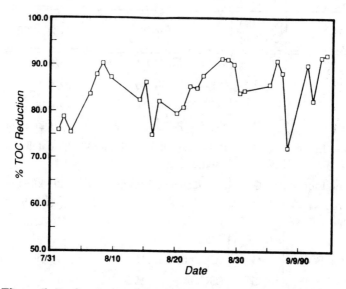

Figure 5 Performance of the SBR-Adsorption leachate treatment system - removal of TOC.

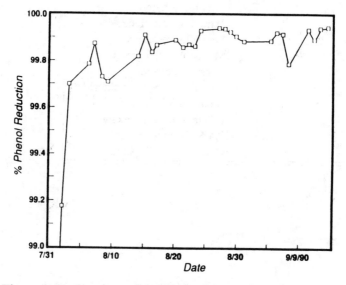

Figure 6 Performance of the SBR-Adsorption leachate treatment system - removal of phenol.

leachate strength has been significantly lower than samples employed in the treatability study, TOC ranging from 300 to 2000 mg/L and phenol from 100 to 800 mg/L. The reduced strength is a result of the fact more leachate is being collected in the expanded collection system as well as natural reduction in leachate strength over time.[20,21,22] Because of the reduced organic loading, only two SBRs, sometime one SBR, are employed for biotreatment when the leachate production rate is reduced. The increase in biomass density has been slow as a result of both the low feed TOC and the long aeration time in the SBRs. The low biomass production rate and the fact a wide MLSS range can be maintained in the SBRs.

Table 16. Performance Data for SBR-Adsorption Treatment System – Removal of TOC and Phenol

Date	SBR Feed		Adsorber Feed		Adsorber Inter Eff		Adsorber Effluent	
	TOC	Phenol	TOC	Phenol	TOC	Phenol	TOC	Phenol
7/27/90	677	156	158	15.9	134	16.3	24	0.08
8/1	585	161	139	7.8	130	0.71	31	0.06
8/6	690	143	112	0.31	124	0.28	35	0.14
8/9	1,380	371	174	1.1	106	0.15	47	0.12
8/14	764	246	131	0.44	109	0.36	51	0.28
8/17	482	119	86	0.15	75	0.10	72	0.07
8/21	459	144	87	0.20	79	0.40	70	0.22
8/24	690	157	85	0.10	60	0.20	64	0.13
8/28	1,089	217	95	0.14	67	0.15	58	0.08
9/4	673	125	90	0.14	122	0.13	50	0.08
9/10	946	153	95	0.10	38	0.19	52	0.11
7/28/91	579	159	39	0.13	31	0.03	28	0.06
8/7	758	199	69	0.17	49	0.05	8	0.05
8/25	492	142	71	0.33	56	0.16	11	0.05
9/7	1,064	282	32	1.60	42	0.16	9	0.01
9/17	1,243	286	57	0.32	74	0.46	7	0.12
9/24	448	128	101	0.68	101	0.51	10	0.02
9/30	247	42	88	0.50	36	0.03	9	0.04
10/5	985	486	77	0.29	30	0.10	14	0.05
10/17	620	150	89	0.44	42	0.05	8	0.05
10/27	976	301	73	0.29	34	0.24	12	0.24
11/3	802	176	115	0.37	46	0.08	10	0.09
11/11	1,146	276	129	0.54	64	0.21	36	0.06
11/19	1,161	234	95	0.29	98	0.31	40	0.04

TOC and phenol concentrations are reported in mg/L.

have significantly reduced the needs for sludge wasting and dewatering. The strong phenolic odor of the leachate has been virtually absent in the treatment building where the pH adjustment tank, clarifier, filters, and carbon adsorbers are located. The vapor phase carbon adsorbers remove all volatile organic compounds present in the exhaust air streams, and the steam regeneration ha been effective in renewing the spent carbon.

During the two years of system operation, only three carbon bed changes were made, one of which was for reasons not associated with organic breakthrough, while previously the leachate treatment carbon adsorption system required two new carbon bed every month. As shown in the operating record sheets (Tables 17 and 18), the SBR-adsorption treatment is treating more than 12 million gal/year of Hyde Park APL leachate and producing an effluent which typically contains less than 50 mg/L in TOC and 0.1 mg/L in phenol.

Table 17. Hyde Park Leachate Inventory

Inventory Balance (gallons)		31 Dec 1991		
		APL	NAPL	Total
On-hand, Beginning of Month		263,209	40,952	304,161
Collection	+	783,977	4,598	788,575
On-hand, End of Month	−	365,586	33,950	399,536
Processed	=	681,600	11,600	693,200

Collection, Shipping & Treatment (gallons/1000)	Current Month		Year-to-Date	
Collection	APL	NAPL	APL	NAPL
OBCS System	483.10	*	7,810.31	*
Bedrock	0.00	*	0.00	*
Source	0.00	*	0.00	*
Treatment Plant	305.47	*	12,736.53	*
Total	783.98	4.60	20,392.92	153.93
Processed	APL	NAPL	APL	NAPL
Shipped	0.00	11.60	0.00	153.61
Treated	681.60	NA	16,028.46	NA
Total	681.60	11.60	16,028.46	153.61

Table 18. Record of Hyde Park Treatment Daily Operations

Inventory (gallons)	APL	NAPL	Total
Leachate 1	86,420	0	86,420
Leachate 2	0	28,409	28,409
Decanter 1	0	0	0
Decanter 2	5,995	5,541	11,536
Decanter 3	0	0	0
SBR Feed	15,188		
SBR 1	102,228		
SBR 2	102,735		
SBR 3	25,954		
SBR Filter	3,129		
Effluent	11,907		
Aeration	4,098		
Sludge	7,932		
Rail Cars	0	0	0
Other On-Hand	0	0	0
Total	365,586	33,950	399,536

Solid Waste	On-Hand	Generation		Disposal	
		Daily	MTD	Daily	MTD
Settler Sludge	35	0	0	0	35
Bio-Sludge	0	0	0	0	0

Materials	On-Hand	Received		Consumed	
		Daily	MTD	Daily	MTD
Caustic	5,483	0	0	11	56
Ammonia (lbs)	3,671	0	0	0	517

SBR Effluent	
TOC	38 ppm
Phenol	0.07 ppm
Operated	12.0 Mrs
Average Flow	GPM
Gallons thru Bed	3,181.8 x 1,000

	Daily	MTD
Overtime	0	18
Downtime	12	196
Operating Hours	12	572
Carbon Changes		0

SUMMARY

Based on results of successful bench- and pilot-Scale treatability studies employing sequencing batch reactor (SBR) biotreatment and pot-treatment by carbon adsorption, a full scale SBR-adsorption system was built to treat leachate from the Hyde Park chemical landfill in Niagara Fall, New York. Since April 1990, up to 100,000 gal/day of leachate, containing a wide variety of organic contaminant, have been treated in three SBR for removal of biodegradable constituents and then in carbon adsorbers for removal of persistent constituent. Treatment performance of two SBR started with biomass grown on the leachate was found to be better than the third SBR started with activated sludge from a local POTW. Excellent treatment results were maintained during build-up of biomass in the SBRs, allowing a very high MLSS (up to 15,000 mg/L) before wasting. Overall, the treatment system has consistently removed more than 95% TOC (from up to 2000 mg/L) and virtually all phenol (from up to 800 mg/L) in the leachate. Because of the very effective biotreatment in the SBR, carbon exhaustion rate has been very low, less than 10% of carbon requirement of the former leachate treatment employing carbon adsorption alone. The treatment system is also equipped with vapor phase adsorption and regeneration as well as sludge thickening and dewatering units to provide complete treatment of leachate in a safe, clean and pleasant environment.

ACKNOWLEDGEMENT

M. Wendell, D. Ernst, G. Moscoto, G. LaLiberty, V. Lloyd, R. Bonk, M. Tucker, S. Sojka, R. Eddy, and E. Dietz contributed significantly to the success of the bench- and pilot-scale SBR treatability studies. J. Updyke and other Occidental Chemical engineers designed the full size SBR-adsorption treatment system.

REFERENCES

1. Ying, W., Dietz, E. A., and Woehr, G. C., "Adsorptive Capacities of Activated Carbon for Organic Constituent of Wastewater," *Environ. Progress*, 9, 1, 1990.

2. Ying, W., et. al., "Biological Treatment of a Landfill Leachate in Sequencing Batch Reactor," *Environ. Progress*, 5, 41, 1986.

3. Ying, W., Bonk, R. R., and Sojka, S. A., "Treatment of a Landfill Leachate in Powdered Activated Carbon Enhanced Sequencing Batch Bioreactors," *Environ. Progress*, 6, 1, 1987.

4. Sojka, S. A. and Ying, W., "Genetic Engineering and Process Technology for Hazardous Waste Control," *Develop. in Industrial Microbiology*, 27, 129, 1987.

5. Tabak, H. H., Quave, S. A., Mashni, C. I., and Barth, E. F., "Biodegradability Studies with Organic Pollutant Compounds," *J. Water Pollu. Control Fed.*, 53, 1503, 1981.

6. Patterson, J. W., and Kodukala, P. S., "Biodegradation of Hazardous Pollutants," *Chem. Engr. Progress*, 77(4), 48, 1981.

7. United States Environmental Protection Agency, *Summary Report - Sequencing Batch Reactors*, Technology Transfer Series, EPA/625/ 8-86/011, 6, 1986.

8. Irvine, R. L., and Busch, A. W., "Sequencing Batch Biological Reactors," *J. Water Pollu. Control Fed.*, 51, 235, 1979.

9. Irvine, R. L., and Letchum, Jr., L. H., "Sequencing Batch Reactor for Biological Wastewater Treatment," *CRC Crit. Rev. Environ. Control*, 18, 255, 1988.

10. "Standard Test Method for Biodegradability of Alkylbenzene Sulfonates," ASTM Method D2667-82, April 1982.

11. Rea, J. E., Jr., "Single Cell Activated Sludge Using Fill and Draw Combined Industrial/Domestic Waste Treatment Plant," *Proc. of the 32nd Purdue Industrial Waste Conf.*, 611, 1978.

12. Herzbrun, P. A., Irvine, R. L., and Malinowski, K. C., "Biological Treatment of Hazardous Wastes in SBR," *J. Water Pollu. Control Fed.*, 57, 1163, 1985.

13. Silverstein, J., and Schroeder, E. D., "Performance of SBR Activated Sludge Processes with Nitrification/Denitrification," *J. Water Pollu. Control Fed.*, 55, 377, 1983.

14. Lurker, P. A., el. al., "Worker Exposure to Chlorinated Organic Compounds from the Activated-sludge Wastewater Treatment Process," *Am. Ind. Hyg. Assoc. J.*, 44, 109, Feb. 1983.

15. Ying, W., Duffy, J.J., and Tucker, M. E., "Removal of Humic Acid and Toxic Organic Compounds by Iron Precipitation," *Environ. Progress*, 7, 262, 1987.

16. Grieves, C. G., et. al., "Powdered Verus Granular Carbon for Oil Refinery Wastewater Treatment," *Jour. Water Pollu. Control Fed.*, 52, 483, 1980.

17. Ying, W., *Investigation and Modeling of Bio-phyicochemical Processes in Activated Carbon Columns*, Ph. D. Dissertation, Univ. of Michigan, 70, 1978.

18. Ying, W., "Integrated Treatment of Hazardous Landfill Leachate," in *Encyclopedia of Environmental Control Technology*, Editor: P. Cheremisinoff, Gulf Publishers, Chapter 18, 1989.

19. Ying, W., Sojka, S. A., and Lloyd, V. J., "Integrated BiologicalAdsorption Process for Treating Waste Water," U. S. Patent 4,755,296, July 5, 1988.

20. Cheremisinoff, P. N., and Gigliello, K. A., Leachate from *Hazardous Waste Sites*, Technomic Publishing Co., Lancaster, PA., Chapter 6, 1986.

21. Kosson, D. S., Dienemann, E. A., and Ahlert, R. C., "Characterization and Treatability Studies of an Industrial Landfill Leachate," Proc. of the 39th Industrial Waste Conf., 329, 1985.

22. Albers H., and Kayer, R., "Two-Stage Biological/Chemical Treatment of Hazardous Waste Landfill Leachate," Proc. of the 42nd Purdue Industrial Waste Conf., 893, 1988.

32

ACHIEVEMENTS IN POLLUTION PREVENTION VIA PROCESS AND PRODUCT MODIFICATION

Kevin F. Gashlin and Daniel J. Watts
Hazardous Substance Management Research Center
New Jersey Institute of Technology
Newark, NJ

INTRODUCTION

Effective process or product design depends upon careful consideration of the product's eventual use, level of technology in use and previous experiences. The increasing emphasis on toxic use reduction/pollution prevention presents additional challenges for product design which can be met, in part, by understanding production and product related issues vis-a-vis proposed process changes. The results of waste reduction audits of selected processes which address the issues of toxic use reduction and the effect of client/supplier relationships will be discussed.

During the course of applying source reduction and recycling principles to business specific circumstances, the management and staff of the New Jersey Technical Assistance Program for Industrial Pollution Prevention (NJTAP) have found that case-study examples frequently lack important insight that the pollution prevention team may have gained during the course of the project. Determining why products are made the way they are, what purpose the product serves and what motivates industry to design products the way they do is a necessary precursor to realistically evaluating pollution prevention options, because such changes frequently affect product performance, cost of production and speed of production, in addition to collateral environmental regulatory issues. Consideration of these issues enhances the credibility of pollution prevention as a viable environmental management option and therefore improves the likelihood of implementing a successful, sustainable pollution prevention program. The following examples selected from NJTAP files examine the product vs. environment conflict and the motivational factors that caused production techniques to be questioned. While many of the issues raised may seem specific to the particular industry studied, they illustrate general principles and challenges which can affect pollution prevention initiatives.

Radiator Repair Shop

Process Description—Automobile and truck radiators are drained, degreased in hot caustic baths and soldered to repair leaks. Repaired units are spray painted prior to shipment.

The business is a fully regulated RCRA generator, discharges to a POTW and has an emissions permit for VOCs and lead fumes.

Product Component/Raw Material

Although manufacture of plastic units for new vehicles has become more common, this paper examines repair of radiators which have been traditionally manufactured of steel and copper, held together with tin/lead solder. As in domestic plumbing, tin/lead solder use has a long history. It is used for two primary reasons; low cost and effectiveness.

Wastes Generated and Recommendations

The following wastes are regulated by USEPA and NJDEPE as RCRA hazardous wastes.

a. **Caustic wastewater contaminated with heavy metals**

Currently sewered. The scouring effect of the caustic removes not only rust, oil and dirt, it also removes zinc, iron, cadmium and lead from the radiator. The discharge is not in compliance for pH or metals. A typical wastewater discharge analysis is provided below.

Metal	Concentration (ppm)	Daily Discharge Limit (ppm)
Zinc	120	2.6
Iron	83	5.0
Cadmium	0.38	0.55
Lead	3000	0.69

*pH = 13.2

Recommended substituting manual cleaning of radiators to reduce volume and toxicity of waste. Another alternative is a "zero discharge" system that will treat wastewater and eliminate the discharge while concentrating metals in sludge, enhancing recovery viability.

b. Sludge contaminated with heavy metals

Change work flow SOP to eliminate the addition of lead/tin solder drippings to the wastewater and sludge.
Substitute lead-free solder; determine feasibility of metal recovery in sludge.

c. Paint cleanup and associated wastes

Potentially a regulatory or emission permit issue to be confronted. Consequences are unclear at this time. Recommended the use of latex based paint.

The following wastes are prohibited from unpermitted discharge by local, state and/or federal law.

d. Antifreeze (local sewer prohibition)

Recommended purchase, an antifreeze recycling unit.

e. CFC-12 - Prohibited from discharge by federal law as of January 1, 1992. Production and use phase-out by 2000. Regulations are pending. Phase-out of HCFCs (CFC substitutes) by 2030. Recommended purchase of an EPA approved Freon reclaiming unit.

Implementation, Results and Consequences:

Process changes are reviewed below.

a. Manual Cleaning

This suggestion was rejected. Though this technique is documented as working successfully elsewhere, the client believes that manual reaming would not be as effective as the current practice and would increase labor cost. Thorough surface cleaning is also needed in order to prepare the entire surface of the unit prior to painting.

Water Treatment System (on order)

Reportedly will reduce water use by 70-80% but will increase amount of RCRA sludge generated. Waste will be disposed through metal reclaimer (secondary smelter) without cost to the generator, providing metal content is maintained.

Capital cost = $11,000, payback period 3-4 years (est.). - RCRA and water discharge compliance problems eliminated.

b. Sludge Contaminated With Heavy Metals

Lead-Free Solder: Reduced lead solders are being evaluated. Strength at high temperature and pressure is questionable. However, reducing or eliminating this source of lead may not render sludge non-hazardous. Much of the lead comes from the caustic bath scouring of the radiator itself. Also, reduced lead in the sludge could reduce its attractiveness to secondary smelters, increasing disposal cost.

To maximize the effectiveness in reducing lead bearing wastes, substitutes must be used by Original Equipment Manufacturers (OEM) so that it is not introduced to the process.

Work Flow/SOP Modification: Similar to the solder issue, the client's position is that reduced lead could actually reduce disposal options. The current system is "convenient" and saves labor costs.

c. Antifreeze

Reclaimer purchased, working "beautifully." Payback undetermined. Discharge eliminated.

d. Solvent Based Paint and Clean-up Solvents

Latex paint is now being used. VOC emissions and permit needs have been eliminated. Based on annual use of 600 gallons, 378 gallons or 2797 pounds of solvent emission has been eliminated. While not in itself significant, when multiplied by the number of radiator shops in the state (250), the industry reduction potential (IRP) for VOC emission reduction is:

$$IRP = 2797 \text{ lbs.} \times 250 = 699,250 \text{ lbs.}$$

However, paint cost = 2 × that of the solvent based paint. Transfer efficiency was initially a problem, but adaptation of a high-volume, low pressure (HVLP) unit has resolved that issue satisfactorily. Adherence is a more persistent problem, requiring even more thorough surface preparation prior to painting.

An unexpected issue is the aesthetic quality of the finish. The latex paint renders a matte finish, not the gloss finish that customers have come to expect. This has caused some problems, but the owner says he is overcoming this concern by explaining the environmental issues to his customers.

e. CFC-12

Reclaimer purchased, emissions eliminated. Payback period undetermined.

Understanding Actions

In this case the shop owner championed the application of innovative technology, even though the payback period for capital costs is uncertain. Regulatory compliance, and altruism appear to be the primary motives. He was less enthusiastic about changing his SOP because his perception was that to do so would increase labor cost, diminish product quality and (ironically) sludge quality. These are bread and butter issues.

Customer expectations and aesthetic preference became an issue to overcome. Who could have expected that people really care how a repaired radiator looks?

It is important to recognize that the job shop owner is limited in his ability to reduce waste because the hazardous constituents enter the shop as part of the radiator, as combustion byproducts or fluids manufactured elsewhere. This problem—lack of control—is a significant one observed across many small industries. The job shop is really an extension of the OEM or customers demand. Materials are frequently dictated by the client or OEM and the jobber is left with the environmental responsibility.

Manufacturing Alternatives

Historically, material choice has been heavily influenced by the sum of the technology + cost + purpose equation. Environmental considerations are relatively new, but are revolutionary in their effect on product design. In this particular case a related environmental/economic issue, the desire to improve fuel economy, resulted in development of lightweight plastic or aluminum radiators. Because they are assembled without using solder a secondary advantage is the source reduction of heavy metal waste.

At this time plastic radiators are not repairable by the traditional industry. A leaking plastic radiator may be recyclable as plastic but not as a radiator. This will cause maintenance costs to rise. As older vehicles are phased out the demand for radiator repair will lessen, eliminating these wastes.

"Progress" of this nature is not without cost. Manufacture of plastic radiators will contribute to other forms of pollution. Those employed in the repair industry and supporting industries will need to be retrained.

Commercial Freezer Manufacturer

Process Description: Sheet metal is shaped to the desired configuration, painted and filled with insulating foam. The units are fitted with appropriate refrigeration hardware constructed elsewhere. The units are then assembled to form commercial freezer units. The business is a fully regulated RCRA generator, has many air permits for VOCs and a discharge to POTW permit.

Product Components/Raw Material

Sheet metal is primed with zinc phosphate and finished with a solvent based paint using an electrostatic spray system. Grandfathered air permits exempt the company from the most stringent controls. Changes to the coating operation would require an application for new permits which would likely be more restrictive than those existing. CFC-12 gives the foam its insulating characteristics. It is mixed with diphenylmethane diisocyanate and blown into a sheet metal mold, forming the walls of the freezer cabinet. CFC-11 is used as the blowing agent. CFC use as a blowing agent or as a component in rigid insulation while being phased out, are unregulated, still inexpensive and provide the best value for this application. Use of these substances in this way will be banned within two years.

Wastes Generated and Recommendations

a. **CFC-11**

About 50% of the blowing agent is lost to the atmosphere during the process.
NJTAP recommended evaluation of two alternative foam manufacturing processes. These processes utilize a less ozone depleting HCFC.

b. **Methylene Chloride (MCF)**

This solvent is scheduled for phase-out by 2002. It is used to clean equipment used in the foam making process. Cleaning must occur after each job because the mixture cures very rapidly (continuous production is not possible given the existing work flow configuration). Neither of the alternative foam blowing processes need to use methylene chloride.

c. **Paint and Paint Related Wastes**

A large volume of solvent waste is generated in part due to the company's desire to meet any color needs their customers may have. Changing color requires the spray paint system to be cleaned, resulting in 2000 - 2500 gallons of RCRA waste being generated.

Overspray is a significant problem. A new, $1.5 million dollar electrostatic system is not working properly. Transfer efficiency is ~50%.

NJTAP recommended consulting the spray system's manufacturer and vendor to improve operating efficiency; evaluating water-based and powder coating systems as replacements; initiating an employee incentive program and purchasing a suitably sized distillation unit to reclaim used solvents the existing system was retaining.

Implementation, Results and Consequences

a. CFC-11 & 12

Alternative HCFC formulas and manufacturing processes have been evaluated. An important issue under review is the effect of a substitute process on the product's R-Factor and cost of production. The company is in the final stages of testing and product redesign, and has begun to accept orders for the CFC-Free product. This does not address the HCFC phase-out scheduled for 2030 nor does it deal in any way with production quotas which will be frozen in 2015. It does give the industry time to develop alternatives.

b. Methylene Chloride

The replacement process eliminates the need for methylene chloride, eliminating regulatory concerns.

c. Paint and Paint Related Wastes

The company prides itself on the ability to provide customers with an almost unlimited choice of colors. This precludes the powder coating alternative at this time. A water-based system is used by others in the industry and remains a possibility. However, the significant and recent investment in their electrostatic system has influenced management to defer the evaluation of other methods. Consultation with the system's manufacturer is a reasonable and low-cost approach that will not effect product quality adversely. Regulatory pressure via the Clean Air Act Amendments may eventually force the issue. The company could be left with an expensive choice either way - a VOC emission treatment system or purchase of a non-solvent coating system, but at this time it is not a high priority.

Understanding Actions

Foam manufacturing is being amended largely due to regulatory intervention coupled with the development of available technology, spurred on by unprecedented global concern. Because the coming CFC phase-out ban affects all manufacturers more or less equally, product redesign to compensate for R-Factor changes puts no one manufacturer at a significant disadvantage. In fact, the leaders in this area are likely to market their product as a "green" one.

The painting issue has been tabled. Certainly the technology exists to reduce raw material costs waste, but cost of existing and candidate technologies coupled with uncertainty over pending Clean Air Act Amendment regulation makes the expenditure unattractive at this time. The company's flexibility fills a niche, and it takes pride in meeting the demands of its customers, a quality that gives their product a competitive edge.

Employee morale was discussed. No one, least of all the spray paint operators, appears to care that the equipment was performing poorly. Recognizing this, the company is initiating a Total Quality Management program, but changing attitudes will take time.

Manufacturing Alternatives

Other options for addressing the VOC/Clean Air issue do exist. The assembly of cabinets manufactured with prepainted sheet metal components or color impregnated plastics may be viable. The first option could merely transfer New Jersey's air problem elsewhere, probably abroad, causing a loss of jobs to compound environmental damage. The second option must be evaluated in terms of customer acceptance, durability, capital and added cost to the product and environmental effect.

Finally, a middle-ground proposal of sorts would have their largest single color (white) coated using powder or waterbased alternatives, thereby moderating the need for clean-up and downtime while maintaining the ability to offer a broad choice of colors.

A Final Note

There are tradeoffs in industry that effect the pollution prevention project manager's ability to effect change. Much depends on what motivates top management. As illustrated here, regulatory issues, capital expense, corporate and personal ego, customer demands, concern for the environment and the tenacity of the corporate decision makers play a role. Resolving one problem may cause another. Understanding the consequences of an action to the degree possible as well as the degree of freedom which one can make changes is essential to developing a productive, sustainable pollution prevention program and to the economic health of the company.

In the legal and regulatory system, OEM and jobber environmental responsibility is typically segmented. Rarely has the manufacturer been assigned liability as a result of wayward or discarded products. This leads to reduced concern about the ripple effects that a product might have on the environment or on the economic liability that supporting industries incur. In their defense, manufacturers cannot possibly foresee all. Ozone depletion is not caused directly by the manufacture of CFCs, freezers and air conditioners, it is caused by the release of CFCs to the atmosphere. Lead poisoning has not been caused by the manufacture of radiators, paint or lead shot, but rather by its ill-advised use and release of lead to the environment. More common sense and technical assessment about the fate of a product's component materials throughout its life-cycle is needed in order to avoid a repetition of such occurrences in the future. Perhaps an examination of applying product liability law to such circumstances would aid in raising industry's consciousness to such issues.

33

PHOTOCATALYTIC OXIDATION OF DICHLORVOS IN WATER USING THIN FILMS OF TITANIUM DIOXIDE

Ming-Chun Lu
Institute of Environmental Engineering,
National Chiao Tung University
Taiwan, R.O.C.

Gwo-Dong Roam
Bureau of Environmental Sanitation and Toxic Substances Control
Environmental Protection Administration
Republic of China

INTRODUCTION

Photocatalytic degradation of organic and inorganic materials on semiconductor powder is a new method for wastewater treatment. Several pollutants in aqueous solution have been reported to be degraded by this method (Al-Ekabi et al., 1988; Matthews, 1987, 1990; Pelizzetti et al., 1990; Terzian et al., 1991). By photocatalytic degradation, the organic pollutants were mineralized. The formation of harmless mineral products is of great importance in water treatment because partial oxidation could conceivably lead to more toxic products than the parent compounds (Matthews, 1990). However, the vast majority of the investigation in this area has employed suspensions of semiconductor. Fixation of the photocatalyst on a stationary support can avoid the need to filter the catalyst from the reaction mixture. The fixed system is stable and efficient for the photodegradation of organic compounds in dilute solutions (Sabate et al., 1991). 2,2-Dichlorovinyl dimethyl phosphate (dichlorvos), a highly hazardous chemical according to WHO rating, is prepared by the reaction of trimethyl phosphate and chloral and is a widely employed insecticide (Fishbein, 1972; Watterson, 1988). It is hardly degraded by biological treatment. Harada et al. (1990) have studied on the photocatalytic degradation of dichlorvos in aqueous titanium dioxide suspensions. It was found that the degradation followed a first-order rate expression. The final degradation products were Cl^-, PO_4^{3-}, H^+ and CO_2. The reaction stoichiometry is:

$$(CH_eO)_2 \text{ POOCHCCl}_2 + 9/2O_2 \rightarrow PO_4^{3-}$$
$$+ 2Cl^- + 4CO_2 + 5H^+ + H_2O$$

In this work, we examined the photocatalytic degradation more further in details. The kinetics of the oxidation of dichlorvos in aerated aqueous media on glass matrix supported TiO_2 in a flow system were reported. In our preliminary research, we have concentrated on the effects of flow rate and temperature on the oxidation rate of dichlorvos (Lu et al., 1991). In this report, therefore, we have studied over a wide range of initial concentrations of dichlorvos, oxygen, electrolytes and hydrogen peroxide. It was found that electrolytes inhibit the degradation rate of dichlorvos, and $NaClO_4$ is the most inert. The influence of oxygen concentration on oxidation rate is obvious. Besides, hydrogen peroxide will compete with dichlorvos for decomposition.

EXPERIMENTAL SECTION

Materials

The dichlorvos was with a 96.33% purity. Titanium dioxide was Degussa P-25 (BET surface area, 50 ± 15 m^2/g; mostly in the anatase form). The water employed in all solution preparations was purified by a Milli-Q/RO system (Millpore) and has a resistivity of $\rho > 18$ MΩcm. The other chemicals and solvents were of reagent grade and used without further purification.

Apparatus

The reactor was composed of a Pyrex glass tube (0.6cm i.d. and 6.6m long) wound 48 times to form a coil which supported 175 mg of TiO_2. The total surface of the coil was 1244 cm^2; TiO_2 was coated on the surface of the coil at 0.14 mg/cm^2. The total volume of the coil was 187 m^3. As shown in Figure 1, a 20-W blacklight fluorescent lamp (Iwsaki Electric Co., LTD), 32 mm diameter, 580 mm long, fitted into the spiral and the unit was mounted in a standard domestic 20-W fluorescent lamp holder. Solutions (1 liter) were circulated through the spiral via reservoir and peristaltic pump in a loop. The reservoir was surrounded by a thermostatically controlled water bath (30^{0}C). Solution pH value and dissolved oxygen were monitored and recorded.

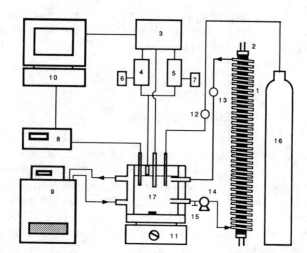

1. reactor (spiral)
2. blacklight fluorcesence lamp
3. pH controller
4. peristaltic pump for acid solution
5. peristaltic pump for base solution
6. acid solution
7. base solution
8. D.O. meter
9. water bath
10. computer
11. magnetic stirrer
12. gas flow meter
13. liquid flow meter
14. peristaltic pump
15. sampling port
16. gas
17. reservoir

Figure 1 Apparatus of photocatalytic system.

Analysis

Dichlorvos were analyzed by gas chromatograph (Hewlett-Packard HP 5890A) with ECD detection (1 μL injection volume). Supelco PTE-5 GC column was used (0.53mm i.d. 15m long). Chloride and phosphate were analyzed by ion chromatograph (Dionex-4500i). Carbon dioxide was measured by the method that the gaseous effluent bubbled through a saturated barium hydroxide solution to trap carbon dioxide.

Actinometry

Potassium ferrioxalate solution was used to measure the photon flux in the coil before it was coated with TiO_2 (Hatchard et al., 1956). Since the maximum emission of the lamp occurred at approximately 360 nm, the quantum yield for ferrous ion formation in the ferrioxalate solution was taken to be 1.28 (Demas et al., 1981).

RESULTS AND DISCUSSIONS

Effect of Initial Concentration

The effect of initial dichlorvos concentration on degradation rate is presented in Figure 2. It is obvious that the lower the initial concentration the higher the efficiency of dichlorvos oxidation and the photocatalytic decomposition of dichlorvos follows the first-order reaction kinetics (Figure 3). The half-life times are compared in Figure 4.

D'Oliveira et al. (1990) reported that initial rate of the photocatalytic transformation of 3-chlorophenol first increases sharply and then progressively levels off with increasing initial concentration. In the case of the effect of solute on the carbon dioxide formation from various solutes in UV-illuminated aqueous suspensions of titanium dioxide powder, it was concluded that the formation rate of carbon dioxide was given by a form of the Langmuir adsorption isotherm (Matthews, 1987). Davis et al.(1990) also showed a Langmuir-type rate expression for the photooxidation of phenol by cadmium sulfide. However, the chlorophenol oxidation by titanium dioxide was not similar to cadmium sulfide (Tseng et al., 1991). They have attributed this relation to the role of adsorption process on overall photocatalytic reaction of cadmium sulfide. In other words, adsorption appears not to be a significant step in chlorophenols oxidation by titanium dioxide.

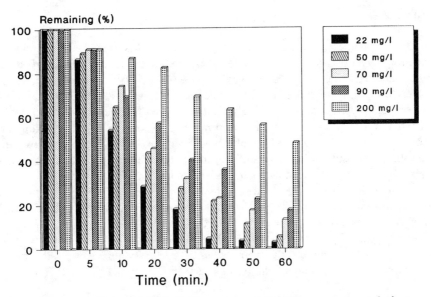

Figure 2 Effect of initial dichlorvos concentration on photocatalytic degradation. Experimental conditions: pH = 4, flow rate = 500ml/min, O_2 purge.

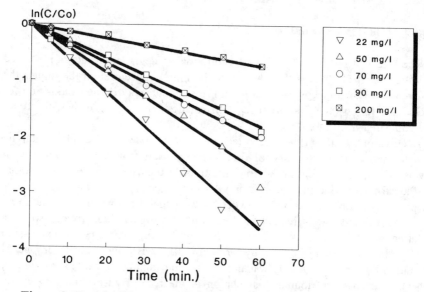

Figure 3 Pseudo-first-order kinetics for dichlorvos in the photocatalytic degradation. Experimental conditions: pH = 4, flow rate = 500 ml/min, O_2 purge.

Effect of Oxygen

During photocatalytic oxidation oxygen is required to trap the photoinduced electrons. The reaction rate is controlled by the quality of oxygen adsorbed on titanium dioxide surface in the Langmuir mode (Ollis, 1989; Okamoto, 1985). In order to investigate the effect of oxygen on the oxidation rate, a series of experiments were conducted using different oxygen/nitrogen ratios to create a different initial dissolved oxygen concentration. As shown in Fig.5, the oxidation rate of 50 mg/l (226 μM) dichlorvos increases with increasing oxygen concentration, and the reaction hardly occurs under a nitrogen gas bubbling. It is clear that oxygen is crucial to photocatalytic oxidation.

From measurements carried out in systems with oxygen and illuminated titanium dioxide, it has been reported that several types of negatively charged adsorbed oxygen species has been derived: $O_2^-, O_3^-, O^-, O_2^=$. The formation of $HO_2^•$ radicals from O_2^- and H^+, giving rise to H_2O_2 has been proposed as another way of producing $OH^•$ radicals.(D'Oliveira et al., 1990). By this viewpoint, it can comprehend that O_2 is significantly important in this process.

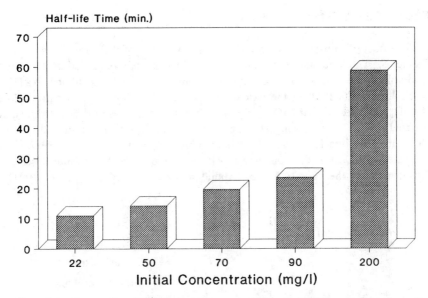

Figure 4 Comparing half-life time among different initial concentration of dichlorvos. Experimental conditions: pH = 4, flow rate = 500 ml/min, O_2, purge.

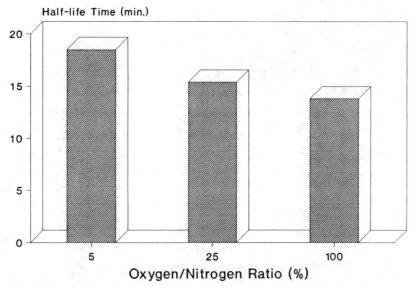

Figure 5 Effect of oxygen concentration on dichlorvos oxidation. Experimental condition: pH = 4, flow rate = 50 ml/min, [dichlorvos] = 500 mg/l, O_2 purge.

Effects of Electrolytes

The effect of inorganic electrolytes on the oxidation of dichlorvos is shown in Figure 6. The rate and the extent of dichlorvos photodegradation are most favorable in the absence of electrolytes. Among the electrolytes studied, $NaClO_4$ is the least influential on the oxidation rate. By changing the concentration of $NaClO_4$, it is found that ClO_4^- only slightly inhibits the reaction rate (Fig.7). It is noted that Cl^- is the strongest inhibitor on photocatalytic oxidation. ClO_4^- and NO_3^- are less interfering agents. The inhibitory effect of Cl^- is attributed to compete with O_2 for electrons and block the active sites of the catalyst surface thus decreasing the oxidation rate. (Abdullah *et al.*, 1991; Tseng *et al.*, 1991).

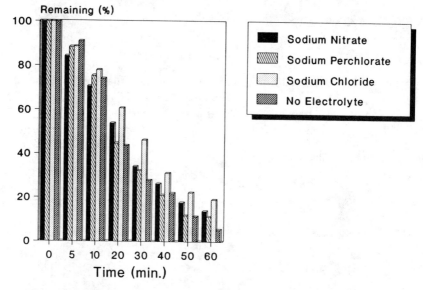

Figure 6 Effect of electrolytes on dichlorvos oxidation. Experimental conditions: Flow rate = 500 mL/min, pH = 4, [Dichlorvos] = 50 mg/l, O_2 purge.

Effect of Hydrogen Peroxide

The addition of H_2O_2 will change the rate of photocatalytic oxidation (Figure 8). In a pervious study, Harada *et al.* (1990) have reported that the addition of 12 mM hydrogen peroxide shortens the half-life by 10-fold during degradation of 1.0 mM dichlorvos under illumination by a 500W super-high pressure mercury lamp. In the absence of either titanium dioxide or hydrogen peroxide, they have found that this effect was not merely the sum of the effect of illuminated hydrogen

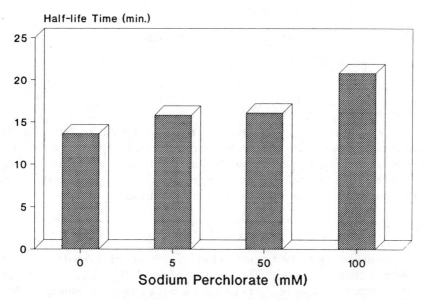

Figure 7 Effect of a NaClO$_4$ on dichlorvos oxidation. Experimental conditions: pH = 4, flow rate = 5 mL/min, [dichlorvis] = 50 mg/l, O$_2$ purge.

peroxide and that of illuminated catalyst. Auguliaro *et al.* (1990) have shown that the rate of phenol photooxidation is independent on the hydrogen peroxide concentration (8.8-20.6 mM), and the two compounds adsorb on sites of different nature. According to their findings, the mechanism of hydrogen peroxide photodecomposition on the titanium dioxide surface, which produces two different radical species able to react with adsorbed organic, is proposed:

$$H_2O_{2(ads)} + h^+ \rightarrow {}^{\bullet}HO_2 + H^+$$

$$H_2O_{2(ads)} + e^+ \rightarrow {}^{\bullet}OH + H^-$$

$$Organics_{(ads)} + \begin{array}{c} {}^{\bullet}HO_2 \\ {}^{\bullet}OH \end{array} \rightarrow CO_2 + H_2O$$

On the other hand, hydrogen peroxide can be photodegraded over illuminated titanium dioxide surface by the followed reaction (Korman *et al.*, 1988; Auguliaro *et al.*, 1990):

$$TiO_2 + h\upsilon \rightarrow TiO_2(h^+ + e^-)$$

$$H_2O_{2(ads)} + 2H^+ \rightarrow O_2 \ 2H^+$$

$$H_2O_{2(ads)} + 2H^+ + 2e^- \rightarrow 2H_2O$$

$$H_2O_{2(ads)} + h\upsilon \rightarrow 2H_2O + O_2$$

From experimental observations, we have found that the oxidation rate of dichlorvos decreases by the addition of hydrogen peroxide. Thus, there may exist a competitive decomposition between dichlorvos and hydrogen peroxide, and the production of radical species from hydrogen peroxide is negligible. However, further study is needed in order to clarify this speculation.

Quantum Yield

The measured rate of ferrous ion formation in 1 liter of 0.006 M potassium ferrioxalate circulated through the reactor, without TiO_2, was 546 μ M/min. Taking quantum yield of Fe^{2+} to be 1.28, a rate of supply of photons, 427 μ M/min is calculated. The quantum yield (QY) of the photocatalytic oxidation can be defined as the ratio of the number of molecules of dichlorvos reacting to the number of the photons produced.

$$QY = [\text{rate of degradation/rate of supply of photons}] \times 100$$

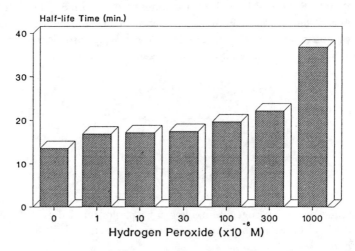

Figure 8 Effect of hydrogen peroxide dichlorvos oxidation. Experimental conditions: pH = 4, flow/rate = 500 mL/min, [dichlorvos] = 50 mg/l, O_2 purge.

Since quantum yield depends on the experimental parameters, it is essential to define the variables before calculating, and assume that all the light is absorbed by the stationary titanium dioxide. Under the experimental conditions (20-W lamp, 30^0C, 500mL/min circulation rate, pH 4, pure oxygen aeration, 226 μ M dichlorvos), the initial degradation rate was found to be 11.39 μ M/min. Therefore, the initial quantum yield for the destruction of dichlorvos is 2.67% (11.39/427 × 100%).

Sabate *et al.* (1991) has shown that when the feed gas is the pure oxygen and the concentration of 3-chlorosalicylic acid in liquid phase is 100 μ M, the apparent quantum yield is 0.12%. Matthews (1987) has found that the quantum yield of salicylic acid degradation at 25^0C is 2.2%. The low quantum yield reflects electron/hole pair recombination as the dominant energy wasting reaction. Consequently, it is necessary to improve the efficiencies of photon utilization in future research.

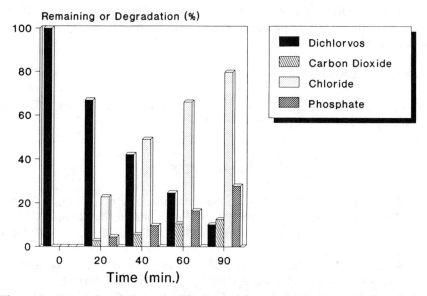

Figure 9 Comparing carbon dioxide formation and dichlorvos oxidation during photocatalytic degradation. Experimental conditions: pH = 4, O_2 purge, flow rate = 500 ml/min, [dichlorvos] = 140 mg/l.

Carbon Dioxide Formation

The evolution of CO_2 during dichlorvos degradation was measured by the total amount of CO_2 determined by trapping in $Ba(OH)_2$ solutions (see Experimental Section). These results show that only ca. 12.5% of expected CO_2 is obtained when dichlorvos is completely destroyed (Fig.9). Comparing the expected formation of CO_2 with those Cl^- and $H_2PO_4^-$, one can find that Cl^- released rate is faster than that of $H_2PO_4^-$. Hence, it appears that chloride-containing intermediates are degraded easier than the phosphate containing compounds. To sum up, the total mineralization requires a much longer illuminated time than the disappearance of dichlorvos (Lu et al., 1991).

CONCLUSION

Photocatalytic oxidation is a promising method for the treatment of hazardous organic compounds. By this process the organics are converted to mineral products. The formation of harmless mineral products is of great importance in water treatment. However, the complete mineralization needs a substantially longer irradiated time, and the initial quantum yields are small. The photocatalytic degradation of dichlorvos obeys the first-order kinetic. Results also indicate that initial concentration and oxygen significantly affect the photooxidation rate. It is generally accepted that H_2O_2 can enhance the photodegradation rate via the formation of $OH^{\bullet}$ radicals. In contrast, as is demonstrated here, dichlorvos competes with H_2O_2 for decomposition, which decreases the degradation rate under pH 4.

ACKNOWLEDGEMENT

We appreciate Professor C.P. Huang, University of Delaware, Newark, DE, USA for his comments and suggestions.

REFERENCES

1. Abdullah M., G. K.-C., Low G., and Matthews, R. W. (1991), "Effects of Common Inorganic Anions on Rates of Photocatalytic Oxidation of Organic Carbon over Illuminated Titanium Dioxide.," *J. Phys. Chem.*, 94, 6820-6825.

2. Al-Ekabi, H. and Serpone, N.(1988), "Kinetic Studied in Heterogeneous Photocatalysis. 1. Photocatalytic Degradation of Chlorinated Phenols in Aerated Aqueous Solutions over TiO_2 Supported on Glass Matrix." *J. Phys. Chem.*, 92, 5726-5731.

3. Auguliaro, V., Davi E. Palmisano, Schiavello, M., and Sclafani, A. (1990), "Influence of Hydrogen Peroxide on the Kinetics of Phenol Photodegradation in Aqueous Titanium Dioxide Dispersion," *Apply Catalysis*, 65, 101-116.

4. Davis, A. P., and Huang, C. P. (1990), "The Removal of Substituted Phenols by A Photocatalytic Oxidation Process with Cadmium Sulfide," *Wat. Res.* 24, 543-550.

5. Demas, J. N., Bowman, W. D., Zalewski, E. F., and Velapoldi, R. A. (1981), "Determination of the Quantum Yield of the Ferrioxalate Actinometer with Electrically Calibrated Radiometers," *J. Phys. Chem.*, 85, 2766-2771.

6. D'Oliveira, J.C., Al-Sayyed, G., and Pichat, P. (1990), "Photodegradation of 2- and 3- Chlorophenol in TiO_2 Aqueous Suspensions," *Environ. Sci. Technol.*, 24, 990-996.

7. Fishbein, L. (1972), "Chromatography of Environmental Hazards, Elsevier Publishing Company," New York, 198-208.

8. Harada, K., Hisanaga, T. and Tanaka, K. (1990), "Photocatalytic Degradation of Organophosphorus Insecticides in Aqueous Semiconductor Suspensions," *Wat. Res.*, 24, 1415-1417.

9. Hatchard, C. G., and Parker, C. A. (1956), "A New Sensitive Chemical Actinometer II. Potassium Ferrioxalate as A Standard Chemical Actinometer," Proc. R. Soc. London, Ser.A, 231, 518-536.

10. Kormann, C., Bahnemann, D. W., and Hoffmann, M. R. (1988), "Photocatalytic Production of H_2O_2 and Organic Peroxides in Aqueous Suspensions of TiO_2, ZnO, and Desert Sand," *Environ. Sci. Technol.*, 22, 798-806.

11. Lu, M.C., Roam, G. D., and Cheng, J. N. (1991), "Photocatalytic of 2,2-Dichlorovinyl Dimethyl Phosphate in Water using Thin Film of Titanium Dioxide," Proc. 16th Conf. on Wastewater Treatment Technology in Republic of China, 807-819.

12. Matthews,R.W. (1987), "Photooxidation of Organic Impurities in Water Using Films of Titanium Dioxide," *J. Phys. Chem.*, 91, 3328-3333.

13. Okamoto, K. I. (1985), "Heterogenous Photocatalytic Decomposition of Phenol over TiO_2 Powder," The Chemical Soc. Jpn., 58, 2015-2022.

14. Ollis, D. F., Pelizzetti, E., and Serpone, N. (1989), "Heterogeneous Photocatalysis in the Environment: Application to Water Purification," *Photocatalysis*, John Wiely & Sons, 603-637.

15. Pelizzetti, E. (1990), "Photocatalytic Degradation of Atrazine and Other s-Triazine Herbicides," *Environ. Sci. Technol.*, 24, 1559-1565.

16. Sabate, J., Anderson, M. A., Kikkawa, H. Edwards, M., and Hill, Jr., C. G. (1991), "A Kinetic Study of the Photocatalytic Degradation of 3-

chlorosalicylic Acid over TiO_2 Membranes Supported on Glass," J. Catal., 127, 167-177.

17. Terzian, R., Serpone, N., Minero, C., and Pelizzetti, E. (1991), "Photocatalyzed Mineralization of Cresols in Aqueous Media with Irradiated Titania," J. Catal., 128,352-365.

18. Tseng, J. M., and Hung, C. P. (1991), "Removal of Chlorphenols from Water by Photocatalytic Oxidation," *Wat. Sci. Tech.*, 23, 377-387.

19. Watterson, A. (1988), *Pesticide Users' Health and Safety Handbook*, Van Nostrand Reinhold, New York, 188-189.

34

USE OF FERRIC CHELATES FOR FENTON (Fe/H_2O_2) TREATMENT OF PESTICIDE CONTAMINATED WATER AND SOIL AT NEUTRAL pH

Katharina Baehr
Universität Hohenheim
Stuttgart, Germany

Joseph Pignatello
Connecticut Agricultural Experimental Station
New Haven, CT

INTRODUCTION

Today there is an urgent need for technologies that can degrade or detoxify organic chemical waste. Since 1930 the production of complex organic compounds has increased. Their global transport in air, water and soil has led to a widespread distribution in additional accumulation in organisms. In many cases it is possible to remove high pollutant concentrations in air and water. There exist, however, few means of detoxification for soil, which itself functions as a filter (Alexander 1991; Ryckman and Ryckman, 1980). From the different physical, biological and chemical possibilities, one of the most promising for soil is a chemical treatment with iron and hydrogen peroxide (Fenton's reagent). Organic compounds are oxidized by hydroxyl radicals and other reactive radicals produced by the reagent (Sun and Pignatello 1992 a, b; Pignatello 1992). The pH optimum of the Fenton Reaction is 2.8 when Fe^{3+} is used, as Fe^{3+} at higher pH precipitates as barely reactive hydrous oxyhydroxide ($Fe_2O_3 \cdot nH_2O$). The pH of soils ranges from 4 to 7. Iron(III) at around neutral pH can be held in solution by chelation (Anderson and Hiller 1975). For successful use, iron chelates should provide the following properties:

1. have catalytic activity towards oxidation of the waste components, i. e. be capable of generating hydroxyl radical and other reactive oxidants from H_2O_2;
2. be resistant to oxidation in the medium;
3. be environmentally safe.

Based on work by Sun and Pignatello (1992 b) out of 40 chelating agents 5 were selected, which fulfilled these conditions in aqueous solution, to use in soil. The growth hormone 2,4-dichlorophenoxyacetic acid (2,4-D) was chosen for degradation in soil, because it breaks down easily with Fenton's reagent in aqueous solution (Sun and Pignatello 1992 a, b' Pignatello 1992). 2,4-D is used as a post emergent herbicide against broad-leafed weeds in agriculture and on lawns. In low concentrations it is easily transformed within days or weeks by microorganisms in the soil (Loos 1975; Sandmann et al., 1988). A concentration of 1000 ppm in the soil, resulting from, say, a spill, depresses the biological activity for long time periods (Ou et al., 1978; Dzantor and Felsot 1991). The goal of this work was to mineralize spill-size concentrations of 2,4-D in the soil by adding iron chelates and hydrogen peroxide without the accumulation of toxic intermediates. The following questions were considered of greatest importance:

1. Can chelating agents be used in the soil suspension to keep iron (as Fe^{3+}) reactive in the Fenton's reaction?
2. What H_2O_2 concentration is optimal for a most complete mineralization of organic wastes?

MATERIAL AND METHODS

Soil

The characteristics of the soil are given in Table 1. It can be described as a coarse loamy, mixed mesic, Typic Dystrochrepts.

Chemicals

The chelating agents hydroxyethyleniminodiacetic acid (HEIDA) and nitrilotriacetic acid (NTA) were obtained from J.T. Baker; picolinic acid, rhodozonic acid and gallic acid were from ICN Biochemicals. All chemicals were used as received. 2,4-D reagent grade was obtained from Sigma and Aldrich in radioactive labeled and unlabeled form, respectively.

Table 1. Selected Properties of Soil

Horizon	Depth (cm)	Sand (%)	Silt (%)	Clay (%)	Texture	C_{org} (%)	pH_{CaCl2}	CEC (mg·kg^{-1})
Ap	0–15	56	36	8	sL	1.57	6.3	94

Procedures

Iron chelate solutions were made by directly mixingg aqueous solutions of Fe^{3+} and the corresponding ligands in a ratio of 1:1. Aqueous Fe^{3+} was prepared fresh daily by dissolving anhydrous $Fe(ClO_4)_3$ (GFS Chemicals) in 0.1M perchloric acid. The iron chelate solutions were adjusted to pH 6 with NaOH. Reactions were carried out in 40ml screw cap vials shaking continuously on a rotary shaker at room temperature. For the chelating agent experiments 2.5 g of air dried soil were mixed with 2.5ml chelating agent solution with or without iron. For the degradation experiment, 1 ml 2,4-D-solution (Na-salt) containing labeled and unlabeled herbicide was added to 2.5g soil. After an incubation period of 24 hours at room temperature, iron chelate solution was added (1 ml) and the Fenton-reaction was initiated with H_2O_2 (0.4 ml). Ratio of soil to solution was 1:1. The initial concentrations in the reaction mixture were 0.001 M Iron(III), 2000ppm = 0.01M 2,4-D, 0.01M chelating agent, H_2O_2 concentrations: I = 0.1 M, II = 0.5 M, III = 1M. No replicates were conducted. Therefore results give only trends, but were not covered by statistical calculations.

Analysis

The pH was measured with a glass electrode. The iron concentration in the supernatant was obtained by an ICP-AES (particle size = diameter ≤ 0.2 µm), 2,4-D and 2,4-dichlorophenol (DCP) were monitored by HPLC on a 25 cm, 5 µm Spherisorb ODS-2 C-18 column (Alltech) using UV detection at 230 nm. The mobile phase (1.5 ml/min) was methanol-water-trifluoroacetic acid (TFA) in the ratio of 70 : 30 : 0.02. The three following fractions of radioactive material were counted in a liquid scintillation counter in 15 ml Opti-Fluor (Packard Instruments Co.) using the external standard method:

1. In the headspace the mineralization product carbon-14 was trapped in a polypropylene center well (Kontes) containing a plug of glass wool impregnated with 0.3 ml of 1 M aqueous ethanolamine. The entire center well was counted after separating the glass wool from the well with forceps.

2. ^{14}C-R (2,4-D and intermediates) in the soil extract (methanol : soil = 10:1, two-hour shake followed by centrifugation) was measured in an 0.5 ml aliquot mixed with the scintillation fluid.

3. Non-extractable ^{14}C-R in the soil was measured by trapping CO_2 in scintillation fluid after combusting 100 mg aliquots of the extracted soil (biological oxidizer, 800° C, 3 min).

RESULTS

Chelate Stability and Ability of Chelating Agents to Complex Soil-borne Iron

Preliminary experiments investigated which chelates were stabile in the soil suspension over a maximum time of 16 to 18 hours. The assumption was made that soil organic matter (SOM) would compete with chelating agent for Fe^{3+}. Five chelating agents were chosen, based on work by Sun and Pignatello (1992 b), which were described as the most reactive ones in the Fenton reaction in aqueous solution. Chelate solution was mixed with air dried soil (ratio 1 : 1) achieving a chelate concentration of 1 and 10 mM at the start of the experiment. Table 2 shows that gallic acid, HEIDA and NTA kept about 1 mM iron in solution over 16 hour time period, if the initial chelate concentration was 10 mM. The contest seemed to be settled during the first hour. These three chelates were chosen for further degradation experiments, at 2000 ppm 2,4-D, 1 mM chelated iron, and various H_2O_2 concentrations.

Table 2. Concentration of Soluble Iron in the Soil Suspension

Initial Concentration Fe (mM)	Time (h)	Percentage Iron Kept Soluble in Soil Suspension				
		Gallic Acid	Picolinic Acid	Rhodozonic Acid	HEIDA[2]	NTA[3]
1	0	100	100	100	100	100
	1	7	0.6	5	5	6
	16	7	0.3	6	6	12
10	0	100	100	100	100	100
	1	12	2	< dl[1]	9	34
	16	20	2	< dl[1]	11	34

Notes:
1. dl = detection limit
2. HEIDA = Hydroxyethyliminodiacetic acid
3. NTA = Nitrilotriacetic acid

A second experiment studied the ability of the three chelating agents to complex soil-borne iron over an 18 hour period. These chelates could be used for the Fenton reaction and could streamline the procedure by just adding chelating agent and hydrogen peroxide to the contaminated soil. Aqueous chelating agent solution was mixed with the soil (ratio 1 : 1) to give an initial concentration of 10 mM and 25 mM. In accordance with an observation of Sun and Pignatello (1992 b) a 3 hour time point was taken. They mentioned that after 3 to 4 hours NTA and HEIDA chelates lost their stability in aqueous solution in the presence of H_2O_2. Iron was released from the complex, precipitated as hydrous oxyhydroxide ($Fe_2O_3 \cdot n\ H_2O$), which did not contribute to the degradation. In an additional replicate, water instead of chelating agent was mixed with the soil.

Table 3 shows the iron concentration in the soil suspension of the different samples. An adequate concentration of 1 mM Fe was only achieved with 25 mM gallic acid. Using 10 mM gallic acid or NTA the concentration was attained after almost 18 hours.

Table 3. Iron Concentration in the Soil Suspension After the Addition of Water or Chelating Agent Respectively

	Iron Concentration in Soil Suspension (mM)			
Time (hours)		0	3	18
H_2O		0.003	0.1	0.3
Gallic Acid (mM)	10	1.8×10^{-3}	0.6	0.8
	25	7.2×10^{-3}	1.0	2.3
HEIDA (mM)	10	1.8×10^{-3}	0.06	0.2
	25	1.8×10^{-3}	0.04	0.2
NTA (mM)	10	5.4×10^{-3}	0.2	0.4
	25	1.1×10^{-3}	0.8	1.0

2,4-D Degradation by the Use of 0.1 M H_2O_2 (Variant I) and Iron Chelates in the Fenton Reaction.

The degradation of 2000 ppm 2,4-D was tested for an 18 hour period using 0.1 M H_2O_2 and 10 mM iron chelate leading to an initial iron concentration of about 1 mM to react in the Fenton-Reaction. Figure 1 shows the distribution of radioactive material (uniformly ring labeled 2,4-D) in the three fractions. As the upper part of the figure shows, about 6 % mineralization occurred for HEIDA and NTA. The reaction seemed to be complete after 3 hours. Using gallic acid the mineralization was lower, but continued after 3 hours. In all samples 80 to 90% radioactivity was found in the soil extract and 10 to 14 % was bound to the soil, i. e., unextractable by methanol used. Differences among the three chelates in the fractions soil extract and soil are small. The total recovery of the radioactivity from ^{14}C-2,4-D in the three fractions was about 100 % in all samples. It is impossible to distinguish between 2,4-D and intermediates in the soil extract fraction by ^{14}C measurement.

Figure 2 compares the 2,4-D concentration in the soil extract (HPLC) and the mineralization rate. Loss of 2,4-D and production of CO_2 are complementary. With all three chelating agents about 50 % of the 2,4-D was lost after 18 hours, but less than 10 % of the transformed 2,4-D was mineralized, 90 % was left as metabolites.

2,4-D Degradation by 0.1 M H_2O_2 (Variant I) and Free Chelating Agents

In an additional experiment free chelating agents (initial concentration 10 mM) instead of iron chelates were added to the reaction mixture. It was thereby tested whether chelating agents complex soil-borne iron in a reactive form and therefore lead to 2,4-D breakdown by the Fenton reaction. Table 4 compares 2,4-D concentrations in the soil extract of samples treated with iron chelates or free chelating agents at 0.1 M H_2O_2. Whereas with the addition of iron about 50 % of the 2,4-D is transformed in 18 hours, no 2,4-D breakdown occurred when no iron was added.

In Table 5 the iron content in the solution is compared in samples with and without the addition of iron. Without the addition of iron the iron concentration is way below 1 mM and corresponds very well with the data of the preliminary experiment. Considering the samples where iron was added, using HEIDA or NTA even after 18 hours, more than 1 mM iron is left in solution. Using gallic acid the iron concentration ranges from a third to half of the adequate concentration, no matter whether iron was added or not.

2,4-D Degradation at Elevated H_2O_2 Concentrations (Variant II = 0.5 M, Variant III = 1 M) and Comparative Breakdown of Ring and Carboxyl labeled 2,4-D

2,4-D contaminated soil was treated over a 3 hour period with 0.5 M or 1 M H_2O_2. The chelating agent concentration was about 10 mM and 1 mM chelated iron was available for the Fenton reaction. In Table 6 2,4-D concentration left in the soil extract is shown compared to the results of the first degradation experiment (Variant I, 3 h, ring). On increasing the H_2O_2 concentration a clear increase of 2,4-D transformation is apparent. After 3 hours there is almost no 2,4-D left in the soil extract using HEIDA or NTA, whereas about 50 % is left using gallic acid. There seems to be no difference in the extent of 2,4-D transformation using 0.5 or 1 M H_2O_2.

In Tables 7, 8 and 9 the results of ^{14}C recovery in the different radioactive fractions are shown. With all three chelating agents, the highest mineralization rate is achieved with 0.5 M H_2O_2. With HEIDA or NTA about 20 % of the ring labeled and 30 % of the carboxyl labeled 2,4-D was mineralized. In the soil extract the same tendency as in the CO_2 fraction is found. If the H_2O_2

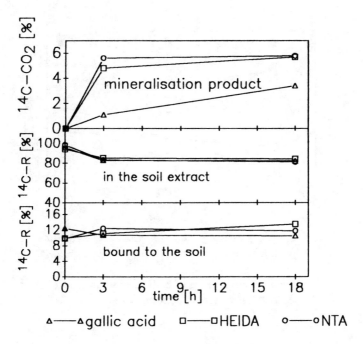

Figure 1 Distribution of Radioactive Material.

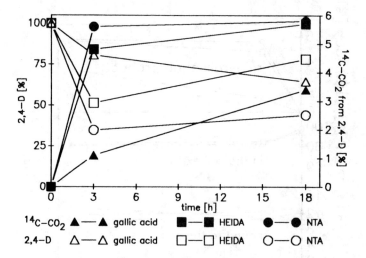

Figure 2 Comparison of 2, 4-D Concentration in Soil Extract and Mineralization Rate.

Table 4. 2,4-D Concentration in the Soil Extract with Iron Chelates or Free Chelating Agents Added

2,4-D Concentration (mM)				
Time (hours)		0	3	18
Gallic Acid (mM)	with Fe	100	8.05	64.4
	without Fe	100	—	97.1
HEIDA (mM)	with Fe	100	51.3	78.2
	without Fe	100	110.0	—
NTA (mM)	with Fe	100	34.7	44.0
	without Fe	100	97.1	—

Table 5. Iron Concentration in the Reaction Mixture

Iron Concentration (mM)				
Time (hours)		0	3	18
Gallic Acid (mM)	with Fe	10	0.4	0.3
	without Fe	0	—	0.5
HEIDA (mM)	with Fe	10	1.5	2.2
	without Fe	0	0.05	—
NTA (mM)	with Fe	10	4.8	2.6
	without Fe	0	0.2	—

concentration increases over 0.5 M, less radioactivity is found in the soil extract. ^{14}C-R bound in the soil was not analyzed due to time constraints.

Summing up the radioactivity from ^{14}C-CO_2 and ^{14}C-R in the soil extract (Table 7 and 8) the percentage of the "bound residues" can be estimated. Assuming the total recovery of radioactivity is about 100 %(as seen in the first degradation experiment) the difference of the data in Table 9 to 100 % shows, that with increasing H_2O_2 concentration up to 50 % "bound residues" (2,4-D and/or intermediates) are formed in the Fenton Reaction. This tendency occurs with all three chelating agents. Comparing 2,4-D breakdown of ring labeled and carboxyl labeled 2,4-D in all samples the tendencies seen with the ring labeled 2,4-D are sharper in the carboxyl labeled samples. The iron concentration was again measured in all samples. After 3 hours in all cases it dropped down to 0.3 or 0.4 mM. Also the pH showed no change over time in any sample. In the controls no 2,4-D degradation due to H_2O_2 treatment alone was observed. This further rules out biodegradation.

Table 6. 2,4-D Concentration in the Soil Extract in the Ratio to the Initial Concentration

Time (hours)	2,4-D in Soil Extract								
	Gallic Acid			HEIDA			NTA		
	Ring	Carboxyl	Mean	Ring	Carboxyl	Mean	Ring	Carboxyl	Mean
0	100	100	100	100	100	100	100	100	100
3 (I)	81	—	81	78	—	78	35	—	35
3 (II)	47	66	56.5	2	3	2.5	2	3	2.5
3 (III)	55	56	55.5	4	2	3	1	2	1.5

Notes:
I = 0.1 M H_2O_2
II = 0.5 M H_2O_2
III = 1 M H_2O_2

DISCUSSION

Background Chemistry

Using Fe/H_2O_2 both Fe^{II} and Fe^{III} react with hydrogen peroxide. The classical "radical mechanism" generates ·OH, which is one of the strongest oxidants known. It reacts nonselectively with organic compounds (Barb et al. 1951, Ingles 1972, Dorfman and Adams 1973, Walling 1975, Buxton et al. 1988). The chain reaction can be described in simple aqueous solution by the following equations (Walling 1975):

(1) $Fe^{2+} + H_2O_2 \longrightarrow Fe^{3+} + HO^- + \cdot OH \quad k_1 = 55\ [M^{-1} \cdot s^{-1}]$

(2) $Fe^{3+} + H_2O_2 \longrightarrow Fe^{2+} + H^+ + HO_2$

(3) $\cdot OH + H_2O_2 \longrightarrow H_2O + HO_2 \cdot \quad k_3 = 4.5 \cdot 10^7\ [M^{-1} \cdot s^{-1}]$

(4) $\cdot OH + Fe^{2+} \longrightarrow Fe^{3+} + OH^- \quad k_4 = 3.0 \cdot 10^8\ [M^{-1} \cdot s^{-1}]$

(5) $\cdot OH + RH \longrightarrow H_2O + R \cdot \quad k_5 =$ substrate specific]

These equations can be used to get an idea which reactions mostly compete in the treatment of the soil with iron (III) and hydrogen peroxide. In the degradation experiments relatively high H_2O_2 concentrations were used. Iron exists mainly as iron III, because reaction (1) is much faster than reaction (2) (Barb et al. 1951). So ·OH generated in the Fenton Reaction reacts with organic compounds (equation 5) or with hydrogenperoxide, if in excess (equation 3). The goal of the

Table 7. ^{14}C from ^{14}C-2,4-D, Found as ^{14}C-CO_2 in the Trap

Time (hours)	Percentage of ^{14}C-CO_2 Mineralization Product						
	Gallic Acid		HEIDA		NTA	
	Ring	Carboxyl	Ring	Carboxyl	Ring	Carboxyl
0	0	0	0	0	0	0
3 (I)	1	—	5	—	6	—
3 (II)	2	12	22	27	18	32
3 (III)	4	8	16	22	15	22

Notes:
I = 0.1 M H_2O_2
II = 0.5 M H_2O_2
III = 1 M H_2O_2

Table 8. ^{14} from ^{14}C-2,4-D, Found in the Soil Extract

Time (hours)	Percentage of ^{14}C-R in Soil Extract					
	Gallic Acid		HEIDA		NTA	
	Ring	Carboxyl	Ring	Carboxyl	Ring	Carboxyl
0	93	96	96	95	94	95
3 (I)	83	—	85	—	83	—
3 (II)	71	72	46	38	36	20
3 (III)	51	47	40	27	36	16

Notes:
I = 0.1 M H_2O_2
II = 0.5 M H_2O_2
III = 1 M H_2O_2

Table 9. Sum of Radioactive Material from ^{14}C-2,4-D Found in the CO_2 Trap and in the Soil Extract

Time (hours)	Percentage of Recovery of Radioactive Material					
	Gallic Acid		HEIDA		NTA	
	Ring	Carboxyl	Ring	Carboxyl	Ring	Carboxyl
0	93	96	96	95	94	95
3 (I)	84	—	90	—	99	—
3 (II)	73	84	68	65	54	52
3 (III)	55	85	56	49	51	38

Notes:
I = 0.1 M H_2O_2
II = 0.5 M H_2O_2
III = 1 M H_2O_2

Fenton treatment is to degrade the organic compound 2,4-D by ·OH. So reaction (3) is undesirable in this treatment, because it diminishes the 2,4-D breakdown. The product of reaction (3) is barely reactive (Bielski et al.). Therefore one of the main questions was: Which H_2O_2 concentration is optimal for a most complete mineralization of organic wastes? Not only H_2O_2 but also other organic compounds compete with 2,4-D for the reaction with ·OH (reaction 5). In the soil suspension there are soluble organic matter and the chelating agents. k_5 is substrate specific and ranges from 10^7 -10^{10} [$M^{-1} \cdot s^{-1}$]. The reactivity for the soluble organic matter or the selected chelating agents is not known. To get an idea of the competition only concentrations can be taken into account. There exists an additional competition between chelating agents and organic matter for the complexation of iron. This competition has a direct influence on the Fenton Reaction itself, because only chelates formed by added chelating agents and iron are probably reactive for the Fenton reaction. As Sun and Pignatello (1992 b) showed, most of the 40 tested chelating agents were not reactive at all in the Fenton Reaction. The competing effects were the reason for the other main question in this work: Can chelating agents be used in the soil suspension to keep iron reactive in the Fenton reaction?

Fate of the Iron Chelates in the Fenton Reaction

Using 0.1 M H_2O_2 Figure 2 showed that in HEIDA and NTA variants 2,4-D transformation and mineralization is complete by the 3 hour time point, whereas with gallic acid transformation and mineralization continue after the 3 hour time point. A possible conclusion for this observation could be a degradation of the chelating agent HEIDA and NTA as described by Sun and Pignatello (1992 b) for aqueous solution. Iron would be released out of the complex, precipitates and does not react in the Fenton Reaction. Looking at iron concentration in Table 5 this explanation does not fit, because using HEIDA or NTA even after 18 hours there is more than 1 mM iron left in solution, whereas using gallic acid already after 3 hours only 0.4 mM iron is left in solution. For these observations the following explanations are possible:

1. HEIDA and NTA were degraded, but iron was kept soluble by soil-borne chelating agents. These chelates were not reactive in the Fenton Reaction.
2. Iron chelates with HEIDA or NTA still existed, but were transformed in a way, that they were not reactive anymore.
3. The gallic acid-chelate was activated, so a much lower iron chelate concentration led to a 2,4-D-transformation over a very long time period.
4. Hydrogenperoxide was already degraded in the first three hours. In the gallic acid variant reactive gallic-acid-ferryl-complexes were formed which lead to oxidation at a slower rate.

For the first two explanations no literature was found probably, because it is very difficult to observe these constitutions as they are very unstable. Several authors observed complexes as described in the third point and explained them by:

a) a ligandfield effect on the redoxproperties of the metal, i. e. an intramolecular oxidation of the chelate by ·OH (Walling and Amarnath 1982) or as found for ethylene-diaminetetraacetate (EDTA) by Rush and Koppenol (1986) by a ferryl group.
b) The lig and favors the complexation of H_2O_2 at a labile coordination site of the metal (Graf et al. 1984).

All authors mentioned worked in aqueous solution without the presence of organic compounds that would interfere. To see, whether the last point could be a true explanation, a higher H_2O_2 concentration was used in the next degradation experiment. Looking at Table 6 the iron concentration in all gallic acid samples seemed to be the same no matter whether iron was added or not. So the question comes up, why 2,4-D is degraded if iron and gallic acid were added to the soil, but no 2,4-D degradation occurred, if gallic acid alone was added. This phenomenon might be explained as follows:

1. Gallic acid complexes soil-borne iron in a nonreactive form.
2. Without the addition of iron it took some time to complex soil-borne iron to an adequate concentration. At the time iron was complexed H_2O_2 was already degraded.

There was no literature found about gallic acid complexes, therefore further experiments are needed to decide these points.

2,4-D's Competition for Reaction with Hydroxyl Radicals

(1) Hydrogenperoxide

In the second degradation experiment the optimum H_2O_2 concentration was determined. Using the "Massenwirkungsgesetz" theoretically a hydrogenperoxide concentration of 1 M would already be too high.

$$\frac{R_3}{R_5} = \frac{k_3 \cdot [H_2 O_2]}{k_5 \cdot [2,4-D]}$$

$k_3 = 4.5 \cdot 10\ 7\ [M^{-1} \cdot s^{-1}]$ (Walling 1975)
$k_5 = 2.7 \cdot 10\ 9\ [M^{-1} \cdot s^{-1}]$ (Pignatello 1992)
$[2,4-D] = 0.01\ M$

Variant II = 0.5 M H_2O_2 leads to: $R_5 = 1.2 \cdot R_3$ That means 2,4-D degradation dominates.

Variant III = 1 M H_2O_2 leads to: $R_3 = 1.7 \cdot R_5$ That means H_2O_2 decomposition dominates.

So it could be assumed, that 2,4-D would be less complete degraded in variant III. As seen in Tables 7, 8 and 9, 2,4-D transformation was the same using 0.5 or 1 M H_2O_2. But with 1 M H_2O_2 2,4-D was less mineralized and more "bound residues" were formed.

(2) Organic compounds: organic matter and chelating agents

This high proportion of "bound residues" is possible due to exchange processes with concurrents in the soil suspension. Concurrents are chelating agents added to the soil and dissolved soilborne organic components. Concentrations of dissolved compounds are decisive, because the reaction with ·OH depends on diffusion constants (Sheldon and Kochi 1981). Less than 50 mg/l soil-borne organic matter is dissolved in the soil suspension (Scheffer and Schachtschabel 1989). So organic carbon dissolved in the soil suspension originates from organic matter, chelating agents and 2,4-D in a ratio of 1 : 16 : 20. This demonstrates that chelating agents and 2,4-D are competing for ·OH. With the proportion of chelating agent and iron in the ratio of 1 : 1 not all the iron is chelated, but iron hydroxide is precipitated. This precipitate does not contribute to the Fenton reaction (Sylva 1972). If less chelating agent would be used, less iron would be available for the Fenton reaction. This indirect factor is as important as the direct factors for the competition. Looking at the ratio of organic C in the soils suspension organic matter competes less with 2,4-D for ·OH. However, organic substances such as fulvic and humic acids have a lot of reactive groups. What kind of exchange processes exist between dissolved and perhaps even solid organic matter, 2,4-D and intermediates (R or R·) can not be said. It seems to be that the development of more and/or different functional groups could have complexed with metabolites, so that they are not extractable. Little work has been done about radical reactions with organic matter or soil. Masten (1990) ozonated chlorinated olefins in the presence of synthetical humic acids. Oxidation was due to ·OH generated in the ozonation. These humic acids seemed to catalyze the degradation of the pollutant and their own aromatic rings were attacked too. In contrast Anderson et al. (1986) observed inhibitory effects of fulvic acids on the ozonation of organic pollutants. Changes in the functional groups of humic or fulvic acids in these experiments were not examined.

Fulvic and humic acids are very heterogenous chemicals and mainly defined by chemical fraction analysis. Further investigations can give information about the reactions that are responsible for the increase of the "bound residues" fraction with increasing H_2O_2 concentration.

2,4-D Transformation and Breakdown

To evaluate the mineralization of the pesticide with Fenton's reagent, the increasing amount of "bound residues" has to be taken into consideration. Comparison of the breakdown of ring labeled and carboxyl labeled 2,4-D gives information about the mineralization path and the effect of different chelates in the Fenton reaction. In all samples carboxyl group of 2,4-D was mineralized by decarboxylation to a greater extent than the aromatic ring. 2,4-Dichlorophenol - a very toxic intermediate- was possibly formed, but was completely transformed in three hours. It could not be detected by HPLC. Comparison of gallic acid and HEIDA or NTA samples respectively shows, that with the optimal H_2O_2 concentration (0.5 M) in gallic acid samples less 2,4-D is transformed (50 % using gallic acid, 95 % using HEIDA or NTA) and less transformed 2,4-D is mineralized. 20 % of the ring and 30 % of the carboxyl group are mineralized using gallic acid. Taking the distribution of C atoms in the 2,4-D molecule into account, 1/6th of transformed 2,4-D is mineralized, 2/3rds of the intermediates are not extractable. Using HEIDA or NTA 25 % of the ring and 35 % of the carboxyl group were mineralized. So a quarter of transformed 2,4-D was mineralized and a third of the metabolites was not extractable from the soil. Further experiments can investigate which mechanisms generate the difference in the degradation by gallic acid and HEIDA or NTA.

CONCLUSIONS

The study shows that the mineralization of high 2,4-D concentration in the soil with iron and H_2O_2 (without accumulation of known toxic intermediates) is the most effective with HEIDA or NTA chelates at a H_2O_2 concentration of 0.5 M. The optimum concentration of H_2O_2 was not determined in this study. An optimum H_2O_2 concentration is found, if maximum mineralization is achieved together with a minimum amount of "bound residues." No degradation was achieved by the addition of free chelating agents and H_2O_2 to the contaminated soil; that is without added Fe^{3+}.

Possibilities for a successful breakdown with free chelating agents do exist and should be investigated in further experiments. Further knowledge about different forms of iron formed in the reaction and exchange processes between iron in the reaction and organic material in the Fenton reaction should be obtained in further experiments. In combination with other methods, i. e. biological processes, the Fenton reaction is promising for the "clean up" of highly contaminated soil material and rinsates on waste disposal sites that contain a mixture of dissolved organic contaminants and soilborne organic matter.

REFERENCES

1. Alexander, M. (1991):"Research Needs in Bioremediation," *Environ. Sci. Technol.*, Vol. 25, pp. 1971-1973.

2. Anderson, L. J., Johnson, J.O., Christman, R.F. (1986): "Extent of Ozone's Reaction with Isolated Aquatic Fulvic Acid," *Environ. Sci. Technol.*, Vol. 20, pp. 739-742.

3. Anderson, W.F., Hiller, M.C. (ed.) (1975): "Development of Iron Chelators for Clinical Use."

4. Barb, W.G., Baxendale, J.H., George, P., Hargrave, R. (1951): "Reactions of Ferrous and Ferric Ions with Hydrogen Peroxide; Part II; The Ferric Ion Reaction," *Trans. Faraday Soc.*, 47, pp. 591-616.

5. Bielski, B.J.H., Cabelli, D.E., Arneli, R.L., Ross, A.B. (1985): "Reactivity of HO_2/O_2-Radicals in Aqueous Solution," *J. Phys. Chem. Ref. Data*, Vol. 14, pp. 1041-1100.

6. Buxton, G.V., Greenstock, C.L., Helman, W.P., Ross, A.B. (1988): "Critical Review of Rate Contants for Reactions of Hydrated Electrons, Hydrogen Atoms and Hydrogen Radicals ($\cdot OH/O\cdot^-$) in Aqueous Solution," *J. Phys. Chem. Ref. Data*, Vol. 17, pp. 513-886

7. Dorfman, L.M., Adams, G.E. (1973): "Reactivity of Hydroxyl Radical," *National Bureau of Standard Rep.*, NSRDS-NBS-46, NBS: Washington, D.C.

8. Dzantor, E.K., Felsot, A.S. (1991): "Microbial Responses to Large Concentration of Herbicides in Soil," *Environ. Toxical. Chem.*, Vol. 10, pp. 649-655.

9. Graff, E., Mahoney, J.R., Bryand, R.G., Eaton, J.W. (1984): "Iron catalyzed Hydroxyl Free Radical Formation," *Journal of Biological Chemistry*, Vol. 259, pp. 3620-3624.

10. Ingles, G.L. (1972): "Studies of Oxidation by Fenton's Reagent Using Redox Titration, I, Oxidation of Organic Compounds," *Aust. Journ. Chem.*, Vol. 25, pp. 87-95.

11. Loos, M.A. (1975): "Phenoxyalkonoic Acids," in: Kearney, P.C., Kaufman, D.D. (ed.) (1975): Herbicides: Chemistry, Degradation and Mode of Action, Vol. 1; Marcel Dekker: New York, pp. 1-128.

12. Masten, S. (1990): "Ozonation of VOC's in the Presence of Humic Acid and Soils," (un-published).

13. Ou, L.T., Rothwell, D.F., Wheeler, W.B., Davidson, J.M. (1978): "The Effect of High 2,4-D-Concentration and Carbon Dioxide Evolution in Soils," *J. Environ. Qual.*, Vol. 7, pp. 241-246

14. Pignatello, J.J. (1992): "Dark and Photoassisted Fe^{3+}-Catalyzed Degradation of Chlorophenoxy Herbicides by Hydrogen Peroxide," *Environ. Sci. Technol.*, (in press).

15. Rush, J.D., Koppenol, W.H. (1986): "Oxidizing Intermediates in the Reaction of Ferrous, EDTA with Hydrogen Peroxide," *J. Biological Chemistry*, Vol. 261, pp. 6730-6733.

16. Ryckman, D.W., Ryckman, M.D. (1980): "Organizing to Cope with Hazardous Material Spills," *J. Am. Water Works Assoc.*, Vol. 72, pp. 196-200.

17. Sandmann, E.R.I.J., Loos, M.A., van Dyck, L.P., (1988): "The Microbial Degradation of 2,4-Dichlorophenoxyacetic Acid in Soil," *Residue Reviews*, Vol. 101, pp. 1-53.

18. Scheffer, F., Schachtschabel, P. (Hrsg.) (1989): "Lehrbuch der Bodenkunde," Ferdinand Enke Verlag, Stuttgart.

19. Scheldon, R.A., Kochi, J.K.I. (1981): "Metal-Catalyzed Reactions of

Organic Compounds: Mechanistic Principles and Synthetic Methodology Including Biochemical Process," *Academic Press*: New York.

20. Sun, Y., Pignatello, J.J. (1992 a): "A Promising Approach to the Chemical Degradation of Pesticide Wastes," *J. Agr. Food Chem.*, (in press).

21. Sun, Y., Pignatello, J.J. (1992 b): "Chemical Treatment of Pesticide Wastes. Evaluation of Fe(III) Chelates for Catalytic Hydrogen Peroxide Oxidation of 2,4-D at Circum Neutral pH," *J. Agr. Food Chem.*, (in press).

22. Sylva, R.N.(1972): "The Hydrolysis of Iron(III)," *Rev. Pure and Appl. Chem.*, Vol. 22, pp. 115-132.

23. Walling, C., Amarnath, K., (1982): "Oxidation of Mandelic Acid by Fenton's Reagent," *J. Am. Chem. Soc.*, Vol. 104, pp. 1185-1189.

INDEX

A

acid extraction 255
acid strength 250
activated carbon 286, 293, 322
activated carbon cartridge 194
adhesive tape 383
adhesive tapes wastewater 375
adsorption leachate treatment 424
adsorption system 409
adsorptive capacities of carbon 426
aerobic bioremediation 363, 364
aerobic treatment 328
air emissions 110
air quality 70
air sample monitoring 74
air stripping 195
air stripping wicks 193
ammonia mobility 343
ammonium nitrate 340, 342
anaerobic biodegradation 293
anaerobic sludge 373
anaerobic sludge bed 376
anaerobic treatment 328
anhydrous ammonia 340, 343
aniline 390
antifreeze 33
artificial intelligence 21
ash processing 121
asphalt concrete plants 70
asphalt plants 79
Atlantic City Medical Center 110
automobile tailpipes 71

B

baseline sampling 354
batch reactor 391
batch tests 121, 128
biodecomposition 390
biodegradation 257, 274, 293, 321, 334
biodegradation pathway 303
biodegradation potential 315
biodegradation rates 344
biokinetic parameters 302
biological oxidation 275
biological oxygen 345
biological soil characteristics 338
biomass density 266
bioremediation 277, 281, 285, 370
biosludge activity 378
biotreatment 413
break through tests 77

C

calcination 225
calorimetric method 390
carbon adsorption 427
carbon dioxide 455
carcinogenic effects 5
cement kiln plant 177
cement kilns 176
chemical soil characteristics 338

chemical treatment 35
Chernobyl explosion 42
chlorinated solvents 37
chrome sludge 225
chromium 247
chromium admixtures 245
chromium contamination 232
chromium extraction 253
citric acid 375
Clean Air Act 71
cleaner wick 194
cleanup levels 1
column tests 121, 126, 132
compressive strength 236
computer programs 19
computer-based models 16
concrete 232
contaminated soils 156
contingency analysis 151
contingency planning 150
control devices 84
covers 83

D

decision diagram 145
deforestation 42
degreasing 37
Delaware 363
Department of Defense 59
deterioration of concrete 233
detoxification 225
disposal facilities 81

E

effect of pH 277
effluent treatment 285
electrolytes 452
electron microscope 236
emissions 72
enclosures 83
endogenous decay 404

energy recovery 170
envirocad 51
environmental disasters 42
Ettringite 234
excavated soils 279
exposure assessment 3
extent of remediation 288

F

ferric chelates 459
field tests 78
fixed film processes 286
fixed-film 321
flow of hazardous material 30
fluid bed reactors 286
fluidized bed 293
fly ashes 131
fuel analysis 173
fuel contamination 333
fuel-contaminated site 332
fuel-contaminated soils 334
fungal biotrap 213
fungal cell walls 204

G

gas chromatographic data 112
graphic display 26
groundwater 274, 308, 333, 335
groundwater flow model 199
growth curves 391

H

hazardous material inventory 32
hazardous waste management 32
heavy metals 213
herbicide 308
highway detention ponds 182, 184
highway runoff 180
hydrogen peroxide 452

I

impact assessment 42
in situ bioremediation 283
in situ treatment 281
incineration 35, 176
industrial boilers 174
industrial emissions 72
input of data 25
iron concentration 463
isotherm study 310

J

jar test 309

K

kinetic measurement 390

L

leachate characteristics 412
leachate constituents 119
leachate treatment 409
leaching procedure 183

M

mass burn 119
matrix solubility 250
medical waste 90
metal sorption 186
microbial populations 334
microorganisms in soils 351
military construction 42
mobility 6
monitoring 335, 354
multispecies biofilm 328

N

National Contingency Plan 2
neutron activation 204
nitrogen addition 332
nitrogen supplementation 340
nutrient addition 340, 342
nutrient requirements 275
nutrient testing 341

O

onsite pretreatment 310
organic destruction 85
organic removal 84
organic waste 321
organic-contaminated wastes 87
oxygen consumption 266
oxygen requirements 276
oxygen tension 257
oxygen utilization 333

P

PCP ammunition boxes 33
pesticide contaminated water 459
petroleum hydrocarbon 351
petroleum-contaminated soils 135
phenol 180
phenol adsorption 190
phenol removal 187
photocatalytic oxidation 446
pollution prevention 437
polycyclic aromatic 257
pretreatment 425
private contractors 67
probability 17
process documentation 19
process simulation 55
project planning 44
pyroxidizer 90

R

raw leachate 425
RDF incinerator 119
reactors 376
recycling 37
regulatory control 82
remedial alternatives 1
remediation conditions 366
remediation times 287
removal of chromium 247
residence time 162
respiration tests 345
reuse 37
rice wine 382
risk assessment 1, 3, 16
rotary kiln 157
rotary kiln thermal desorber 158

S

sampling 357
sampling procedures 120, 342
saturated soil 283
saturated zone 285
scrap tires 170
screening method 136
screw feeder 160
sensitive populations 326
sequential extractions 122
site characterization 335
slag characteristics 248
smog 71
soil biodegradation 348
soil cell 281
soil cell treatment 280
soil conditions 336
soil remediation 135
soil suspension 463
soils bioventing 332
solid waste amendments 29
solvent cleaning 37

solvent extractions 161
standard gas 77
stream properties 56
structure of cement 234
structure of concrete 233
substrate inhibition 403
superfund 1

T

temperature profiles 165
Tenax cartridges 73
thermal desorption 135, 138, 157, 166
thermal vapor incineration 85
thin films 446
threshold criteria 4
titanium dioxide 446
toxic metals 180
toxic substrates 324
toxicity 6, 277
treatability studies 412, 419
treatment of soils 274
treatment system selection 150

U

unsaturated soil 257, 282
used oil 33
utility industries 72

V

vadose zone 339
VOC emissions 81
volatile organic compound 81

W

washing 247
waste batteries 33

waste classification   139, 144
waste cleanup   59
waste disposal   55
waste minimization   29
waste oil   38
waste recovery   50, 55
waste solvents   33, 37, 87
waste treatment   55, 81, 83
wastewater treatment   373
wastewaters   213
water treatment   286
wick flow tests   199
winery plant   379, 382